Lecture Notes in Mathematics

Edited by A. Dold and B. Eckmann

1388

J. Shinoda T.A. Slaman T. Tugué (Eds.)

Mathematical Logic and Applications

Proceedings of the Logic Meeting held in Kyoto, 1987

Springer-Verlag

Berlin Heidelberg New York London Paris Tokyo Hong Kong

Editors

Juichi Shinoda
Tosiyuki Tugué
Department of Mathematics, College of General Education
Nagoya University, Nagoya 464-01, JAPAN

Theodore A. Slaman
University of Chicago, Department of Mathematics
Chicago, IL 60637, USA

Mathematics Subject Classification (1980): 03 B xx, 03 D xx, 03 E xx, 03 F xx, 03 H xx.

ISBN 3-540-51527-5 Springer-Verlag Berlin Heidelberg New York
ISBN 0-387-51527-5 Springer-Verlag New York Berlin Heidelberg

© Springer-Verlag Berlin Heidelberg 1989
Printed in Germany

Printing and binding: Druckhaus Beltz, Hemsbach/Bergstr.
2146/3140-543210

PREFACE

The '87th Meeting on Mathematical Logic and its Applications was held at the Research Institute of Mathematical Sciences (RIMS) of Kyoto University during August 3–6, 1987. The present volume represents the proceedings of the meeting, which also includes some contributed papers of the participants.

The editors acknowledge with their gratitude that the meeting was sponsored by the RIMS, and was partly supported by Grant-in-Aid for Co-operative research project No. 61302010, (the head investigator is Professor T. Uesu, Tokyo Science University), Ministry of Education, Science and Culture. We should also notify that the large number of participants from the United States got their travel expenses through NSF.

The second editor visited Japan, mainly in Nagoya, from March through September in 1987 as a visiting professor on a grant from the Nihon Gakujutsu Shinkokai (the Japan Society for the Promotion of Science) through the first editor. Let him take this opportunity to express his heartfelt thanks to institutions and logicians in Japan concerned.

Juichi Shinoda (Nagoya)
Theodore A. Slaman (Chicago)
Tosiyuki Tugué (Nagoya)

Table of Contents

Recursively Enumerable Sets in Models of Σ_2 Collection*

C. T. Chong
National University of Singapore

This paper studies a particular aspect of ***reverse recursion theory***: the problem of the existence and non-existence of recursively enumerable (r.e.) sets under the Σ_2 collection scheme. While investigations on the subject of reverse recursion theory began only recently, the results obtained so far have pointed to the central importance of Σ_2 induction and Σ_2 collection schema in the construction of r.e. sets. Previous works (Mytilinaios and Slaman [12], Mytilinaios [12], Groszek and Slaman [7], Slaman and Woodin [18], Chong [1 and 2]) have shown that practically all r.e. sets constructed using the finite injury priority method, and many of those obtained via the $0''$-priority method, can be proved to exist under the hypothesis of Σ_2 induction. The situation becomes quite different with the weaker hypothesis of $B\Sigma_2$ induction. In this case counterexamples have been found. Our interest here lies in obtaining a necessary and sufficient condition in the hierarchy of Peano axioms for the existence of r.e. sets of a given specified property. In particular, since Σ_2 collection is not strong enough to guarantee the existence of r.e. sets of a certain type, will their existence, under Σ_2 collection, then imply Σ_2 induction ? An affirmative answer to this question will provide a nice characterization of the relative strength of different classes of r.e. sets vis à vis the Peano axioms, and identify the existence of each class with a particular induction scheme, over a base theory. Two classes of r.e. sets are singled out for special attention: the class of maximal sets and the class of incomplete non-low r.e. sets. We prove that in a model of Σ_2 collection, the existence of any of these two classes is equivalent to the hypothesis that the Σ_2 induction scheme holds. We also show that the $\forall\exists$ theory of the r.e. degrees (in the language involving ordering and join) under Σ_2 collection is different from that under the full Peano axioms. This is done by producing a model of Σ_2 collection in which Lachlan [9]'s non-splitting result is false. A brief survey of some results related to the main theme is given along the way.

Despite the subject's modest history, answers obtained from the investigations have been interesting, not only because they provide a better understanding of the fundamental constructions in recursion theory, but also because many of the techniques used to obtain the answers were inspired by those introduced in α recursion theory. Indeed in many cases the original techniques appear to fit nicely into the new situation, giving the impression of a technical development that is historically correct. We will attempt to point out this connection at various places.

---

*The author is grateful to Ted Slaman for the willingness to discuss and share his expertise on the subject matter of this paper. In particular, Theorems 3 and 5 owe their final version to observations made by Slaman.

We begin by fixing the notations. Let P^- be the set of axioms of Peano arithmetic minus the induction scheme. The language used includes function symbols $+$, $\cdot$, successor $'$, and exponentiation, and the constant symbol 0. The axioms consist of the usual axioms for successor, addition, multiplication, and the commutativity, associativity, and distributivity of numbers under the four arithmetical operations. We also include axioms for exponentiation:

$$x^{y'} = x^y \cdot x,$$

$$x^0 = 0'.$$

The induction scheme is arranged into a hierarchy of increasing complexity strength. For each $n < \omega$, let $I\Sigma_n$ be the *induction scheme* for Σ_n formulas. We then have Peano Arithmetic to be equal to $P^- + \{I\Sigma_n | n < \omega\}$.

A scheme which is closely related to the Σ_n induction scheme is the Σ_n *least member scheme* ($L\Sigma_n$). This states that every nonempty Σ_n definable set has a least element. We also have the Σ_n *collection scheme* ($B\Sigma_n$): If φ is Σ_n, then

$$(\forall y < x)(\exists w)\varphi(y, w)$$

implies there is a b such that

$$(\forall y < x)(\exists w < b)\varphi(y, w).$$

Thus a Σ_n function on a proper initial segment is bounded.

Define $B\Pi_n$, $I\Pi_n$ and $L\Pi_n$ similarly for Π_n formulas. The relative strength of these schema are established by the next result:

PROPOSITION 1 (PARIS AND KIRBY [14]). *In every model of $P^- + I\Sigma_0$, we have*

$$I\Sigma_{n+1} \rightarrow B\Sigma_{n+1} \rightarrow I\Sigma_n$$

$$I\Sigma_n \leftrightarrow I\Pi_n \leftrightarrow L\Sigma_n \leftrightarrow L\Pi_n$$

$$B\Pi_n \leftrightarrow B\Sigma_{n+1}$$

Within the system of $P^- + I\Sigma_0$, all the basic notions of recursion theory can be formalized. For example, using exponentiation one may code n-tuples of numbers by single elements in models of $P^- + I\Sigma_0$. In general, given $\mathcal{M} \models P^- + I\Sigma_0$, one has the following definition:

DEFINITION. *$H \subset \mathcal{M}$ is $\mathcal{M}$-finite if H has a code in $\mathcal{M}$.*

In particular, we have many more $\mathcal{M}$-finite sets than just finite sets. Indeed, the bounded Δ_0 sets in a model of $P^- + I\Sigma_0$ are all $\mathcal{M}$-finite. Given such an $\mathcal{M}$, a set is defined to be recursively enumerable (r.e.) if it is $\Sigma_1(\mathcal{M})$, and is recursive if its complement is r.e. as well. There is an effective Gödel numbering of r.e. sets and $\mathcal{M}$-finite sets. We let W_e and K_e to be the eth r.e. set and $\mathcal{M}$-finite set respectively under this numbering. The notion of reduction may now be introduced:

DEFINITION. *Let X and Y be subsets of $\mathcal{M} \models P^- + I\Sigma_0$. X is r.e. in Y if there is an r.e. set W_e of quadruples such that for all x,*

$$x \in X \longleftrightarrow (\exists H)(\exists K)[(x,1,H,K) \in W_e \ \& \ H \subset Y \ \& \\ K \cap Y = \emptyset],$$

where H, K are $\mathcal{M}$-finite sets.

The notation $X = W_e^Y$ is used to express the fact that X is r.e. in Y via the set W_e. If X and $\bar{X}$ are both r.e. in Y, then we say that X is *pointwise recursive in* Y, or *weakly recursive in* Y. This is written $X \leq_w Y$. We give $X <_w Y$ the obvious meaning. It is not difficult to see that if $\mathcal{M}$ is the standard model of arithmetic, then $\leq_w$ is a transitive relation. In general, however, the transitivity of $\leq_w$ is not automatic (Groszek and Slaman [7] proved that transitivity of $\leq_w$ is a theorem of $P^- + I\Sigma_2$ but not of $P^- + I\Sigma_1$).

An important notion in the enumeratiion of r.e. sets, widely studied in α recursion theory, is that of regularity. This is defined as follows:

DEFINITION. *Let $\mathcal{M}$ be a model of $P^- + I\Sigma_0$. A set $A \subset \mathcal{M}$ is regular if its intersection with every proper initial segment is $\mathcal{M}$-finite.*

PROPOSITION 2. *Every r.e. set in a model of $P^- + I\Sigma_1$ is regular.*

This result is fairly well-known, and is attributed to H. Friedman. A proof of it appears in Mytilinaios [12].

Recursion-theoretic results which were proved using finite or infinite injury methods have been the principal objects of study in reverse recursion theory. We give a summary below, starting with r.e. degrees:

THEOREM 1 (MYTILINAIOS [12], SLAMAN AND WOODIN [18]). *Every finite injury construction can be carried out within the system $P^- + I\Sigma_1$.*

Thus the Friedberg-Muchnik Theorem and the Sacks Splitting Theorem are all provable in $P^- + I\Sigma_1$. The complete r.e set $\emptyset'$ is defined to be $\{e|e \in W_e\}$. An r.e. set A is incomplete if $\emptyset'$ is not pointwise recursive in it. It is low if its jump $A' = \{e|e \in W_e^A\}$ is pointwise recursive in $\emptyset'$. The Sacks Jump Inversion Theorem states that every set above $\emptyset'$ and r.e. in $\emptyset'$ is the degree of the jump of an r.e. set. In particular, there exists (in the standard model of arithmetic) an incomplete non-low r.e. set. The method used to prove this result is an instance of infinite injury priority argument.

THEOREM 2 (GROSZEK-MYTILINAIOS [6], MYTILINAIOS-SLAMAN [13]). *The existence of an incomplete non-low r.e. set is a theorem of $P^- + I\Sigma_2$, but not of $P^- + B\Sigma_2$.*

We now turn to the lattice of r.e. sets. An r.e. set in a model $\mathcal{M}$ of $P^- + I\Sigma_0$ is *maximal* if its complement is not $\mathcal{M}$-finite, and cannot be split by an r.e. set into two non-$\mathcal{M}$-finite parts. Maximal sets were first constructed by Friedberg [4] for the standard model $\mathcal{N}$. It has since become a subject of intense study for recursion theorists (see Soare [20] for an exposition). The next result, together with Theorem 1, imply that there is no finite injury construction of a maximal set.

THEOREM 3 (CHONG [1]. *(a) There is a maximal set in every model of $P^- + I\Sigma_2$.*
(b) There is a model of $P^- + B\Sigma_2 + \neg I\Sigma_2$ with no maximal set.

The proof of Theorem 3 (b) examines the Mytinaios-Slaman model $\mathcal{M}_0$, and shows that it has no maximal sets. $\mathcal{M}_0$ is an uncountable model of $P^- + B\Sigma_2 + \neg I\Sigma_2$, with standard system equal to 2^ω (i.e. every set of natural numbers is the standard part of an $\mathcal{M}_0$-finite set), such that there is a Σ_2 cofinal function f defined on $\mathcal{N}$. This is a candidate for an 'arithmetical analog' of $\aleph_\omega^L$. It is endowed with properties reminiscent of the ordinal $\aleph_\omega^L$. In Lerman and Simpson [11], these properties of $\aleph_\omega^L$ were sufficient to show that no maximal sets exist. The idea was to split $\aleph_\omega^L$ recursively into the union $\{A_n\}$ of ω many pairwise disjoint simultaneous r.e . sets. By choosing those n's for which A_n has nonempty intersection with a given Π_1 set X, one gets an $\aleph_\omega^L$-finite subset K of ω, with the propertry that $X \cap A_n \neq \emptyset$ for each $n \in K$. One can now easily split K into two disjoint infinite $\aleph_\omega^L$-finite sets K_1 and K_2, so that the corresponding r.e. sets $\cup\{A_n|n \in K_1\}$ and $\cup\{A_n|n \in K_2\}$ split X into two non-$\aleph_\omega^L$-finite pieces.

Now for models of fragments of arithmetic such as $\mathcal{M}_0$, a recursive splitting of the universe into ω pieces is not possible (by the Overspill Lemma), and so a different strategy is required. The intuition remains the same: Given a Π_1 set X, devise a method of recursively guessing (correctly) ω many elements of X, without 'touching'

ω many other elements of X. Here ideas from Chong and Lerman [3] developed for hyperhypersimple sets turn out to be relevant (see the proof of Theorem 5).

An r.e. set is hyperhypersimple (hh-simple) if it is not recursive and the lattice of its r.e. supersets forms a Boolean algebra. Post [15] originally defined hh-simple sets in terms of recursive arrays of pairwise disjoint r.e. sets. The aim was to show that hh-simple sets had complements sufficiently thin to guarantee an incomplete degree. Lachlan [8] showed that over the standard model, the original defintion was equivalent to the lattice-theoretic definition given above. This paved the way for the study of hh-simple sets from the lattice-theoretic point of view, leading to various results on the $\forall\exists$ theory of r.e. sets.

THEOREM 4 (CHONG [2]). *The existence of an hh-simple set can be proved within the system $P^- + B\Sigma_2$.*

Theorems 3 and 4 show that the difference in terms of the proof-theoretic strength of maximal sets and hh-simple sets appear at the level $P^- + B\Sigma_2$.

Arising from Theorem 3 is the question of an exact classification within the hierarchy of induction scheme for the existence of a maximal set. A related question, in view of Theorem 2, is a similar classification for the existence of a non-low incomplete r.e. set. The answers for these questions are provided by the next result:

THEOREM 5. *Let $\mathcal{M}$ be a model of $P^- + B\Sigma_2$. Then each of the following is equivalent to $I\Sigma_2$:*

(a) There is a maximal set in $\mathcal{M}$;
(b) There is an incomplete non-low r.e. set.

We give a proof of this theorem. We begin with a notion which is a generalization of that of a standard system in models of arithmetic. We say that a set X is bounded in a model $\mathcal{M}$ of $P^- + I\Sigma_0$ if there is a $b \in \mathcal{M}$ which is greater than every member of X.

Recall that a cut in a model $\mathcal{M}$ is a set that is closed downwards and has no end point. The set ω is a cut in every nonstandard model of $P^- + I\Sigma_0$. If X is a cut, and $Y \subset \mathcal{M}$, we say that $Y \cap X$ is the X-standard part of Y.

DEFINITION. *Let X be a cut in a model $\mathcal{M}$ of $P^- + I\Sigma_0$. A set $Y \subset X$ is $\Delta_n(X)$ if there exist Σ_n and Π_n formulas (with parameters) φ and ψ, respectively, such that for $z \in X$,*

$$z \in Y \longleftrightarrow \mathcal{M} \models \varphi(z)$$
$$\longleftrightarrow \mathcal{M} \models \psi(z).$$

DEFINITION. *Let $\mathcal{M} \models P^- + I\Sigma_0$ and let X be a cut in $\mathcal{M}$. Then $\mathcal{M}$ admits a Δ_2 system for X if every $Y \subset X$ which is $\Delta_2(X)$ is the X-standard part of an $\mathcal{M}$-finite set.*

If X is a cut in a model $\mathcal{M}$ which admit a Δ_2 system for X, then X behaves like an ordinal 'below the Δ_2 projectum' of $\mathcal{M}$ (at least in one form of the definition). For ease of discussion, let us recall the following notions from Jensen's fine structure theory: An ordinal α is Σ_n admissible if it satisfies the Σ_n collection scheme. The Δ_n projectum of α, written $\delta np(\alpha)$, is the least ordinal less than or equal to α with a Δ_n subset which is not α-finite. Fine structure theory of the constructible universe L then implies that $\delta np(\alpha)$ is precisely the least ordinal ν for which there is a Σ_n function mapping ν onto α. Models of $P^- + I\Sigma_2$ admit Δ_2 systems on every cut, and therefore correspond to ordinals α with $\delta 2p(\alpha) = \alpha$. For models which satisfy only $B\Sigma_2$ and not $I\Sigma_2$, the appropriate analogs are Σ_1 admissible ordinals α with $\delta 2p(\alpha) < \alpha$. It should be noted that the various equivalent definitions for $\delta np(\alpha)$ do not apply in the arithmetical case. In particular, even when $\mathcal{M}$ does not admit a Δ_2 system for X, there is still no Δ_2 projection from X onto the model under $B\Sigma_2$. As a footnote, we are interested only in Δ_2 systems and not in Σ_2 systems because we will be considering Σ_2 functions defined on cuts, and the graphs of these objects are automatically Δ_2 (over the cut in question).

LEMMA 1 (CHONG [2]. *Let $\mathcal{M}$ be a model of $P^- + B\Sigma_2$. If $\mathcal{M}$ is not a model of $I\Sigma_2$, then there is a cut X, and a Σ_2 function f total on X, such that the range of f on X is cofinal in $\mathcal{M}$.*

LEMMA 2 (CHONG [2]). *Let X be a cut in a countable model $\mathcal{M}$ of $P^- + B\Sigma_2$ satisfying the conclusion of Lemma 1. Then $\mathcal{M}$ admits a Δ_2 system for X.*

We begin the proof of Theorem 5 by first considering countable models.

Proof that existence of a maximal set in a countable model of $P^- + B\Sigma_2$ implies $I\Sigma_2$

Let $\mathcal{M}$ be a countable model of $P^- + B\Sigma_2$. If $\mathcal{M}$ does not satisfy $I\Sigma_2$, then by Lemma 1 there is a cofinal Σ_2 function f from a cut X of $\mathcal{M}$ into $\mathcal{M}$. By Lemma 2 $\mathcal{M}$ admits a Δ_2 system for X. We may assume that f is bounded on every set bounded in X. The details are worked out in [2], and we shall not repeat them here.

LEMMA 3. *There is a recursive approximation f' of f defined on X such that*
(a) *$\lim_s f'(s,i) = f(i)$ for each $i \in X$;*
(b) *$f'(s,j) \leq f'(t,i)$ for all $s \leq t$ and $j \leq i$.*

The existence of f' can be extracted from the Σ_2-ness of f. It is a familiar procedure and we shall omit the details (see Chong [1] for the case when X is ω).

Let A be an r.e. set in $\mathcal{M}$ which is not recursive. By the Proposition, A is regular. Hence neither A nor its complement is bounded. We show that A is not maximal by splitting its complement into two unbounded parts, using an r.e. set.

Let $a(0)$ be the least element of $\bar{A}$. The regularity of A ensures that $a(0)$ exists. If $i \in X$ and $a(i) \in \bar{A}$ is defined, let k^* be the least $k > i$ such that $A^{f(k)}|a(i) = A|a(i)$ (A^s is the $\mathcal{M}$-finite set of elements of A enumerated by stage s). The regularity of A and $B\Sigma_2$ ensure that k^* exists. Let $a(i+1)$ be the least member of $\bar{A}$ greater then $f(k^*)$. Then the set $\{(i, a(i)\}_{i \in X}$ is $\Delta_2(X)$ and, since $\mathcal{M}$ admits a Δ_2 system, is the X standard part of an $\mathcal{M}$-finite set K. We fix K as a parameter. Using K, there is a recursive approximation $a'(s,i)$ of the $a(i)$ at stage s, with the property that $a'(s,j) \leq a'(t,i)$ for $s \leq t$ and $j \leq i$. We leave it to the reader to work out the details.

We note the following facts:

(1) If $i \notin X$, then $lim_s f'(s,i)$ and $lim_s a'(s,i)$ do not exist.

(2) For $j < i$ in X, for each s, if $a'(s,i) = a(i)$ then $a'(s,j) = a(j)$.

By the Overspill Lemma, the functions f' and a' are defined at many values of $i \notin X$, but none of these has a limit. Hence (1) is non-vacuous. The choice of $a(i+1)$ to be greater than $f(k^*)$ implies that (2) holds.

Define a function $S' : \mathcal{M} \times \mathcal{M} \times \mathcal{M} \to 2$ as follows: Given s and $j < i$, let $t(s)$ be the least t such that $a'(t,i) = a'(s,i)$. If there is no $u < t(s)$ such that $a'(u,i) = a'(t(s),j)$, let $S'(s,j,i) = 0$. Otherwise let $u(s)$ be the least such u and set $S'(s,j,i) = 1 - S'(u(s),j,i)$. A simple argument using $I\Sigma_1$ shows that $S'(s,j,i) < 2$ for all s, j and i. We also have

(3) If $j < i$ and $j \in X$, then $lim_s S'(s,j,i) = S(j,i)$ exists and is less than 2.

The next lemma says that there is a recursive way of avoiding specific elements of $\bar{A}$ by not making wrong guesses. It is proved as in Chong [1, Lemma 5].

LEMMA 3. *Let $j \in X$ and suppose that $S'(s,j,i) = S(j,i)$ where $j < i$. Then $a'(s,i) \neq a(j)$.*

We need the following result of Paris and Kirby [14]:

LEMMA 4. *A countable model of $P^- + B\Sigma_2$ has a proper Σ_2 elementary end extension.*

Now choose b to be an upper bound for X and set

$$D = \{((j,i),r) | j < i \in X \ \& \ i \text{ is even} \ \& \ j \text{ is odd} \ \& \\ (\exists s)(\forall t \geq s)(S'(s,j,i) = r)\}.$$

By (3), D is $\Delta_2(X)$. Let $\mathcal{M}'$ be a proper Σ_2 elementary end extension guaranteed by Lemma 4. Choose c be an upper bound of $\mathcal{M}$ in $\mathcal{M}'$. Then D is contained in the following set

$$Z = \{((j,i),r) < b | i \text{ even} \ \& \ j \text{ odd} \ \& \mathcal{M}' \models j < i \ \& \ r < 2 \ \& \\ (\forall t \geq c)[S'(t,j,i) = r]\}.$$

Furthermore, for $j \in X$ and $j < i$, $((j,i),r) \in Z$ if and only if $r = S(j,i)$. On the other hand, if $((j,i),r) \in Z$ then it implies that $lim_s S'(s,j,i)$ exists in $\mathcal{M}'$. Σ_2 elementariness then implies that $lim_s S'(s,j,i)$ exists in $\mathcal{M}$ as well. It follows that $Z \cap X = D$, and that for all odd $j \in X$, $(j,i,S(j,i)) \in Z$ whenever i is even and satisfies $j < i < b$. For the same (j,i), we also have $((j,i),r) \in Z$ if and only if

$$\mathcal{M}' \models (\exists t \geq c)[S'(t,j,i) = r].$$

Now for $x = ((j,i),r)$ with $j < i$ and j odd, i even, and $r < 2$, define

$$\Phi(x,y,t) \longleftrightarrow \mathcal{M}' \models y \geq t \ \& \ S'(y,j,i) \neq r \ \vee \\ y \geq t \ \& \ S'(y,j,i) = r.$$

For each t and each $x = ((j,i),r)$ of the form j odd, i even and $r < 2$, there is a y such that $\Phi(x,y,t)$ holds in $\mathcal{M}'$. Furthermore, if $t \in \mathcal{M}' \setminus \mathcal{M}$, then $((j,i),r) \in Z$, with $j \in X$, if and only if

$$\Phi(x,y,t) \longleftrightarrow y \geq t \ \& \ S'(y,j,i) = r.$$

Thus we have $\mathcal{M}' \models (\forall t)(\forall x < b)(\exists y)\Phi(x,y,t)$ (strictly speaking we are considering only those x of the form $((j,i),r) < b$). The same sentence is true in $\mathcal{M}$ and so by $B\Sigma_2$ in $\mathcal{M}$ we have

$$\mathcal{M} \models (\forall t)(\exists w)(\forall x < b)(\exists y < w)\Phi(x,y,t).$$

By Σ_2 transfer, $\mathcal{M}'$ satisfies this condition on uniform upper bound, so that if, in particular, we let t be the c considered earlier, and choose a corresponding upper bound w, then we get

$$Z^* = \{((i,j),r) < b | j < i \ \& \ j \text{ odd} \ \& \ i \text{ even} \ \& \\ \mathcal{M}' \models (\exists y < w)[y \geq c \ \& \ S'(y,j,i) = r\}$$

to be $\mathcal{M}'$-finite. Then Z^* is $\mathcal{M}$-finite as well since $\mathcal{M}'$ is an end extension of $\mathcal{M}$. Now if $j \in X$, and $i > j$, then $((j,i),r) \in Z^*$ if and only if $S(j,i) = r$. Define

$$B = A \cup \{a'(s,i) | (\forall j < i)(\forall r)[(j,i,r) \in Z^* \rightarrow S'(s,j,i) = r]\}.$$

Then $B \supset A$ and for each $i \in X$ even, $a(i)$ belongs to B. On the other hand, if $j \in X$ is odd, then Lemma 3 ensures that $a(j)$ is never enumerated in B by error. Thus

B splits the complement of A into two non-$\mathcal{M}$-finite parts. We conclude that A is not maximal.

Proof that existence of an incomplete non-low r.e. set in a countable model of $P^- + B\Sigma_2$ implies $I\Sigma_2$

Let $\mathcal{M}$ be countable as above. We follow the same notations. We shall prove that in $\mathcal{M}$, every incomplete r.e. set A is low, i.e. the jump A' of A is recursive in $\emptyset'$.

DEFINITION. *A set A is hyperregular if whenever W_e^A is a function with bounded domain, it has a bounded range.*

LEMMA 5 (MYTILINAIOS AND SLAMAN [13]). *Let $\mathcal{M}$ be as above and let A be r.e. in $\mathcal{M}$. The following are equivalent:*
(a) A is hyperregular;
(b) $\mathcal{M}$ is a model of $I\Sigma_1^A$ (Σ_1 formulas with extra predicate for A);
(c) W_e^A is regular for each e.

Let A be an incomplete r.e. set.

Claim 1. A is hyperregular.

If not, let $g = W_e^A$ be a function with bounded domain and unbounded range in $\mathcal{M}$. Applying $B\Sigma_2$ we may assume that the domain of g is a cut X in $\mathcal{M}$ and that for $a \in X$, the range of g on the initial segment a is bounded in $\mathcal{M}$. We describe an algorithm to compute $\emptyset'$. Set

$$(j,i) \in Y \longleftrightarrow j \in X \ \&\ i \in X\ \&$$
$$i = \mu y[(\emptyset')^{g(y)}|g(j) = \emptyset'|g(j)],$$

where $(\emptyset')^{g(y)}$ is $\emptyset'$ computed $g(y)$ steps. Y is $\Delta_2(X)$ and so since $\mathcal{M}$ admits a Δ_2 system on X (Lemma 2), Y is the X-standard part of an $\mathcal{M}$-finite set K. This is then an $\mathcal{M}$-finite partial function defined for every $j \in X$.

Now given a, find the least j such that $g(j) > a$. Use K to find the corresponding i, and compute $(\emptyset')^{g(i)}$. Then $a \in \emptyset'$ if and only if it is an element of $(\emptyset')^{g(i)}$. Hence A is a complete r.e. set, a contradiction.

Claim 2. If A is hyperregular, then A is low.

By Lemma 5, $\mathcal{M}$ is a model of $I\Sigma_1^A$. By Lemma 5 (c), every W_e^A (in particular A') is regular. Let

$$(j,i) \in V \longleftrightarrow i = \mu y[A'|f(j) = (A')^{f(y)}|f(j)],$$

where $(A')^{f(i)}$ is A' computed $f(i)$ steps using A as an oracle. Now V is a $\Delta_2(X)$ set, and so is the X-standard part of an $\mathcal{M}$-finite set K. Note that the function f is recursive in $\emptyset'$: For $j \in X$ and s given, $f'(s,j) = f(j) = w$ if $(\forall t \geq s)(f'(t,j) = w)$. Thus we may use $\emptyset'$ as an oracle to compute A', namely, to decide if $x \in A'$, find the least j such that $x < f(j)$. Then $j \in X$ and so choose the least i such that $(j,i) \in K$. Then $x \in A'$ if and only if it is so by stage $f(i)$. Now $(A')^{f(i)}$ can be pointwise recursively computed from A and hence from $\emptyset'$, since A is hyperregular. This proves Claim 2.

Thus every incomplete r.e. set in $\mathcal{M}$ is low. It follows that if $\mathcal{M}$ is countable and models $P^- + B\Sigma_2$, then $\mathcal{M}$ satisfies $I\Sigma_2$ if and only if $\mathcal{M}$ has an incomplete non-low r.e. set.

To prove Theorem 5, it is now sufficient to consider uncountable models of $P^- + B\Sigma_2$. Let J be such a model, and let $\mathcal{M}$ be a countable elementary substructure. Then $\mathcal{M}$ is a model of $P^- + B\Sigma_2$ as well. If J does not satisfy $I\Sigma_2$, then neither does $\mathcal{M}$. In this case, by the above arguments we see that $\mathcal{M}$ has no maximal set and no incomplete non-low r.e. set. These facts can be expressed in a first order sentence in the language of arithmetic. It follows that J satisfies the same sentence.

The proof of Theorem 5 is complete.

In each of the results discussed above, ideas and techniques from α recursion theory were applied in an essential way. In particular, models of $P^- + I\Sigma_2$ behave much like Σ_2 admissible ordinals, except that in the former there is the additional feature that it is closed under the exponential function. On the other hand, by Lemma 1 models of $P^- + B\Sigma_2 + \neg I\Sigma_2$ bear a striking resemblence to ordinals which are not Σ_2 admissible. Furthermore, if we restrict ourselves to countable models of $P^- + B\Sigma_2 + \neg I\Sigma_2$, then Lemma 2 reminds one of Σ_2 inadmissible ordinals α whose Σ_2 cofinality is less than the Δ_2 projectum, such as $\aleph_\omega^L$. As noted earlier, the uncountable model $\mathcal{M}_0$ also share these properties. Whether this is more than a mere coincidence is not clear. I have been given to understand, however, that applications to recursion theory on non-standard models of arithmetic were far from the minds of the pioneers of higher recursion theory.

One should hasten to point out that the analogies, strong as they are, are not perfect, due basically to the existence of non-standard elements. A good example is that while Lerman [10] has identified a particular class of countable ordinals as those and only those for which maximal sets exist, Theorem 3 (a) places no restriction on the cardinality of the model.

The structure $\mathcal{M}_0$ is of great interest since it is an archetypical example of a model of $P^- + B\Sigma_2 + \neg I\Sigma_2$ with special properties, e.g. being Σ_2 cofinal with ω and admitting a Δ_2 system for ω. It is a rich source of problems and conjectures in reverse recursion theory. We give one more illustration of this point by proving the result that in $\mathcal{M}_0$, every r.e. set splits over all lower r.e. sets. Shore [16] showed that this was true for $\aleph_\omega^L$, establishing a difference between ω and $\aleph_\omega^L$ in the first order theory of r.e. degrees (in the language consisting of $\leq$, 0, and $\vee$ as special symbols). A corollary of the theorem that we prove here is that the first order theory of r.e degrees under $P^- + B\Sigma_2$ is not the same as that of the full Peano arithmetic. The 'blocking technique' of Shore is used. However, the presence of the cofinal function f is exploited to group the requirements into ω many blocks in the limit. This is an approach which is closer to the original one developed by Shore [16] than that used in Mytilinaios [12], where in the absence of a cofinal function (in general), a different device is needed to form the blocking.

THEOREM 6. *Let $A <_w B$ be r.e. sets $\mathcal{M}_0$. There exist r.e. sets C_1 and C_2 such that $C_1 \oplus C_2 = B$, and $C_1 \oplus A$ and $C_2 \oplus A$ have incomparable degrees.*

PROOF: For $i < 2$, consider requirements are of the type

$$R_e: \qquad \{e\}^{C_i \oplus A} \neq B,$$

where $\{e\}^X$ is W_e^X if W_e^X is looked upon as a characteristic function. Using the same notations as before, we let $f : \omega \to \mathcal{M}_0$ be a Σ_2 cofinal function and let f' be its recursive, non-decreasing approximation. R_e is said to be in block k at stage s if k is the least element such that $f'(s,k) > e$. Thus every requirement eventually falls into a kth block for some finite k.

Requirements in the same block are given equal priority. This means that we do not injure one computation associated with a requirement for the sake of satisfying another in the same block. At stage s, the length of agreement for $C_i^s \oplus A^s$ with respect to the eth requirement is

$$L_i(s,e) = \mu x[\{e\}_s^{C_i^s \oplus A^s}(x) \downarrow \longrightarrow \{e\}_s^{C_i \oplus A^s}(x) \neq B^s(x)],$$

where $\{e\}_s^X(x)$ means the result of s steps of computation with input x, whereas $\{e\}^X(x) \downarrow$ means that the computations gives an output. The length of agreement for block k at stage s is defined to be

$$\mathcal{L}_i(s,k) = sup_{e<f'(s,k)} L_i(s,e).$$

The use function $U_i(s,e)$ for requirement e is defined to be

$$\mu y[(\forall x < L_i(s,e))(\{e\}^{(C_i^s \oplus A^s)|y}(x) = \{e\}^{C_i^s \oplus A^s}(x)].$$

And we set the use function $\mathcal{U}_i(s,k)$ for block k at stage s to be $sup_{e<f'(s,k)} U_i(s,e)$.

We begin with empty sets at stage 0, setting the lengths of agreement to be 0. At stage $s+1$, let b_s be the element of B enumerated at stage s. Find the least (k,i) for which $\mathcal{U}_i(s,k) > b_s$ for some $i < 2$. Enumerate b_s in C_1 if $i = 0$. Otherwise enumerate it in C_0. If such a k does not exist, enumerate b_s in C_1.

By Claims 1 and 2 of Theorem 5, we see that A is hyperregular and low. This implies that we have a recursive function g' such that $lim_s g'(s,x) = A'(x)$ for each x. This follows from the fact that $A' \leq_w \emptyset'$ and so (using the regularity of A and $\emptyset'$) the function used to compute A' from $\emptyset'$ can be recursively approximated (cf. Chong [1] for a proof of a special case).

LEMMA 6. *R_e is satisfied for each e.*

PROOF: We do this by induction on blocks $k \in \omega$. Suppose that for all $n < k$, if $e < f(n)$ then R_e is satisfied. The function g' then implies that for each $e < f(n)$, $lim_s L_i(s,e) = L_i(e)$ exists. We claim that $\mathcal{L}_i(n) = lim_s \mathcal{L}_i(s,n)$ is bounded for $n < k$. If not, pick the least counterexample n, called n_0 and assume that $i = 0$.

Consider the set

$$K = \{e | e < f(n) \ \& \ (\exists x)[\{e\}^{C_0 \oplus A}(x) = 0 \neq B(x) = 1\},$$

where we adopt the convention that $X(x) = 1 \leftrightarrow x \in X$. Now K is bounded in $\mathcal{M}_0$, and $\mathcal{M}_0$ is a Σ_1 elementary end-substructure of a saturated model $\mathcal{M}'$ of Peano arithmetic (Mytilinaios and Slaman [13]). Since C_0, A and B are Σ_1, the computation resulting in the set K persist in $\mathcal{M}'$ and is in fact $\mathcal{M}'$-finite. Since $\mathcal{M}_0$ is end extended by $\mathcal{M}'$, K is $\mathcal{M}_0$-finite as well.

Choose s_0 such that: (i) $f'(s, n_0) = f(n_0)$ for all $s \geq s_0$, (ii) $\mathcal{U}_i(n) = \mathcal{U}_i(s,n)$ for $n < n_0$ and $s \geq s_0$, and (iii) all members of B less than $sup\ \mathcal{U}_i(n)$, $n < n_0$, have been enumerated (regularity of B and the boundedness of $\mathcal{L}_i(n)$ ensure that this is possible). After stage s_0, the requirements in block n_0 have attained their final priority. and so using $f(n_0)$ as a parameter, we may argue that $B \leq_w A$ as follows: To compute $B(x)$, go to an $s > s_0$ and an $e \notin K$ in block n_0 such that $\{e\}_s^{|C_s \oplus A^s}(x) = B^s(x)$, and such that $A^s|y = A|y$ where y is the length of information about A^s used in the computation. This computation gives the correct computation on B since no injury on the computation is allowed by construction (any $b_s < \mathcal{U}_0(s, n_0)$ automatically go to C_1). This contradiction implies that $\mathcal{L}_0(n_0)$ is bounded. We argue as above that $\mathcal{L}_1(n_0)$ is bounded in a similar fashion, using the least stage $s > s_0$ (in place of s_0) such that $\mathcal{U}_0(s, n_0)$ attains its final value.

Using the bounds on $\mathcal{L}_i(n)$, $n < k$, one argues firstly that R_e, for e in block k, is satisfied (else $B \leq_w A$), and then using the function g' concludes that $lim_s L_i(s,e)$ exists. We leave it to the reader for filling in the details.

Thus each R_e is satisfied, and a straight forward argument implies Theorem 6.

COROLLARY. *The $\forall\exists$ theory of r.e. degrees for $P^- + B\Sigma_2$ is not equivalent to that for full Peano arithmetic.*

PROOF: The sentence

$$(\forall a < b)(\exists c_0, c_1)[a < c_0 < b \ \& \ a < c_1 < b \ \& \ c_0 \vee c_1 = b]$$

is a Π_2 sentence true in $\mathcal{M}_0$ but false in any model of full Peano arithmetic.

Finally, we consider some problems in degree theory. The basic questions here are of the Kleene-Post type: Do there exist two incomparable degrees ? This is quickly settled by Theorem 1 under $P^- + I\Sigma_1$. The next question is a relativized version of the above, and seeks to decide the statement:

(*): There exists an incomparable pair of degrees above $0'$.

Now it is not difficult to show that the answer is affirmative if we assume $P^- + I\Sigma_2$. This follows roughly from the observation that under $I\Sigma_2$, constructions r.e. in $\emptyset'$ are just like r.e. constructions under $I\Sigma_1$. The problem becomes challenging with the weaker assumption of $P^- + B\Sigma_2$. Here one runs into the difficulty of satisfying requirements of the type

$$\{e\}^{A \oplus \emptyset'} \neq B$$

and

$$\{e\}^{B \oplus \emptyset'} \neq A.$$

Specifically, according to Lemma 1 there is a function pointwise recursive in $\emptyset'$ with domain a cut and range cofinal. In the language of ordinal recursion theory and that of Mytilinaios and Slaman [13], sets above $\emptyset'$ are then not hyperregular. For Σ_2 inadmissible cardinals, this forces the breakdown of the stage by stage construction at limit ordinal levels, and in fact in the case of $\aleph^L_{\omega_1}$ it results in a negative solution to the problem (Friedman [6]). In $\mathcal{M}_0$ one again encounters the difficulty of meeting the requirements given above, in view of the presence of the function f.

In a similar vein, one may consider the question of the existence of a minimal degree, i.e. a non-recursive degree that does not bound any strictly lower non-recursive degree. It is not difficult to show that the existence of a set of minimal degree follows from $P^- + I\Sigma_2$. At the $P^- + B\Sigma_2$ level, one is confronted with the problem of meeting requirements of the type

$$\{e\}^A <_w A \rightarrow \{e\}^A \equiv_w \emptyset,$$

where $A \equiv_w B$ means $A \leq_w B$ and $B \leq_w A$. For the model $\mathcal{M}_0$ we have a situation that is not unlike that of $\aleph^L_\omega$: If one attempts the construction of a set of minimal degree via the method of splitting trees and full trees, then a moment's reflection shows that the process breaks down at stages which are non-standard numbers (recursive trees need not be closed under ω intersections). Since the problem of the existence of a set of minimal degree for $\aleph^L_\omega$ is a major unsolved problem in recursion theory, the corresponding one for $\mathcal{M}_0$, bearing numerous similarities with the former one, becomes especially interesting. It is even tempting to speculate that the solution of one problem would throw light onto the solution of the other.

We end this paper with some questions:

(a) It is proved in Chong [1] that the existence of maximal sets does not require any assumption stronger than $P^- + I\Sigma_0$, provided that the underlying universe is carefully chosen. In Chong [1], the model chosen has the property that there is a Σ_2 map from

ω onto the whole universe. Do all models of $P^- + I\Sigma_0 + \neg I\Sigma_1$ with maximal sets have this property ?

(b) What is the complexity, in the hierarchy of fragments of Peano arithmetic, of various theorems on maximal sets ? In particular, is Soare's theorem (Soare [19]) on the automorphisms of the lattice of r.e. sets sending maximal sets to maximal sets provable in $P^- + I\Sigma_2$?

(c) Does $P^- + I\Sigma_2$ prove Lachlan's non-splitting theorem [9] ? Theorem 6 implies that it is not provable in $P^- + B\Sigma_2$. Since the proof of Lachlan's theorem uses a $\emptyset'''$ priority argument, we are also asking if such strategies can be carried out using Σ_2 induction alone.

REFERENCES

[1] C. T. Chong, Maximal sets and fragments of Peano arithmetic, *Nagoya Math. J., to appear*

[2] C. T. Chong, Hyperhypersimple sets and Δ_2 systems, *Annals of Pure and Applied Logic, to appear*

[3] C. T. Chong and M. Lerman, Hyperhypersimple α r.e. sets, *Annals of Math. Logic* **9** (1976), 1–48

[4] R. M. Friedberg; Three theorems on recursive enumeration: I. Decomposition, II. Maximal set, III. Enumeration without duplication, *J. Symbolic Logic* **23** (1957), 309–316

[5] S. D. Friedman, Negative solutions to Post's problem, II., *Annals Math.*, **113** (1981), 25–43

[6] M. Groszek and M. Mytilinaios, Σ_2 induction and the construction of a high degrees, *to appear*

[7] M. Groszek and T. A. Slaman, On turing reducibility, *to appear*

[8] A. H. Lachlan, On the lattice of recursively enumerable sets, *Trans. Amer. Math. Soc.* **35** (1968), 123–146

[9] A. H. Lachlan, A recursively enumerable set which will not split over all lesser ones, *Annals Math. Logic* **9** (1975), 307–365

[10] M. Lerman, Maximal α-r.e. sets, *Trans. Amer. Math. Soc*, **188** (1974), 341–386

[11] M. Lerman asnd S. G. Simpson, Maximal sets in α recursion theory, *Israel J. Math.* **4** (1973), 236–247

[12] M. Mytilinaios, Finite injury and Σ_1 induction, *to appear*

[13] M. Mytilinaios and T. A. Slaman, Σ_2 collection and the infinite injury priority method, *J. Symbolic Logic, to appear*

[14] J. B. Paris and L. A. Kirby, Σ_n collection schemas in arithmetic, in: *Logic Colloquium '77*, North-Holland, 1978

[15] E. L. Post, Recursively enumerable sets of integers and their decision problems, *Bull. Amer. Math. Soc.*, **50** (1944), 284–316

[16] R. A. Shore, The $\forall\exists$ theory of r.e. sets, in: *Generalized Recursion Theory II*, North-Holland, 1978

[17] R. A. Shore, On the jump of an α-recursively enumerable set, *Trans. Amer. Math. Soc.* **217** (1978), 351–363

[18] T. A. Slaman and H. Woodin, Σ_1 collection and the finite injury priority method, *to appear*

[19] R. I. Soare, Automorphisms of the latttice of recursively enumerable sets, Part I: Maximal sets, *Annals of Math.* (2) **100** (1974), 80–120

[20] R. I. Soare, *Recursively Enumerable Sets and Degrees*, Ω Series, Springer Verlag, 1987

The Role of a Filter Quantifier in Set Theory

Yuzuru Kakuda*
Kobe University

If we reconsider the axiom of **ZF** through the concept of *small* or *large*, or in the other words, through the concept of ideals or filters, then we might notice some aspects of the axioms of **ZF**. Here, by the axioms of **ZF** we mean the follwing:

Extensionality:	$(\exists z)(z \in x \leftrightarrow z \in y) \rightarrow x = y$;
Regularity:	$x \neq 0 \rightarrow (\exists y \in x)(\forall z \in x)(z \notin y)$;
Pair:	$(\exists z)(x \in z \wedge y \in z)$;
Union:	$(\exists z)(\forall y \in x)(y \subseteq z)$;
Collection:	$(\forall x \in z)(\exists y)\varphi \rightarrow (\exists u)(\forall x \in z)(\exists y \in u)\varphi$, where u dose not occur free in φ;
Separation:	$(\exists z)(\forall x)(x \in z \leftrightarrow x \in y \wedge \varphi)$, where z does not occur free in φ;
Infinity:	$(\exists x)(0 \in x \wedge (\forall y \in x)(y \cup \{y\} \in x))$;
Power Sets:	$(\exists z)(\forall y)(y \subseteq x \rightarrow y \in z)$.

First, let us examine the intuitive idea that a class contained in a set is regarded as to be small. In order to express this idea in the language of **ZF**, we define an auxiliary quantifier *set* by:

$$(\mathit{set}\,x)\varphi \equiv (\exists y)(\forall x)(\varphi \rightarrow x \in y), \text{ where } u \text{ dose not occur free in } \varphi.$$

Then, the following can be proved in **ZF**:

(i) $(\forall x)(\varphi \rightarrow \psi) \rightarrow ((\mathit{set}\,x)\psi \rightarrow (\mathit{set}\,x)\varphi)$,
(ii) $(\mathit{set}\,x)\varphi \wedge (\mathit{set}\,x)\psi \rightarrow (\mathit{set}\,x)(\varphi \vee \psi)$,
(iii) $\neg(\mathit{set}\,x)(x = x)$,
(iv) $(\forall x \in z)(\mathit{set}\,y)\varphi \rightarrow (\mathit{set}\,y)(\exists x \in z)\varphi$,
(v′) $(\forall y)(\mathit{set}\,x)(y = x)$.

* This work was partially supported by JSPS under a grant of Japan-U.S.A. Cooparative reserach.

(i)–(iii) and (v′) says that hte collection of ***small*** classes in the first sence forms a proper non-trivial ideal on the universe of sets. (iv) says that the union of ***set-size-wisely*** many ***small***, in other words, theideal of ***small*** classes is ***set-size-wisely*** complete. (iv) is an important property for set theory. In fact, Collection, together with Union, is equovalent to (iv) under the other axioms of **ZF**. If we want to stress ***large*** instead of ***small***, we define the following quantifier aa which corresponds to the dual filter:

$$(\mathit{aa}\,x)\varphi \equiv (\mathit{set}\,x)\neg\varphi.$$

If we restate (i)–(iv′) by using aa, then we have the follwing:

(i) $(\forall x)(\varphi \to \psi) \to ((\mathit{aa}\,x)\varphi \to (\mathit{aa}\,x)\psi)$,
(ii) $(\mathit{aa}\,x)\varphi \wedge (\mathit{aa}\,x)\psi \to (\mathit{aa}\,x)(\varphi \wedge \psi)$,
(iii) $\neg(\mathit{aa}\,x)(x \neq x)$,
(iv) $(\forall x \in z)(\mathit{aa}\,y)\varphi \to (\mathit{aa}\,y)(\forall x \in z)\varphi$.

Secondly, we consider the idea that a class containing a final ***segment***, that is, containing a class of the form $\{\, x \mid x \text{ is a set and } y \subseteq x \,\}$ is regarde as to be ***large***. As in the first case, we define a quantifier aa by:

$$(\mathit{aa}\,x)\varphi \equiv (\exists y)(\forall x)(y \subseteq x \to \varphi), \text{ where } y \text{ does not occur free in } \varphi.$$

In **ZF**, we can prove the same schemes (i)–(iv) as in the first case, together with

(v) $(\forall y)(\mathit{aa}\,x)(y \in x)$.

In fact, (ii)–(v) are logically equivalent to Pair, Union and Collection. We can say much more: If we extend the language $\mathcal{L}$ of **ZF** by adding a new quantifier aa and formalize a theory on the extended language $\mathcal{L}(\mathit{aa})$ with

(o) $(\mathit{aa}\,x)\varphi(x) \leftrightarrow (\mathit{aa}\,y)\varphi(y)$, where y does not occur in $\varphi(x)$.

and (i)–(v) as its own axioms, then this theory is just a conservative extension of the first-order theory on the language of **ZF** of which axioms are Pair, Union and Collection. (See §2.) Even if we formalize a set otheory on the language $\mathcal{L}(\mathit{aa})$ with (o)–(v), Extensionality, Regularity, Infinity, Power Set and Separation (ranging over all formulas of $\mathcal{L}(\mathit{aa})$) as its own axioms, this set theory is a conservative extension of **ZF**. (See §4.) Let us continue our attitude further, A large class in the third sense is a class containing a $\in$-closed-unbounded class. (For the definition of $\in$-closed-unbounded class, see Definition 3.2, in §3.) This time, however, we encounter the essential difficulty if we want ot express this ides in the language $\mathcal{L}$ of **ZF**. For, we cannno avoid using the existential quantifier over classes to define large classes in the third sence. However, if we introduce a new qauntifier aa into the language $\mathcal{L}$ of **ZF** and choose certain statements for the axioms about aa which are intuitively true whenever $(\mathit{aa}\,x)\varphi$ means that the extension $\{\, x \mid \varphi(x) \,\}$ is large in the third sence, then we might have an interesting set theory on the new extended language $\mathcal{L}(\mathit{aa})$. By a sort of completeness theorem (Theorem 3.11), (o)–(ii) and ((v), together with

(vi) $(\forall x)(\mathit{aa}\,y)(x \subseteq y)$,

and

(vii) $(\forall x)(\mathit{aa}\, y)\varphi\ to (\mathit{aa}\, y)(\forall x \in y)\varphi$,

are appropriate to the axioms for aa in the intended meaning. (vii) saya that the collection of large classes in third sense if closed under diagonal intersections. As we shall see later, the theory on the language $\mathcal{L}(\mathit{aa})$ wiht (o)–(iii) as its own axioms is a conservative extension of the first-order theory with the following reflection scheme (*) as its own axioms:

$$(*) \qquad (\exists u)(u \text{ is transitive} \wedge (\forall x_1 \cdots x_m \in u) \bigwedge_{k=1}^{n} (\varphi_k \leftrightarrow \varphi_k^{(u)})),$$

where each φ_k is a formula of $\mathcal{L}$ with free variables among $x_1, \ldots, x_m$.

Since this reflection scheme is provable in **ZF**, the separation scheme ranging over all formulas of $\mathcal{L}(\mathit{aa})$, together with (vii), may have the decisive power to go beyond **ZF**. (The collection scheme ranging over all formuala of $\mathcal{L}(\mathit{aa})$ is deducible from (o) to (v).) Actually, we shall see later that theset theory on the language $\mathcal{L}(\mathit{aa})$ with (o)–(iii), (v)–(vii), Extensionality, Regularity, Power set and Separation Scheme (ranging over all formulas of $\mathcal{L}(\mathit{aa})$) as its axioms, which we shall call **ZF**(aa), is essentially stronger than **ZF**.

The set theory **ZF**(aa) started to be investigated by the present author and Matt Kaufmann independently in 1980. A part of the present author's work about **ZF**(aa) in 1980 was announced in [7].[1] This paper was rewritten by adding to some results to the present author's unpublished paper which had been prepared while visiting the Pennsylvania State University during the fall and winter term of 1981–82. The author would like to thank Thomas Jech for valuable comments on the subject of this work.

---

[1] For another aspect of **ZF**(aa), see the recent work of the present author in [8].

Part I

§1. Language with a generalized quantifier and their structures

In this section, we shall state some general facts about language with a generalized first-order quantifier and their structures.

Let $\mathcal{L}$ be a first-order language with equality. We form the language $\mathcal{L}(Q)$ by adding to $\mathcal{L}$ a new quantifier Q. Formulas of $\mathcal{L}(Q)$ are defined as usual except with the additional clause: if φ is a formula then so is $(Qx)\varphi$.

The logical axioms for $\mathcal{L}(Q)$ and the rules of inferences are given as follows:

Ax.0 All the usual first-order logical axiom schemes of formulas in $\mathcal{L}(Q)$, including the equality axioms;

Ax.1 $(Qx)\varphi(x) \leftrightarrow (Qy)\varphi(y)$, where $\varphi(x)$ is a formula of $\mathcal{L}(Q)$ in which y does not occur;

Ax.2 $(\forall x)(\varphi \leftrightarrow \psi) \rightarrow ((Qx)\varphi \leftrightarrow (Qx)\psi)$;

Inference Rules: Modus Ponens and Generalization.

Following standard usage, $\vdash \varphi$ means that φ is provable in $\mathcal{L}(Q)$. If Σ is a set of sentences of $\mathcal{L}(Q)$, then $\Sigma \vdash \varphi$ means that φ is deducible from Σ.

By a structure of $\mathcal{L}(Q)$, we mean a pair $(\mathcal{A}, q)$ of a structure $\mathcal{A}$ of $\mathcal{L}$ and a subset q of $P(A)$, where A is the universe of $\mathcal{A}$. If $(\mathcal{A}, q)$ is a structure of $\mathcal{L}(Q)$ and $\varphi(x_1 \dots x_n)$ is a formula of $\mathcal{L}(Q)$, we define $(\mathcal{A}, q) \models \varphi[a_1 \dots a_n]$ as usual by induction on the complexity of φ except with the following clause:

$(\mathcal{A}, q) \models \varphi[a_1 \dots a_n] \leftrightarrow \{\, a \in A \mid (\mathcal{A}, q) \models \psi[aa_1 \dots a_n] \,\} \in q$, where $\varphi(x_1 \dots x_n)$ is a formula of $\mathcal{L}(Q)$ of the form $(Qx)\psi(xx_1 \dots x_n)$, and $a_1, \dots, a_n \in A$.

Let Σ be a set of sentences of $\mathcal{L}(Q)$. We say that a structure $(\mathcal{A}, q)$ is a model of Σ if $(\mathcal{A}, q) \models \sigma$ for all $\sigma \in \Sigma$. By the standard Henkin style argument, we have the following lemma. (For the proof, see **[12]**.)

1.1 Lemma. *Let Σ be a set of sentences of $\mathcal{L}(Q)$. Σ is consistent iff Σ has a model. Moreover, if Σ is consistent and $\overline{\overline{\Sigma}} \leq \kappa$, then Σ has a model of power at most κ, where κ is an infinite cardinal.* $\dashv$

The following lemma can be proved in the same way as in first-order model theory.

1.2 Lemma. *Assume that $\mathcal{L}$ is countable. Let T be a consistent theory in $\mathcal{L}(Q)$, and $\Sigma(x)$ be a set of formulas with at most x free variable. If, for any formula $\varphi(x)$ of $\mathcal{L}(Q)$ consistent with T, there exists a $\sigma(x)$ in $\Sigma(x)$ such that $(\exists x)(\varphi(x) \wedge \neg\sigma(x))$ is consistent with T, then there exists a countable model $(\mathcal{A}, q)$ of T which omits Σ.* $\dashv$

Let $(\mathcal{A}, q)$ and $(\mathcal{B}, r)$ be structures of $\mathcal{L}(Q)$. We say that $(\mathcal{A}, q)$ is an elementary substructure of $(\mathcal{B}, r)$, in symbol, $(\mathcal{A}, q) \prec (\mathcal{B}, r)$ if for any formula $\varphi(x_1 \dots x_n)$ of $\mathcal{L}(Q)$ and any $a_1 \dots a_n$ in$A$,

$$(\mathcal{A}, q) \models \varphi[a_1 \dots a_n] \quad \text{iff} \quad (\mathcal{B}, r) \models \varphi[a_1 \dots a_n].$$

By an elementary chain of structure of $\mathcal{L}(Q)$, we mean a sequence $(\mathcal{A}_\alpha, q_\alpha)_{\alpha<\lambda}$ of structure of $\mathcal{L}(Q)$ such that $(\mathcal{A}_\alpha, q_\alpha) \prec (\mathcal{A}_\beta, q_\beta)$ for all $\alpha < \beta < \lambda$.

The union of an elementary chain $(\mathcal{A}_\alpha, q_\alpha)_{\alpha<\lambda}$ is the structure $(\mathcal{A}, q)$ such that:

(i) $\mathcal{A} = \bigcup_{\alpha<\lambda} \mathcal{A}_\alpha$, and
(ii) $q = \{ S \subseteq A \mid (\exists\beta < \lambda)(\forall\alpha < \lambda)(\beta \leq \alpha \to S \cap A_\alpha \in q_\alpha) \}$.

The proof of the following lemma can be found in [**12**, Lemma 2.5].

1.3 Lemma. *Let $(\mathcal{A}_\alpha, q_\alpha)_{\alpha<\lambda}$ be an elementary chain. Let $(\mathcal{A}, q) = \bigcup_{\alpha<\lambda}(\mathcal{A}_\alpha, q_\alpha)$. Then, for all $\alpha < \lambda$, $(\mathcal{A}_\alpha, q_\alpha) \prec (\mathcal{A}, q)$.*

1.4 Definition. The theory **FL** (relative to a language $\mathcal{L}$) is a theory formulated on the language $\mathcal{L}(Q)$, of which axioms are the universal closure of the following formulas:

FL1. $(\forall x)(\varphi \to \psi) \to ((Qx)\varphi \to (Qx)\psi)$;
FL2. $(Qx)\varphi \wedge (Qx)\psi \to (Qx)(\varphi \wedge \psi)$;
FL3. $\neg(Qx)(x \neq x)$;
FL4. $(Qx)(x = x)$.

Remark. Ax.0 for $\mathcal{L}(Q)$ is a consequence of **FL1**. Therefore, **Ax.2** is superfluous when we consider the theory **FL**.

A quantifier satisfying **FL1** to **4** is said to be a filter quantifier. Whenever we consider a filter quantifier, we shall use the symbol $\mathfrak{a}$, instead of Q.

We list formal consequences of **FL**.

1.5 Lemma. *The following are provable in* **FL**.

(i) $(\forall x)\varphi \to (\mathfrak{a}\, x)\varphi$.
(ii) $(\mathfrak{a}\, x)(\forall y)(\mathfrak{a}\, x)\varphi \to (\forall y)(\mathfrak{a}\, x)\varphi$.
(iii) $(\exists x)(\mathfrak{a}\, y)\varphi \to (\mathfrak{a}\, y)(\exists x)\varphi$.
(iv) $(\mathfrak{a}\, x)(\varphi \to \psi) \to ((\mathfrak{a}\, x)\varphi \to (\mathfrak{a}\, x)\psi)$.
(v) $\varphi \leftrightarrow (\mathfrak{a}\, x)\varphi$, *where x does not occur free in φ.*
(vi) $\varphi \vee (\mathfrak{a}\, x)\psi \leftrightarrow (\mathfrak{a}\, x)(\varphi \vee \psi)$.

Proof. For sample, we shall prove (v). The remainder will be left to the reader. By predicate logic,

$$\varphi \to (\forall x)\varphi$$

By (i) and propositional logic,

$$\varphi \to (\mathfrak{a}\, x)\varphi.$$

For the converse, by propositional logic,

$$\neg\varphi \to (\varphi \to x \neq x),$$

so by predicate logic,

$$\neg\varphi \to (\forall x)(\varphi \to x \neq x).$$

By **FL3** and propositional logic,

$$(aa\,x)\varphi \to \varphi. \dashv$$

Lemma 1.5 shows that a filter quantifier *aa* behaves like an universal quantifier. So we shall define the dual quantifier "*stat*" of *aa* which behaves like an existential quantifier on the other hand.

1.6 Definition. We define a quantifier *stat* by:

$$(stat\,x)\varphi \overset{\text{def}}{\leftrightarrow} \neg(aa\,x)\neg\varphi$$

1.7 Lemma. *The following are provable in* **FL**.

(o) $(\forall x)(\varphi \to \psi) \to ((stat\,x)\varphi \to ((stat\,x)\psi)$.

(i) $(stat\,x)\varphi \to (\exists x)\varphi$.

(ii) $(\exists x)(stat\,y)\varphi \to (stat\,y)(\exists x)\varphi$.

(iii) $(stat\,x)(\forall y)\varphi \to (\forall y)(stat\,x)\varphi$.

(iv) $(aa\,x)\varphi \wedge (stat\,x)\psi \to (stat\,x)(\varphi \wedge \psi)$.

(v) $\varphi \leftrightarrow (stat\,x)\varphi$, *where x does not occur free in φ.*

(vi) $\varphi \wedge (stat\,x)\psi \leftrightarrow (stat\,x)(\varphi \wedge \psi)$, *where x does not occur free in φ.*

(vii) $(stat\,x)\varphi \vee (stat\,x)\psi \leftrightarrow (stat\,x)(\varphi \vee \psi)$.

(viii) $(aa\,x)\varphi \to (stat\,x)\varphi$.

PROOF. (o) is just the restatement of **FL1**, and (i) to (ii) are also the restatement of (i) to (vi) in Lemma 1.5 respectively. For (viii), ,by (iv), we have

$$(aa\,x)\varphi \to (stat\,x)(\varphi \wedge x = x).$$

By **FL3** and propositional logic,

(1)
$$(aa\,x)\varphi \to (stat\,x)(\varphi \wedge x = x).$$

But we have

$$(\forall x)(\varphi \wedge x = x \to \varphi).$$

By (0),

(2)
$$(stat\,x)\varphi \wedge x = x) \to (stat\,x)\varphi.$$

By (1) and (2),

$$(aa\,x)\varphi \to (stat\,x)\varphi. \dashv$$

§2. Theories UB and ST on $\mathcal{L}(aa)$

From now on, we shall be concerned only with language containing a binary predicate symbol. We always assume that $\mathcal{L}$ is a first-order language with equality and a binary predicate symbol $\in$ (possibly with other non-logical symbols).

2.1 Definition. (1) The theory **UB** (relative to $\mathcal{L}$) is a theory formulated on the language $\mathcal{L}(aa)$, of which axioms are **FL1**, **FL2**, **FL3** and the universal closure of the following formulas:

UB1. $(\forall x)(aa\, y)(x \in y)$;

UB2. $(\forall x \in z)(aa\, y)\varphi \rightarrow (aa\, y)(\forall x \in z)\varphi$.

(2) The theory **ST** (relative to $\mathcal{L}$) is a theory formulated on the language $\mathcal{L}(aa)$, of which axioms are **FL1**, **FL2**, **FL3**, **UB1** and the universal closure of the following formulas:

ST1. $(\forall x)(aa\, y)(x \subseteq y)$;

ST2. (the axiom of diagonal intersection). $(\forall x)(aa\, y)\varphi \rightarrow (aa\,()\forall y)(\forall x \in y)\varphi$.

Since **FL4** is deducible from **FL1** and **UB1** in $\mathcal{L}(aa)$, we see that **UB** and **ST** are extensions of **FL**.

2.2 Lemma. *The following are provable in* **UB**.

(i) $(\forall x)(aa\, y)(x \subseteq y)$.

(ii) $(\forall x \in z)(aa\, y)\varphi \leftrightarrow (aa\, y)(\forall x \in z)\varphi$.

(iii) $(\exists y)(\forall y)(\varphi \rightarrow x \in y) \rightarrow (aa\, y)(\forall x)(\varphi \rightarrow x \in y)$, *where* y *does not occur free in* φ.

PROOF. (i): By **UB1**, we have

$$(\forall x \in x)(aa\, y))(z \in y).$$

By **UB2**, we have

$$(aa\, y)(\forall z \in x)(z \in y).$$

That is,

$$(aa\, y)(x \subseteq y).$$

(ii): It suffices to show that

$$(aa\, y)(\forall x \in z)\varphi \rightarrow (\forall x \in z)(aa\, y)\varphi.$$

So by (vi) in Lemma 1.5,

$$(aa\, y)(\forall x \in z)\varphi \rightarrow (\forall x \in z)(aa\, y)\varphi.$$

(ii): Let z be a new variable. Then, we have

$$(\forall z)[y \subseteq z \land (\forall x)(\varphi \rightarrow x \in y) \rightarrow (\forall x)(\varphi \rightarrow x \in z)],$$

so by **FL1**,

$$(aa\, z)(y \subseteq z \land (\forall x)(\varphi \rightarrow x \in y)) \rightarrow (aa\, z)(\forall x)(\varphi \rightarrow x \in z).$$

By **FL1**, (i) and (v) in Lemma 1.5,

$$(\forall x)(\varphi \rightarrow x \in y) \rightarrow ((\mathit{aa}\, z)(\forall x)(\varphi \rightarrow x \in z).$$

By predicate logic,

$$(\exists y)(\forall x)(\varphi \rightarrow x \in y) \rightarrow (\mathit{aa}\, z)(\forall x)(\varphi \rightarrow x \in z),$$

so by **A0** for $\mathcal{L}(\mathit{aa})$,

$$(\exists y)(\forall x)(\varphi \rightarrow x \in y) \rightarrow (\mathit{aa}\, y)(\forall x)(\varphi \rightarrow x \in y). \dashv$$

2.3 Definition. We define a quantifier "*set*" by:

$$(\mathit{set}\, x)\varphi \overset{\text{def}}{\leftrightarrow} (\exists y)(\forall x)(\varphi \rightarrow x \in y), \quad \text{where } y \text{ does not occur free in } \varphi.$$

2.4 Lemma. *The following are provable in* **UB**.

(i) $(\mathit{set}\, x)\varphi \leftrightarrow (\mathit{stat}\, y)(\forall x)(\varphi \rightarrow x \in y) \leftrightarrow (\mathit{aa}\, y)(\forall x)(\varphi \rightarrow x \in y)$, *where y does not occur free in φ.*

(ii) $(\forall x \in z)(\mathit{set}\, y)\varphi \rightarrow (\mathit{set}\, y)(\exists x \in z)\varphi$.

PROOF. (i): This is clear from (iii) in Lemma 2.2, (viii) and (i) in Lemma 1.7.
(ii): We have

$$(\forall w)((\forall x \in z)(\forall y)(\varphi \rightarrow y \in w) \rightarrow (\forall y)((\exists x \in z)\varphi \rightarrow y \in w)),$$

where w is a new variable. By **FL1**,

$$(\mathit{aa}\, w)(\forall x \in z)(\forall y)(\varphi \rightarrow y \in w) \rightarrow (\mathit{aa}\, w)(\forall y)((\exists x \in z)\varphi \rightarrow y \in w).$$

By **UB1**,

$$(\forall x \in z)(\mathit{aa}\, w)(\forall y)(\varphi \rightarrow y \in w) \rightarrow (\mathit{aa}\, w)(\forall y)((\exists x \in z)\varphi \rightarrow y \in w).$$

By (i),

$$(\forall x \in z)(\mathit{set}\, y)\varphi \rightarrow (\mathit{set}\, y)(\exists x \in z)\varphi. \dashv$$

2.5 Lemma. *Union, Pair and Collection Scheme (ranging over all formulas in $\mathcal{L}(\mathit{aa})$) are provable in* **UB**.

PROOF. For Union, we have

$$(\forall x \in z)(\mathit{set}\, y)(y \in x).$$

By (ii) in Lemma 2.4,

$$(\mathit{set}\, y)(\exists x \in z)(y \in x).$$

That is,

$$(\exists w)(\forall y)((\exists x \in z)(y \in x) \rightarrow y \in w).$$

For Pair, by **UB1**, we have

$$(\mathit{aa}\, z)(x \in z), \quad (\mathit{aa}\, z)(y \in z).$$

By **FL2**,

$$(\text{aa}\, z)(x \in z \wedge y \in z).$$

It follows that

$$(\exists z)(x \in z \wedge y \in z).$$

For Collection, we must prove

$$(\forall x \in z)(\exists y)\varphi \rightarrow (\exists w)(\forall x \in z)(\exists y \in w)\varphi,$$

where φ is a formula of $\mathcal{L}(\text{aa})$ in which w does not occur free.

By **UB1**, we have

$$\varphi \rightarrow \varphi \wedge (\text{aa}\, w)(y \in w).$$

By predicate logic,

$$(\exists y)\varphi \rightarrow (\exists y)(\varphi \wedge (\text{aa}\, w)(y \in w)).$$

Since w does not occur free in φ, we have

$$(\exists y)\varphi \rightarrow (\exists y)(\text{aa}\, w)(\varphi \wedge y \in w).$$

It follows (from (iii) in Lemma 1.5) that

$$(\exists y)\varphi \rightarrow (\text{aa}\, w)(\exists y)\varphi$$

By predicate logic,

$$(\forall x \in z)(\exists y)\varphi \rightarrow (\forall x \in z(\text{aa}\, w)(\exists y \in w)\varphi.$$

By **UB2**,

$$(\forall x \in z)(\exists y)\varphi \rightarrow (\text{aa}\, w)(\forall x \in z)(\exists y \in w)\varphi.$$

It follows that

$$(\forall x \in z)((\exists y)\varphi \rightarrow (\exists w)(\forall x \in z)(\exists y \in w)\varphi. \dashv$$

We can now show that **UB** is a conservative extension of the first-order theory $\mathbf{T}^{\mathbf{UB}}$ on $\mathcal{L}$ which is defined in the following definition.

2.6 Definition. $\mathbf{T}^{\mathbf{UB}}$ is a first-order theory on $\mathcal{L}$, of which axioms are Pair, Union and Collection Scheme (ranging over all formulas in $\mathcal{L}$).

By the immediate consequence of Lemma 2.5, we have the following lemma.

2.7 Lemma. *Let Σ be a set of sentences in $\mathcal{L}$, and φ be a formula of $\mathcal{L}$. If φ is deducible from Σ in the theory* $\mathbf{T^{UB}}$, *then φ is deducible from Σ in the theory* **UB**.

2.8 Definition. Let φ be a formula of $\mathcal{L}(aa)$. We define a formula φ^* of$\mathcal{L}$ as follows:
(1) If φ is an atomic formula, then φ^* is φ.
(2) $(\neg\varphi)^*$ is $\neg\varphi^*$.
(3) $(\varphi \wedge \psi)^*$ is $\varphi^* \wedge \psi^*$.
(4) $((\forall x)\varphi)^*$ is $(\forall x)\varphi^*$.
(5) $((aa\, x)\varphi)^*$ is $(\exists y)(\forall x)(y \subseteq x \to \varphi^*)$, where y does not occur in φ^*.

Let Σ be a set of formulas of $\mathcal{L}(aa)$. By Σ^*, we mean the set $\{\varphi^* \mid \varphi \in \Sigma\}$ of formulas of $\mathcal{L}$.

2.9 Lemma. *Let Σ be a set of sentences in $\mathcal{L}(aa)$, and φ be a formula of $\mathcal{L}(aa)$. If φ is deducible from Σ in* **UB**, *then φ^* is deducible from Σ^* in* $\mathbf{T^{UB}}$.

PROOF. The lemma is proved by induction on the length of given deduction from Σ to φ in **UB**. It is enough to show that φ^* is a theorem of $\mathbf{T^{UB}}$ if φ is one of the axioms of **UB**. For sample, we shall show that $\mathbf{T^{UB}} \vdash \varphi^*$ if φ is **UB2**. The remainder is left to the reader. Now, let φ be of the form

$$(\forall x \in z)(aa\, y)\psi \to (aa\, y)(\forall x \in z)\psi.$$

Then, φ^* is of the form

$$(\forall x \in z)(\exists u)(\forall y)(u \subseteq y \to \psi^*) \to (\exists v)(\forall y)(v \subseteq y \to (\forall x \in z)\psi^*).$$

To prove this, assume that

$$(\forall x \in z)(\exists u)(\forall y)(u \subseteq y \to \psi^*).$$

By Collection, for some w,

$$(\forall x \in z)(\exists u \in w)(\forall y)(u \subseteq y \to \psi^*).$$

By Union, there is a v such that

$$(\forall u \in w)(u \subseteq v).$$

Then, clearly,

$$(\forall y)(v \subseteq y \to (\forall x \in z)\psi^*). \dashv$$

2.10 Corollary. *Let Σ be a set of sentences of $\mathcal{L}(aa)$ such that $\Sigma^* \subseteq \Sigma$. Let φ be a formula of $\mathcal{L}$. Then, φ is deducible from Σ in* **UB** *iff φ is deducible from Σ^* in* $\mathbf{T^{UB}}$. $\dashv$

As an immediate consequence of Corollary 2.10, we have the following theorem.

2.11 Theorem. **UB** *is a conservative extension of* $\mathbf{T^{UB}}$.

Our next task is to study the relationship between **UB** and **ST**. First, we list some formal consequences of **ST**.

2.12 Lemma. *The following are provable in* **ST**.

(i) $(\forall x)(\mathsf{aa}\, y)\varphi \leftrightarrow (\mathsf{aa}\, y)(\forall x \in y)\varphi$.

(ii) $(\forall x \in z)(\mathsf{aa}\, y)\varphi \rightarrow (\mathsf{aa}\, y)(\forall x \in z)\varphi$.

(iii) $(Q_1 x_1)\ldots(Q_n x_n)\varphi \leftrightarrow (\mathsf{aa}\, z)(Q_1 x_1 \in z)\ldots(Q_n x_n \in z)\varphi$
$\leftrightarrow (\mathsf{stat}\, z)(Q_1 x_1 \in z)\ldots(Q_n x_n \in z)\varphi$,
where each Q_i is either $\exists$ or $\forall$ and φ is a formula of $\mathcal{L}(\mathsf{aa})$ in which z does not occur free.

(iv) $(\mathsf{aa}\, x)(x$ is transitive$)$.

(v) $(\exists z)(\forall y)(y \in z) \leftrightarrow (\mathsf{aa}\, x)(x \in x) \leftrightarrow (\mathsf{stat}\, x)(x \in x)$.

PROOF. (i): $(\forall x)(\mathsf{aa}\, y)\varphi \rightarrow (\mathsf{aa}\, y)(\forall x \in y)\varphi$ is **ST2** itself. For the converse, by predicate logic

$$(\forall x)(x \in y \rightarrow \varphi) \wedge x \in y \rightarrow \varphi.$$

By **FL1** and **FL2**,

$$(\mathsf{aa}\, y)(\forall x \in y)\varphi \wedge (\mathsf{aa}\, y)(x \in y) \rightarrow (\mathsf{aa}\, y)\varphi.$$

By **UB1**,

$$(\mathsf{aa}\, y)(\forall x \in y)\varphi \rightarrow (\mathsf{aa}\, y)\varphi.$$

By predicate logic,

$$(\mathsf{aa}\, y)(\forall x \in y)\varphi \rightarrow (\forall x)(\mathsf{aa}\, y)\varphi.$$

(ii): By (vi) in Lemma 1.7,

$$(x \in z \rightarrow (\mathsf{aa}\, y)\varphi) \rightarrow (\mathsf{aa}\, y)(x \in z \rightarrow \varphi).$$

By predicate logic,

$$(\forall x \in z)(\mathsf{aa}\, y)\varphi \rightarrow (\forall x)(\mathsf{aa}\, y)(x \in z \rightarrow \varphi).$$

By **ST2**,

$$(\forall x \in z)(\mathsf{aa}\, y)\varphi \rightarrow (\mathsf{aa}\, y)(\forall x \in y)(x \in z \rightarrow \varphi).$$

By **ST1**,

$$(\forall x \in z)(\mathsf{aa}\, y)\varphi \rightarrow (\mathsf{aa}\, y)(\forall x \in y)(x \in z \rightarrow \varphi) \wedge (\mathsf{aa}\, y)(z \subseteq y).$$

By **FL2**, we have

$$(\forall x \in z)(\mathsf{aa}\, y)\varphi \rightarrow (\mathsf{aa}\, y)((\forall x \in y)(x \in z \rightarrow \varphi) \wedge z \subseteq y). \qquad (1)$$

On the other hand, by predicate logic,

$$(\forall x \in y)(x \in z \rightarrow \varphi) \wedge z \subseteq y \rightarrow (\forall x \in z)\varphi,$$

so by **FL1**,

$$(\mathsf{aa}\, y)((\forall x \in y)(x \in z \rightarrow \varphi) \wedge z \subseteq y) \rightarrow (\mathsf{aa}\, y)(\forall x \in z)\varphi. \qquad (2)$$

By (1) and (2), we have

$$(\forall x \in z)(\mathsf{aa}\, y)\varphi \rightarrow (\mathsf{aa}\, y)(\forall x \in z)\varphi.$$

(iii): First, we prove

$$(\mathit{aa}\,z)(Q_1x_1 \in z)\dots(Q_nx_n \in z)\varphi \leftrightarrow (\mathit{stat}\,z)(Q_1x_1 \in z)\dots(Q_nx_n \in z)\varphi$$

by induction on n.
For $n = 0$, it is clear since z does not occur free in φ. Suppose that $n > 0$. It is enough to show the case that $Q_1 = \exists$. By (i),

$$(\mathit{stat}\,z)(\exists x_1 \in z)\dots(Q_nx_n \in z)\varphi \leftrightarrow (\exists x_1)(\mathit{stat}\,z)(Q_2x_2 \in z)\dots(Q_nx_n \in z)\varphi. \quad (1)$$

By the induction hypothesis,

$$(\exists x_1)(\mathit{stat}\,z)(Q_2x_2 \in z)\dots(Q_nx_n \in z)\varphi \leftrightarrow (\exists x_1)(\mathit{aa}\,z)(Q_2x_2 \in z)\dots(Q_nx_n \in z)\varphi.$$

By **UB1**,

$$(\exists x_1)(\mathit{stat}\,z)(Q_2x_2 \in z)\dots(Q_nx_n \in z)\varphi$$
$$\leftrightarrow (\exists x_1)(\mathit{aa}\,z)(x_1 \in z) \wedge ((\exists x_1)(\mathit{aa}\,z)(Q_2x_2 \in z \dots (Q_nx_n \in z)\varphi.$$

It follows that

$$(\exists x_1)(\mathit{stat}\,z)(Q_2x_2 \in z)\dots(Q_nx_n \in z)\varphi$$
$$\leftrightarrow (\exists x_1)(\mathit{aa}\,z)(x_1 \in z \wedge (Q_2x_2 \in z)\dots(Q_nx_n \in z)\varphi),$$

and so we have

$$(\exists x_1)(\mathit{stat}\,z)(Q_2x_2 \in z)\dots(Q_nx_n \in z)\varphi$$
$$\rightarrow (\mathit{aa}\,z)(\exists x_1 \in z)(Q_2x_2 \in z)\dots(Q_nx_n \in z)\varphi. \quad (2)$$

By (1) and (2), we have

$$(\mathit{stat}\,z)(\exists x_1 \in z)(Q_2x_2 \in z)\dots(Q_nx_n \in z)\varphi$$
$$\rightarrow (\mathit{aa}\,z)(\exists x_1 \in z)(Q_2x_2 \in z)\dots(Q_nx_n \in z)\varphi.$$

Now we prove

$$(Q_1x_1)\dots(Q_nx_n)\varphi \leftrightarrow (\mathit{aa}\,z)(Q_1x_1 \in z)\dots(Q_nx_n \in z)\varphi$$

by induction on n.
For $n = 0$, it is clear. Suppose that $n > 0$. By the former result it is enough to show the case that $Q_1 = \forall$. By (i), we have

$$(\mathit{aa}\,z)(\forall x_1 \in z)(Q_2x_2 \in z)\dots(Q_nx_n \in z)\varphi$$
$$\leftrightarrow (\forall x_1)(\mathit{aa}\,z)(Q_2x_2 \in z)\dots(Q_nx_n \in z)\varphi.$$

By the induction hypothesis, we have

$$(\mathit{aa}\,z)(\forall x_1 \in z)(Q_2x_2 \in z)\dots(Q_nx_n \in z)\varphi \leftrightarrow (\forall x_1)(Q_2x_2)\dots(Q_nx_n)\varphi.$$

(iv): By **ST1**, we have

$$(\forall y)(\mathit{aa}\,x)(y \subseteq x).$$

By using **ST2**, we have

$$(\mathit{aa}\,x)(\forall y \in x)(y \subseteq x).$$

That is,

$$(\mathit{aa}\,x)(x \text{ is transitive}).$$

(v): It is enough to show that

$$\text{(a) } (\mathit{stat}\,x)(x \in x) \to (\exists z)(\forall y)(y \in z),$$

and (b) $(\exists z)(\forall y)(y \in z) \to (\mathit{aa}\,x)(x \in x).$

For (a), by predicate logic,

$$x \in x \to (\exists z \in x)(\forall y \in x)(y \in z).$$

It follows that

$$(\mathit{stat}\,x)(x \in x) \to (\mathit{stat}\,x)(\exists z \in x)(\forall y \in x)(y \in z).$$

By using (iii), we have

$$(\mathit{stat}\,x)(x \in x) \to (\exists z)(\forall y)(y \in z).$$

For (b), by predicate logic,

$$(\forall y)(y \in x) \to x \in x.$$

It follows that

$$(\mathit{aa}\,x)(\forall y)(\in x) \to (\mathit{aa}\,x)(x \in x).$$

On the other hand, by (i) of Lemma 2.4,

$$(\exists z)(\forall y)(y \in z) \leftrightarrow (\mathit{aa}\,x)(\forall y)(y \in x).$$

Thus, we have

$$(\exists z)(\forall y)(y \in z) \to (\mathit{aa}\,x)(x \in x). \dashv$$

2.13 Lemma. **ST** *is an extension of* **UB**. *Moreover, if* $\mathbf{T}^{\mathbf{UB}}$ *is consistent with the following statement*

$$(\forall x)(\exists y)(x \subseteq y \wedge x \in y \wedge (\forall x \in y)(x \notin z)),$$

then **ST** *is a proper extension of* **UB**.

PROOF. From (ii) in Lemma 2.12, it is clear that **ST** is an extension of **UB**. Since $(\forall x)(\exists y)(x \in y)$ is provable in **UB**, *a fortiori* in **ST**,

$$(\mathit{aa}\,z)(\forall x \in z)(\exists y \in z)(x \in y)$$

is provable in **ST** by (iii) in Lemma 2.12. Suppose that

$$(\mathit{aa}\,z)(\forall x \in z)(\exists y \in z)(x \in y)$$

is provable in **UB**. Then, by Lemma 2.9,

$$((\mathit{aa}\,z)(\forall x \in z)(\exists y \in z)(x \in y))^*$$

is provable in $\mathbf{T}^{\mathbf{UB}}$, that is,

$$(\exists w)(\forall z)(w \subseteq z \to (\forall x \in z)(\exists y \in z)(x \in y))$$

is provable in $\mathbf{T^{UB}}$. Therefore,

$$(\forall w)(\exists z)(w \subseteq z \wedge (\exists x \in z)(\forall y \in z)(x \notin y))$$

is inconsistent with $\mathbf{T^{UB}}$. Since

$$(\forall x)(\exists y)(x \subseteq y \wedge x \in y \wedge (\forall z \in y)(x \notin z)) \\ \to (\forall w)(\exists z)(w \subseteq z \wedge (\exists x \in z)(\forall y \in z)(x \notin y)),$$

$\mathbf{T^{UB}}$ is consistent with

$$(\forall x)(\exists y)(x \subseteq y \wedge x \in y \wedge (\forall z \in y)(x \notin z)). \dashv$$

By the proof of Lemma 2.13, we can see that **ST** is still a proper extension of the theory obtained by adding to **UB** the set

$$\Sigma = \{\varphi \mid \varphi \text{ is a sentence of } \mathcal{L} \text{ such that } \mathbf{ST} \vdash \varphi\}.$$

However, we can have the analogous theorem of Theorem 2.11. Before stating this, we shall state *Reflection Principle* in **ST**.

2.14 Definition. Let φ be a formula of $\mathcal{L}(aa)$, and u be a variable not occurring in φ. The relativization $\varphi^{(u)}$ of φ to u is defined by induction on the complexity of φ as follows:

(1) If φ is an atomic formula, then $\varphi^{(u)}$ is φ.
(2) $(\neg\varphi)^{(u)}$ is $\neg\varphi^{(u)}$.
(3) $(\varphi \wedge \chi)^{(u)}$ is $\varphi^{(u)} \wedge \chi^{(u)}$.
(4) $((\forall x)\varphi)^{(u)}$ is $(\forall x \in u)\varphi^{(u)}$.
(5) $((aa\, x)\varphi)^{(u)}$ is $(aa\, x)\varphi$.

2.15 Lemma. *The following are provable in* **ST**.

(i) $\varphi \leftrightarrow (aa\, u)\varphi^{(u)} \leftrightarrow (stat\, u)\varphi^{(u)}$.
(ii) $(\exists u)(u \textit{ is transitive} \wedge (\forall x_1 \ldots x_n \in u) \bigwedge_{i=1}^{l}(\varphi_i \leftrightarrow \varphi_i^{(u)}))$, *where* $\varphi_1, \ldots, \varphi_l$ *are formulas of* $\mathcal{L}(aa)$ *with free variables among* $x_1, \ldots, x_n$.

PROOF. For (i), we can prove it just like as in (ii) of Lemma 2.12 by induction on the complexity of φ. For (ii), by (i), we have, for each i $(i = 1, \ldots, l)$,

$$(aa\, u)(\varphi_i \leftrightarrow \varphi_i^{(u)}).$$

It follows from **FL2** that

$$(aa\, u) \bigwedge_{i=1}^{l} (\varphi_i \leftrightarrow \varphi_i^{(u)}),$$

and so,

$$(\forall x_1 \ldots x_n)(aa\, u) \bigwedge_{i=1}^{l} (\varphi_i \leftrightarrow \varphi_i^{(u)}).$$

By **ST2**,

$$(\text{aa}\, u)(\forall x_1 \ldots x_n \in u) \bigwedge_{i=1}^{l} (\varphi_i \leftrightarrow \varphi_i^{(u)}).$$

By (iV) of Lemma 2.12 and **FL2**,

$$(\text{aa}\, u)(u \text{ is transitive} \wedge (\forall x_1 \ldots x_n \in u) \bigwedge_{i=1}^{l} (\varphi_i \leftrightarrow \varphi_i^{(u)}).$$

By (iV) of Lemma 2.12 and **FL2**,

$$(\text{aa}\, u)(u \text{ is transitive} \wedge (\forall x_1 \ldots x_n \in u) \bigwedge_{i=1}^{l} (\varphi_i \leftrightarrow \varphi_i^{(u)}).$$

It follows that

$$(\exists u)(u \text{ is transitive} \wedge (\forall x_1 \ldots x_n \in u) \bigwedge_{i=1}^{l} (\varphi_i \leftrightarrow \varphi_i^{(u)}). \dashv$$

2.16 Definition. The theory $\mathbf{T}^{\mathbf{ST}}$ (relative to $\mathcal{L}$) is a first-order theory on $\mathcal{L}$, of which axioms are the following reflection scheme:

$$(\exists u)(u \text{ is transitive} \wedge (\forall x_1 \ldots x_n \in u) \bigwedge_{i=1}^{l} (\varphi_i \leftrightarrow \varphi_i^{(u)}),$$

where each φ_i is a formula of $\mathcal{L}$ with free variables among $x_1, \ldots, x_n$.

A model-theoretic proof of the following theorem will be given in the end of §3. At present, we do not know any finitary proof for the theorem.

2.17 Theorem. **ST** *is a conservative extension of* $\mathbf{T}^{\mathbf{ST}}$.

We close this section for giving a lemma which will be frequently used latter.

2.18 Lemma. *Let $\varphi(x)$ and $\theta(x, y)$ be formulas of $\mathcal{L}(\text{aa})$ in which z does nit occur. Then, the following are provable in* **ST**:

$$(\forall x)\exists y(\theta(x,y) \wedge \varphi(y)) \wedge (\text{aa}\, z)(((\forall x \in z)(\exists y \in z)(\theta(x,y) \wedge \varphi(y)) \rightarrow \varphi(z)) \rightarrow (\text{aa}\, z)\varphi(z).$$

PROOF. By (iv) of Lemma 1.5, we have

$$(\text{aa}\, z)((\forall x \in z)(\exists y \in z)(\theta(x,y) \wedge \varphi(y)) \rightarrow \varphi(z)) \rightarrow ((\text{aa}\, z)(\forall x \in z)(\exists y \in z)(\theta(x,y) \wedge \varphi(y)) \rightarrow (\text{aa}\, z)\varphi(z)). \quad (1)$$

By (iii) of Lemma 2.12,

$$(\forall x)(\exists y)(\theta(x,y) \wedge \varphi(y)) \rightarrow (\text{aa}\, z)(\forall x \in z)(\exists y \in z)(\theta(x,y) \wedge \varphi(y)). \quad (2)$$

Thus, by (1) and (2),

$$(\forall x)(\exists y)(\theta(x,y)\wedge\varphi(y))\wedge(\mathrm{aa}\,z)((\forall x\in z)(\exists y\in z)(\theta(x,y)\wedge\varphi(y))\rightarrow\varphi(z))\rightarrow(\mathrm{aa}\,z)\varphi(z).\ \dashv$$

§3. Standard models of ST

The aim of this section is to get a sort of completeness theorem for **ST**. For this purpose, we must first see what a kind model of **ST** should be standard.

3.1 Definition. Let A be a non-empty set, and $\leq$ be a partial order on A. Let C be a subset of A. C is said to be ***unbounded with respect to*** $\leq$ if for every $x \in A$ there exists a $y \in C$ such that $x \leq y$. C is said to be ***closed with respect to*** $\leq$ if whenever $y = supD$, $D \subseteq C$ and D is linearly ordered by $\leq$, then $y \in C$.

Let A be a non-empty set and E be a binary relation on A. The ***E-extension*** a_E of element a of A is the set $\{b \in A \mid b\,E\,a\}$. We write $a \subseteq_E b$ if $a_E \subseteq b_E$. E is said to be ***extensional*** if $a_E = b_E$ implies $a = b$ for all $a, b \in A$. We should note that the relation $\subseteq_E$ on A is a partial order on A if E is extensional on A.

3.2 Definition. Let A be a non-empty set, and E be an extensional binary relation on A. Let C be a subset of A. C is said to be ***E-unbounded*** if C is unbounded with respect to $\subseteq_E$. C is said to be ***E-closed*** if whenever $b_E = \bigcup\{a_E \mid a \in D\}$, $D \subseteq C$ and D is linearly ordered by $\subseteq_E$, then $b \in C$. If C is E-closed and E-unbounded, then we say that C is ***closed unbounded*** or shortly ***E-cub***. We denote by $cub_{(A;E)}$ the set $\{X \subseteq A \mid X$ contains an E-cub subset of $A\}$.

3.3 Definition. Let A be a non-empty set, and E be a binary relation on A. A proper filter $\mathcal{F}$ over A is said to be ***E-normal*** if $\mathcal{F}$ satisfies the following conditions.

(i) $\{b \in A \mid a\,E\,b\}$ belongs to $\mathcal{F}$ for each $a \in A$.

(ii) $\{b \in A \mid a \subseteq_E b\}$ belongs to $\mathcal{F}$ for each $a \in A$.

(iii) Let $\langle X_a : a \in A\rangle$ be such that $X_a \in \mathcal{F}$ for each $a \in A$. Then, the ***diagonal intersection*** $\Delta X_a = \{b \in A \mid (\forall a)(a\,E\,b \rightarrow b \in X_a)\}$ belongs to $\mathcal{F}$.

In general, $cub_{(A;E)}$ is not an E-normal filter. However, under certain conditions, $cub_{(A;E)}$ can be an E-normal filter. The following are typical examples that $cub_{(A;E)}$ are E-normal filters.

EXAMPLE 1. Let κ be a regular uncountable cardinal. Then, $cub_{(\kappa;<)}$ is a $<$-normal filter. In fact, $cub_{(\kappa;<)}$ is the normal filter generated by closed unbounded subset of κ in the usual sense.

EXAMPLE 2. Let κ be a regular uncountable cardinal. Let H_κ be the set of all sets hereditrarily of cardinality less than κ. Then, $cub_{(H_\kappa;\in)}$ is an $\in$-normal filter.

Examples 1 and 2 suggest to us that there are two ways to naturally make $cub_{(A;E)}$ an E-normal filter. First, we shall consider the method containing Example 1 as a special case.

3.4 Definition. Let A be a non-empty set, and E be a binary relation on A. Let κ be an infinite cardinal. A map $e : \kappa \to A$ is said to be a *κ-cumulation of* $(A; E)$ if e satisfies the following conditions.

(i) $\overline{\overline{(e(\alpha)_E)}} < \kappa$ for all $\alpha < \kappa$.
(ii) $\alpha < \beta < \kappa$ implies $e(\alpha)\ E\ e(\beta)$.
(iii) Each $e(\alpha)$ is an E-segment of A, that is, $a\ E\ e(\alpha)$ and $b\ E\ a$ imply $b\ E\ e(\alpha)$.
(iv) e is continuous, that is, $e(\mu)_E = \bigcup_{\alpha<\mu} e(\alpha)_E$ for a limit ordinal $\mu < \kappa$.
(v) For any element a of A, there exists an $\alpha < \kappa$ such that $a\ E\ e(\alpha)$.

If κ is a regular uncountable cardinal, and e is a κ-cumulation of $(A; E)$, we denote by $\mathcal{F}_e$ the set $\{ X \subseteq A \mid e^{-1}(X)$ contains a closed unbounded subset of $\kappa \}$.

3.5 Lemma. *Let κ be a regular uncountable cardinal, and e be a κ-cumulation of $(A; E)$. Then, $\mathcal{F}_e$ is an E-normal filter over A. Moreover, if E is extensional on A, then $\mathcal{F}_e = cub_{(A;E)}$ for any κ-cumulation e of $(A; E)$.*

PROOF. It is clear that $\mathcal{F}_e$ is a κ-complete filter such that $\{ b \in A \mid a\ E\ b \}$ and $\{ b \in A \mid a \subseteq_E b \}$ belong to $\mathcal{F}_e$ for all $a \in A$. To see that $\mathcal{F}_e$ is closed under diagonal intersections, let $\langle X_a : a \in A \rangle$ be such that $X_a \in \mathcal{F}_e$ for all $a \in A$. For each $a \in A$, choose a closed unbounded subset K_a of κ so that $K_a \subseteq e^{-1}(X_a)$. By using the fact that $\mathcal{F}_e$ is κ-complete and that $\overline{\overline{(e(\alpha)_E)}} < \kappa$ for any $\alpha < \kappa$, we can see that $K = \{ \alpha < \kappa \mid \alpha \in \bigcap_{aEe(\alpha)} K_a \}$ is closed unbounded in κ. It is clear that $K \subseteq e^{-1}(\Delta X_a)$. For the latter part of the lemma, we should note that the following remarks holds under the assumption that E is extensional on A:

(1) If K is a closed unbounded subset of κ, then $e''K$ is E-cub.
(2) If C is an E-cub subset of A, then $e^{-1}(C)$ is closed unbounded in κ.

From these remarks, we can easily see that $\mathcal{F}_e = cub_{(A;E)}$ for any κ-cumulation e of $(A; E)$, providing that E is extensional.

3.6 Definition. A subset K of $P_\kappa(A)$ is said to be *closed unbounded* if K is closed unbounded with respect to $\subseteq$.

It is well-known that the filter generated by closed unbounded subsets of $P_\kappa(A)$ is a κ-complete normal filter. (See [5].)

3.7 Definition. Let A be a non-empty set, and E be a binary relation on A. Let κ be a regular uncountable cardinal. A subset D of A is said to be a *κ-base of* $(A; E)$ if it satisfies the following conditions.

(i) $\overline{\overline{(a_E)}} < \kappa$ for every $a \in D$.
(ii) $\{ a_E \mid a \in D \}$ is a closed unbounded subset of $P_\kappa(A)$.
(iii) For every $b \in A$, there exists an $a \in D$ such that $b \subseteq_E a$.

If D is a κ-base of $(A; E)$, we denote by $\mathcal{F}_D$ the set $\{ X \subseteq A \mid \{ a \in D \mid a_E \in H \} \subseteq X$ for some closed unbounded subset H of $P_\kappa(A) \}$.

3.8 Lemma. *Let κ be a regular uncountable cardinal, and D be a κ-base of $(A;E)$. then, $\mathcal{F}_D$ is an E-normal filter over A. Moreover, if E is extensional on A, then $\mathcal{F}_D = cub_{(A;E)}$ for any κ-base D of $(A;E)$.* ⊣

The next lemma clarifies the connection between κ-cumulations and κ-bases of $(A;E)$.

3.9 Lemma. *Let κ be a regular uncountable cardinal, and e be a κ-cumulation of $(A;E)$. Then, $\{e(\alpha) \mid \alpha < \kappa\}$ is a κ-base of $(A;E)$.*

PROOF. It is enough to show that $\{e(\alpha)_E \mid \alpha < \kappa\}$ is closed unbounded in $P_\kappa(A)$. Let $S \subseteq \{e(\alpha)_E \mid \alpha < \kappa\}$ be such that S is linearly ordered by $\subseteq$ and $x = \bigcup S \in P_\kappa(A)$. For each $a \in x$, let $\beta_a < \kappa$ be the least ordinal such that $a \mathrel{E} e(\beta_a)$ and $e(\beta_a)_E \in S$. Clearly, $x = \bigcup\{e(\beta_a)_E \mid a \in x\}$. Since $\overline{\overline{x}} < \kappa$, $\alpha = \sup_{a\in x} \beta_a < \kappa$. Since e is continuous, $e(\alpha)_E = \bigcup\{e(\beta_a)_E \mid a \in x\}$. It follows that $x = e(\alpha)_E$, that is, $\bigcup S \in \{e(\alpha)_E \mid \alpha < \kappa\}$. Thus, $\{e(\alpha)_E \mid \alpha < \kappa\}$ is closed. To prove that $\{e(\alpha) \mid \alpha < \kappa\}$ is unbounded in $P_\kappa(A)$, let $x \in P_\kappa(A)$ be arbitrary. For each $a \in x$, let β_a be the least ordinal such that $a \mathrel{E} e(\beta_a)$. Let $\alpha = \sup_{a\in x} \beta_a$. By the continuity of e, $e(\alpha)_E = \bigcup_{a\in x} e(\beta_a)_E$. It follows that $e(\alpha)_E \supseteq x$. Thus, $\{e(\alpha) \mid \alpha < \kappa\}$. Thus, $\{e(\alpha) \mid \alpha < \kappa\}$ is unbounded in $P_\kappa(A)$. ⊣

3.10 Definition. Let $\mathcal{L}$ be a first-order language with at least one binary predicate symbol $\in$. A structure $\mathcal{A} = (A; E, \ldots)$ of $\mathcal{L}$ is said to be a *standard model of* **ST** if E is extensional on A and $cub_{(A;E)}$ is an E-normal filter over A.

If $\mathcal{A}$ is a standard model of **ST**, it is clear that $(\mathcal{A}, cub_{(A;E)})$ is a model of **ST**. If $\varphi(x_1 \ldots x_n)$ is a formula of $\mathcal{L}(aa)$, we shall write $\mathcal{A} \models \varphi[a_1 \ldots a_n]$ instead of $(\mathcal{A}, cub_{(A;E)}) \models \varphi[a_1 \ldots a_n]$ whenever $\mathcal{A}$ is standard.

Now, our completeness theorem for **ST** can be described as the following theorem.

3.11 Theorem. *Let $\mathcal{L}$ be a countable language with at least one binary predicate symbol $\in$. Let Σ be a set of sentences of $\mathcal{L}(aa)$. Σ has a standard model iff Σ is consistent with* **ST** *and the axiom of extensionality.*

Theorem 3.11 is an immediate consequence of the following lemma.

3.12 Lemma. *Let $\mathcal{L}$ be as in Theorem 3.11. Let $(\mathcal{A}, q)$ be a countable model of* **ST**. *Then, there exist a model $\mathcal{B} = (B; F, \ldots)$ of $\mathcal{L}$ and an ω_1-cumulation e such that $(\mathcal{A}, q) \prec (\mathcal{B}, \mathcal{F}_e)$.*

Lemma 3.12 can be proved just like as the completeness theorem for $\mathcal{L}(Q)$ in [12] and the completeness theorem for the stationary logic in [1]. We shall here give a brief sketch of the proof. To the end of the proof of Lemma 3.12, we assume that $\mathcal{L}$ is a countable language with at least one binary predicate symbol $\in$.

3.13 Definition. Let $\mathcal{A} = (A; E, \ldots)$ and $\mathcal{B} = (B; F, \ldots)$ be structures of $\mathcal{L}$ such that $\mathcal{A} \subseteq \mathcal{B}$.

(i) $\mathcal{B}$ is said to be an end extension of $\mathcal{A}$ if $b\ F\ a$ implies $b \in A$ whenever $a \in A$ and $b \in B$,.

(ii) $\mathcal{B}$ is said to be a blunt end extension of $\mathcal{A}$ if $\mathcal{B}$ is an end extension of $\mathcal{A}$ and $A = b_F$ for some $b \in B$.

Let $\mathcal{A}$ be a structure of $\mathcal{L}$. By $\mathcal{L}^*$, we denote the language obtained by adding a new constant $\boldsymbol{a}$ for each $a \in A$. We also denote by $\mathcal{A}^* = (\mathcal{A}, a)_{a \in A}$ the corresponding expansion of $\mathcal{A}$.

As in [12], the following lemma is a key for the proof of 3.12.

3.14 Lemma. *Let $(\mathcal{A}, q)$ be a countable model of* **ST**, *and $\varphi(c)$ be a formula of $\mathcal{L}^*(aa)$ (with at most one free variable x) such that $(\mathcal{A}^*, q) \models (\mathit{stat}\, x)\varphi$. Then, there exists a countable structure $(\mathcal{B}, r)$ and an element $b \in B$ such that:*

(i) $(\mathcal{A}, q) \prec (\mathcal{B}, r)$.

(ii) $\mathcal{B} = (B; F, \ldots)$ *is an end extension of* $\mathcal{A} = (A; E, \ldots)$.

(iii) $(\mathcal{B}^*, r) \models \varphi[b]$.

(iv) *For any formula $\psi(y)$ of $\mathcal{L}^*(aa)$ (with at most one free variable y) such that* $(\mathcal{A}^*, q) \models (aa\, y)\psi(y)$, $(\mathcal{B}^*, r) \models \psi[b]$.

(v) $A = \{ a \in B \mid a\ F\ b \}$.

PROOF. Let $\mathcal{L}'$ be the language obtained by adding a new constant c to $\mathcal{L}^*$. Let Γ be the following set of sentences of $\mathcal{L}'(aa)$:

(a) All sentences of $\mathcal{L}^*(aa)$ which are true in $(\mathcal{A}^*, q)$,

(b) $\varphi(c)$,

(c) $\psi(c)$ for each formula $\psi(y)$ of $\mathcal{L}^*(aa)$ such that $(\mathcal{A}^*, q) \models ((aa\, y)\psi(y)$,

(d) $\boldsymbol{a} \in c$ for each $a \in A$.

We can prove the following claim as in [12, p.19, Lemma 2.7].

CLAIM 1. Let $\theta(y)$ be a formula of $\mathcal{L}^*(aa)$ with at most one free variable y. Then, $\theta(c)$ is consistent with Γ iff $(\mathcal{A}^*, q) \models (\mathit{stat}\, u)(\theta(u) \wedge \varphi(u))$.

By Claim 1, it is clear that Γ is consistent. (Take $y = Y$ for $\theta(y)$.) Now, let $\Sigma(x) = \{ x \in c \} \cup \{ x \neq \boldsymbol{a} \mid a \in A \}$. By using the axiom of diagonal intersections and Claim 1, we can prove the following claim.

CLAIM 2. Let $\theta(xc)$ be a formula of $\mathcal{L}'(aa)$ such that $(\exists x)\theta(xc)$ is consistent with Γ. Then, there exists a $\sigma(x) \in \Sigma(x)$ such that $(\exists x)(\theta \wedge \neg\sigma)$ is consistent with Γ.

By Claim 2 and Lemma 1.2, Γ has a countable model which omits $\Sigma(x)$. Let $(\mathcal{B}', r)$ be such a model. We may assume that for each $a \in A$ the constant $\boldsymbol{a}$ is interpreted in $\mathcal{B}'$ by the element a. Let b be the elements of the universe of $\mathcal{B}'$ which is the interpretation of c in $\mathcal{B}'$. Let $\mathcal{B}$ be the reduct of $\mathcal{B}'$ to $\mathcal{L}$. Noting that $\mathbf{ST} \vdash (aa\, y)(\forall z \in y)(z \subseteq y)$, and so $(\mathcal{B}', r) \models (\forall z \in c)(z \subseteq c)$, we can see that $(\mathcal{B}, r)$ is a required model.

3.15 Proof of Lemma 3.12. Let $(\mathcal{A}, q)$ be a countable model of **ST**. Let C be a set of new constants with $\overline{\overline{C}} = \omega_1$. Let $\mathcal{L}'$ be the language obtained by adding C to $\mathcal{L}$. Let

$(\varphi_\zeta)_{\zeta<\omega_1}$ be an enumeration of all sentences of $\mathcal{L}'(aa)$ with form $(stat\,x)\psi$. Let $(S_\zeta)_{\zeta<\omega_1}$ be a partition of ω_1 such that each S_ζ is a stationary subset of ω_1. We define an elementary chain $(\mathcal{A}_\alpha, q_\alpha)_{\alpha<\omega_1}$ of countable models, a monotone sequence $(f_\alpha)_{\alpha<\omega_1}$ of injections, $f_\alpha : A_\alpha \to C$, and a sequence $(a_\alpha)_{\alpha<\omega_1}$, $a_\alpha \in A_{\alpha+1}$ as follows:

(1) $(\mathcal{A}_0, q_0) = (\mathcal{A}, q)$. Choose an injection $f_0 : A_0 \to C$, and we consider $f_0(a)$ as a name of a for each $a \in A_0$.

(2) $(\mathcal{A}_\alpha, q_\alpha) = \bigcup_{\beta<\alpha}(\mathcal{A}_\beta, q_\beta)$ for a limit ordinal α. Set $f_\alpha = \bigcup_{\beta<\alpha} f_\alpha$, and we consider $f_\alpha(a)$ as a name of a for each $a \in A_\alpha$.

(3) Let ζ be the unique ζ such that $\alpha \in S_\zeta$. Let φ_ζ be $(stat\,x)\psi(x)$. We consider two cases.

Case 1. φ_ζ is a sentence of the language of $\mathcal{A}^*_\alpha$ and $(\mathcal{A}^*_\alpha, q_\alpha) \models \varphi_\zeta$:

Then, we apply Lemma 3.14. We can choose a countable model $(\mathcal{A}_{\alpha+1}, q_{\alpha+1})$ and $a_\alpha \in A_{\alpha+1}$ so that:

(i) $(\mathcal{A}_\alpha, q_\alpha) \prec (\mathcal{A}_{\alpha+1}, q_{\alpha+1})$.

(ii) $\mathcal{A}_{\alpha+1} = (A_{\alpha+1}; E_{\alpha+1}, \ldots)$ is an end extension of $\mathcal{A}_\alpha = (A_\alpha; E_\alpha, \ldots)$.

(iii) $(\mathcal{A}^*_{\alpha+1}, q_{\alpha+1}) \models \psi_\zeta[a_\alpha]$.

(iv) For any formula $\varphi(y)$ of the language of $\mathcal{A}^*_\alpha$ such that
$$(\mathcal{A}^*_\alpha, q_\alpha) \models (aa\,y)\varphi(y) \text{ and } (\mathcal{A}^*_{\alpha+1}, q_{\alpha+1}) \models \varphi[a_\alpha].$$

(v) $A_\alpha = \{\, a \in A_{\alpha+1} \mid a\ E_{\alpha+1}\ a_0 \,\}$.

Choose an injection $f_{\alpha+1}$ is an extension of f_α, and we consider $f_{\alpha+1}(a)$ as a name of a for each $a \in A_{\alpha+1}$.

Case 2. Otherwise. We choose a countable model $(\mathcal{A}_{\alpha+1}, q_{\alpha+1})$ and an $a_\alpha \in A_{\alpha+1}$ satisfying (i), (ii), (iv) and (v) in the case 1. $f_{\alpha+1} : A_{\alpha+1} \to C$ is chosen as in the case 1.

Now, we set $(\mathcal{B}, r) = \bigcup_{\alpha<\omega_1}(\mathcal{A}_\alpha, q_\alpha)$. If we define $e : \omega_1 \to B$ so that $e(\alpha) = a_\alpha$ for $\alpha < \omega_1$, then, clearly, e is an ω_1-cumulation of $(B; F)$, where $calB = (B; F, \ldots)$. As in [1], we can prove that, for any formula $\varphi(x_1 \ldots x_n)$ of $\mathcal{L}(aa)$, and any $b_1 \ldots b_n \in B$,

$$(\mathcal{B}, r) \models \varphi[b_1 \ldots b_n] \quad \text{iff} \quad (\mathcal{B}, \mathcal{F}_e) \models \varphi[b_1 \ldots b_n].$$

Thus, $(\mathcal{A}, q) \prec (\mathcal{B}, \mathcal{F}_e)$. ⊣

3.16 Proof of Theorem 3.11. It is clear that Σ is consistent with **ST** and the axiom of extensionality if Σ is consistent with **ST** and the axiom of extensionality. then, by Lemma 1.1, there is a countable model $(\mathcal{A}, q)$ of Σ, **ST** and the axiom of extensionality. By Lemma 3.12, there exist a model $\mathcal{B}$ and an ω_1-cumulation e of $(B; F)$ such that $(\mathcal{A}, q) \prec (\mathcal{B}, \mathcal{F}_e)$. Since the axiom of extensionality holds in $\mathcal{A}$, F is extensional on B. It follows from 3.5 that $\mathcal{F}_e = cub_{(B;F)}$. Thus, $\mathcal{B}$ is a standard model of Σ. ⊣

The model $\mathcal{B} = (B; F, \ldots)$ constructed in the proof of Theorem 3.11 has much more information than required in the theorem. $(B; F)$ has an ω_1-base $D = \{\, a_\alpha \mid \alpha < \kappa \,\}$. Since F is extensional on B, $\mathcal{F}_D = cub_{(B;F)}$ (by Lemma 3.8). This fact may clarify the connection between **ST** and the stationary logic of Barwise-Kaufmann-Makkai. Now, let $\mathcal{L}$ be a first-order language with at least one binary predicate symbol $\in$ ($\mathcal{L}$ is not necessarily countable). We consider the stationary logic $\mathcal{L}$(**aa**) relative to $\mathcal{L}$. (We use a symbol **aa** for the "for almost all" second-order quantifier in the stationary logic). In $\mathcal{L}(\mathbf{aa})$, we define a new first-order quantifier aa as follows:

$$(\mathit{aa}\, x)\varphi \leftrightarrow (\mathbf{aa}s)(\exists x)((\forall y)(y \in x \leftrightarrow s(y)) \wedge \varphi).$$

Then, we can have the following theorem:

3.17 Theorem. *Let φ be a formula of Laa. Then, φ is deducible from the axiom of extensionality in* **ST** *iff φ is deducible from the axiom of extensionality and $(\mathbf{aa}s)(\exists x)(\forall y)(y \in x \leftrightarrow s(y))$ in the stationary logic $\mathcal{L}(\mathbf{aa})$.*

PROOF. By the definition of aa in $\mathcal{L}(\mathbf{aa})$, it is clear that all axioms of **ST** are deducible from the axiom of extensionality and $(\mathbf{aa}s)(\exists x)(\forall y)(y \in x \leftrightarrow s(y))$ in $\mathcal{L}(\mathbf{aa})$. It follows that φ is deducible from the axiom of extensionality and $(\mathbf{aa}s)(\exists x)(\forall y)(y \in x \leftrightarrow s(y))$ in $\mathcal{L}\mathbf{aa}$) if φ is deducible from **ST** and the axiom of extensionality. For the converse, we suppose that there is a formula φ of $\mathcal{L}(\mathit{aa})$ such that φ is deducible from the axiom of extensionality and $(\mathbf{aa}s)(\exists x)(\forall y)(y \in x \leftrightarrow s(y))$ in $\mathcal{L}(\mathbf{aa})$, but φ is not deducible from the axiom of extensionality in **ST**. Let σ be the universal closure of φ. Let $\mathcal{L}_0$ be the countable sublanguage of $\mathcal{L}$ such that all nono-logical symbols appearing in the deduction of σ from the axiom of extensionality and $(\mathbf{aa}s)(\exists x)(\forall y)(y \in x \leftrightarrow s(y))$ appear in $\mathcal{L}_0$. So σ is deducible from the axiom of extensionality and **ST**(relative to $\mathcal{L}_0$). By the proof of Theorem 3.17, there is a standard model $\mathcal{B}$ of σ such that $(B; F)$ has an ω_1-base D. Since $\{\, b_F \in P_{\omega_1}(B) \mid b \in D \,\}$ is closed unbounded subset of $P_{\omega_1}(B)$, $(\mathbf{aa}s)(\exists x)(\forall y)(y \in x \leftrightarrow s(y))$ holds in $(\mathcal{B}, cub_{P_{\omega_1}(B)})$, where $cub_{P_{\omega_1}(B)}$ is the filter generated by closed unbounded subsets of $P_{\omega_1}(B)$. Thus, $(\mathcal{B}, cub_{P_{\omega_1}(B)}) = \sigma$. Since $(\mathcal{B}, cub_{(B;F)}) \models \sigma$, we reach the contradiction if we prove the following:
For any formula $\psi(x_1 \dots x_n)$ of $\mathcal{L}(\mathit{aa})$ and any $b_1, \dots, b_n \in B$,

$$(\mathcal{B}, cub_{P_{\omega_1}(B)}) = \psi[b_1 \dots b_n] \quad \text{iff} \quad (\mathcal{B}, cub_{(B;F)}) \models \psi[b_1 \dots b_n].$$

This can be easily proved by induction on the complexity of $\psi(x_1 \dots x_n)$.

In general, a κ-cumulation e of $(A; E)$ does not give an isomorphism between $(\kappa; <)$ and $(\{\, e(\alpha) \mid \alpha < \kappa \,\}; E)$. Thus, we have the following definition.

3.18 Definition. A κ-cumulation e of $(A; E)$ is faithful if e gives an isomorphism between $(\kappa; <)$ and $(\{\, e(\alpha) \mid \alpha < \kappa \,\}; E)$, in other words, $\{\, e(\alpha) \mid \alpha < \kappa \,\}$ is linearly ordered by E.

The following lemma is clear from the proof of Lemma 3.17 and of Lemma 2.13.

3.19 Lemma. *Let $\mathcal{L}$ be a countable language with at least one binary predicate symbol $\in$. Let Σ be a set of sentences of $\mathcal{L}(\mathit{aa})$. Then, Σ has a model $\mathcal{A} = (A; E, \dots)$ such that $(A; E)$ has a faithful ω_1-cumulation if and only if Σ is consistent with* **ST** *and* $\neg(\exists y)(\forall x)(x \in y)$. ⊣

We conclude this section to give a proof of Theorem 2.17. The proof is essentially due to the observation of Schmerl about the axiomatization of the class of ω_1-like models which embeds ω_1. (See [**18**].)

3.20 Proof of Theorem 2.17. It is clear that **ST** is an extension of $\mathbf{T}^{\mathbf{ST}}$ by Lemma 2.15 Suppose that **ST** is not a conservative extension of $\mathbf{T}^{\mathbf{ST}}$. Then, there exists a sentence σ of the language $\mathcal{L}$ such that $\mathbf{ST} \vdash \sigma$ and $\mathbf{T}^{\mathbf{ST}} \not\vdash \sigma$. By compactness, we may assume that $\mathcal{L}$ has only finitely many non-logical symbols. By using the following facts, we can have two countable models $\mathcal{A}$ and $\mathcal{B}$ such that $\mathcal{A}$ is a blunt end elementary extension of $\mathcal{B}$ and $\mathcal{A} \models \neg\sigma$.

Fact. Let T_c be the theory on the language obtained by adding a new constant c to$\mathcal{L}$. The axioms of T_c are:

(i) c is transitive,

(ii) $(\forall x_1 \ldots x_n \in c)(\varphi \leftrightarrow \varphi^{(c)})$, where φ is a formula of $\mathcal{L}$ with free variables among $x_1 \ldots x_n$ and $\varphi^{(c)}$ is the relativization of φ to c.

Then, T_c is a conservative extension of $\mathbf{T}^{\mathbf{ST}}$.

Now, let $\mathcal{A}_0$ and $\mathcal{B}_0$ be countably homogeneous models such that $\mathcal{A}_0 \simeq \mathcal{B}_0$ and $(\mathcal{A}, B) \prec (\mathcal{A}_0, B_0)$, where B and B_0 are the universes of $\mathcal{B}$ and $\mathcal{B}_0$ respectively. ([3, Proposition 3.2.10].) Then, $\mathcal{A}_0$ is a blunt end elementary extension of $\mathcal{B}_0$. By the proof of Vaught's two cardinal theorem, we can construct an elementary chain $(\mathcal{B}_\xi)_{\xi<\omega_1}$ of countable models such that:

(i) $\mathcal{B}_{\xi+1}$ is a blunt end elementary extension of $\mathcal{B}_\xi$ for all $\xi < \omega_1$,

(ii) $\mathcal{B}_\eta = \bigcup_{\xi<\eta} \mathcal{B}_\xi$ for a limit ordinal $\eta < \omega_1$.

Put $\mathcal{C} = \bigcup_{\xi<\omega_1} \mathcal{B}_\xi$. Define $e : \omega_1 \to C$ so that $e(\xi)_E = B_\xi$, where $\mathcal{C} = (C; E, \ldots)$ and B_ξ is the universe of $\mathcal{B}_\xi$. Clearly, e is an ω_1=cumulation of $(C; E)$. Thus, $(\mathcal{C}, \mathcal{F}_e)$ is a model of **ST**, and so $\mathcal{C} \models \sigma$. This is a contradiction because $\mathcal{B}_0 \models \neg\sigma$ and $\mathcal{B}_0 \prec \mathcal{B}$. $\dashv$

§4. Formulation of ZF($\mathfrak{a}\mathfrak{a}$)

In this section and the following sections in Part I, we we always assume that $\mathcal{L}$ is the language of **ZF**, that is, $\mathcal{L}$ is a first-order language with equality and binary predicate symbol $\in$. However, with a little modification, all results are still valid for languages with other non-logical symbols.

4.1 Definition. **ZF**($\mathfrak{a}\mathfrak{a}$) is a theory formulated on the language $\mathcal{L}(\mathfrak{a}\mathfrak{a})$, of which axioms are:

(1) all axioms of **ST** (relative to $\mathcal{L}$),
(2) (i) Extensionality,
  (ii) Regularity,
  (iii) Separation Schemes ranging over all formulas of $\mathcal{L}(\mathfrak{a}\mathfrak{a})$,
  (iv) Power Set.

4.2 Lemma. **ZF**($\mathfrak{a}\mathfrak{a}$) *extends* **ZF**.

PROOF. By the results of §2, we see that Pair, Union, Collection Scheme (ranging over all formulas of $\mathcal{L}(\mathfrak{a}\mathfrak{a})$) are provable in **ZF**($\mathfrak{a}\mathfrak{a}$). So only we must do is to show that Infinity is provable in **ZF**($\mathfrak{a}\mathfrak{a}$).
By using **UB**1, we have

$$(\forall y)(\mathfrak{a}\mathfrak{a}\, x)(y \cup \{y\} \in x), \quad (\mathfrak{a}\mathfrak{a}\, x)(0 \in x).$$

By using **ST**2, we have

$$(\mathfrak{a}\mathfrak{a}\, x)(\forall y \in x)(y \cup \{y\} \in x).$$

Then, by **FL**2, we have

$$(\mathfrak{a}\mathfrak{a}\, x)(0 \in x \wedge (\forall y \in x)(y \cup \{y\} \in x)),$$

and so,

$$(\exists x)(0 \in x \wedge (\forall y \in x)(y \cup \{y\} \in x)). \quad \dashv$$

The following lemma is an immediate consequence of Corollary 2.10.

4.3 Lemma. *Let* $\mathbf{ZF}^{\mathbf{UB}}$ *be a theory formulated on the language* $\mathcal{L}(\mathfrak{a}\mathfrak{a})$, *of which axioms are:*

(1) *all axioms of* **UB** *(relative to* $\mathcal{L}$*),*
(2) (i) *Extensionality,*
  (ii) *Regularity,*
  (iii) *Infinity,*
  (iv) *Separation Scheme (ranging over all formulas of* $\mathcal{L}(\mathfrak{a}\mathfrak{a})$*),*
  (v) *Power Set.*

Then, $\mathbf{ZF}^{\mathbf{UB}}$ *is a conservative extension of* **ZF**. $\dashv$

The following lemma, which is an immediate consequence of Lemma 2.17, shows that the separation axiom ranging over all formulas of $\mathcal{L}(\mathfrak{a}\mathfrak{a})$ plays an essential role in **ZF**($\mathfrak{a}\mathfrak{a}$).

4.4 Lemma ([**10**, prop. 1.8]). *Let* $\mathbf{ZF}^W(aa)$ *be the theory obtained by weakening* $\mathbf{ZF}(aa)$ *so that the separation scheme is restricted to formulas of* $\mathcal{L}$. *Then,* $\mathbf{ZF}^W(aa)$ *is a conservative extension of* $\mathbf{ZF}$. ⊣

Kaufmann (in [**10**]) stands on a little different point of view to formulate $\mathbf{ZF}(aa)$. His intended meaning of $(aa\,x)\varphi(x)$ is that $\{\,\alpha \mid \alpha(\alpha)\,\}$ contains a closed unbounded class of *ordinals*. To distinguish his quantifier aa from ours, we shall use a symbol "ao" for his quantifier, and call his system $\mathbf{ZF}(ao)$ for a while.

4.5 Definition. $\mathbf{ZF}(ao)$ is a theory formulated on the language $\mathcal{L}(aa)$, of which axioms are:

(1) all axioms of **ZF** (with the collection axiom and separation axiom ranging over all formulas of $\mathcal{L}(ao)$),

(2) (i) $(aox)(x \text{ is an ordinal})$,

(ii) $(\forall\alpha)((aox)(\alpha \in x)$,

(iii) $(\forall x)(\varphi \to \psi) \to ((aox)\varphi \to (aox)\psi)$,

(iv) $(aox)\varphi \wedge (aox)\psi x \to aox(\varphi \wedge \psi)$,

(v) $(\forall x)(ao\alpha)\varphi \to (ao\alpha)(\forall x \in R_\alpha)\varphi$, where $(ao\alpha)\varphi$ is an abbreviation for $(ao\alpha)(\alpha$ is an ordinal $\to \varphi)$ and R_α $(\alpha \in On)$ is the usual cumulative hierarchy,

(vi) $\neg(aox)(x \neq x)$,

However, Shizuo Kamo ([9]) observed that $\mathbf{ZF}(aa)$ and $\mathbf{ZF}(ao)$ are the essentially the same.

4.6 Definition. In $\mathbf{ZF}(aa)$, we define a quantifier ao by:

$$(aox)\varphi \leftrightarrow (aa\,u)(\exists x)(x \text{ is an ordinal} \wedge u = R_x \wedge \varphi),$$

where u does not occur free in φ.

4.7 Lemma ([9]). *All axioms of* $\mathbf{ZF}(ao)$ *are provable in* $\mathbf{ZF}(aa)$ *if* ao *is defined as in* Def.4.6.

4.8 Definition. In $\mathbf{ZF}(ao)$, we define a quantifier aa by:

$$(aa\,x)\varphi \leftrightarrow (ao\alpha)(\exists x)(x = R_\alpha \wedge \varphi), \quad \text{where } \alpha \text{ does not occur free in } \varphi.$$

4.9 Lemma([9]). *All axioms of* $\mathbf{ZF}(aa)$ *are provable in* $\mathbf{ZF}(ao)$ *if* aa *is defined as in* Def. 4.8.

Both Lemma 4.7 and 4.9 are easily proved by using the following lemma.

4.10 Lemma. $\mathbf{ZF}(\mathit{aa}) \vdash (\mathit{aa}\, z)(\exists\alpha)(z = R_\alpha)$.

PROOF. Let $\varphi(y)$ and $\theta(x, y)$ be $((\exists\alpha))(y = R_\alpha)$ and $x \in y$ respectively. It is enough to show that the following are provable in $\mathbf{ZF}(\mathit{aa})$:

(i) $(\forall x)(\exists y)(\theta(x, y) \wedge \varphi(y))$.
(ii) $(\mathit{aa}\, z)[(\forall x \in z)(\exists y \in z)(\theta(x, y) \wedge \varphi(y)) \to \varphi(z)]$.

Then, by Lemma 2.18, we have $(\mathit{aa}\, z)\varphi(z)$, that is, $(\mathit{aa}\, z)(\exists\alpha)(z = R_\alpha)$ is provable in $\mathbf{ZF}(\mathit{aa})$. (i) is clear. (ii) is from that $(\mathit{aa}\, z)(z$ is transitive). ⊣

From now on, if we write $(\mathit{aa}\,\alpha)$, it always means $(\mathit{aa}\, x)(\exists\alpha)(x = R_\alpha \wedge \varphi)$ whenever any confusion does not occur.

§5. Definability of satisfaction predicate

In this section, we shall show that there is a binary predicated St which constitute a definition for first-order formulas over the universe.

Since $\mathbf{ZF}(\mathit{aa})$ extends $\mathbf{ZF}$, we can formalize basic syntactical notions and model theory. Therefor, we ca define a set Form(aa) of formulas of $\mathcal{L}(\mathit{aa})$ in $\mathbf{ZF}(\mathit{aa})$ so that every metaformula φ of $\mathcal{L}(\mathit{aa})$ corresponds to a certain chosen term $\ulcorner\varphi\urcorner$ such that $\ulcorner\varphi\urcorner \in$ Form(aa). ($\ulcorner\varphi\urcorner$ is called the Gödel set of φ.) By Form, we mean the subset $\{\varphi \in \text{Form}(\mathit{aa}) \mid \varphi$ is a first-order formula of $\mathcal{L}\}$ of Form(aa).

5.1 Definition. Let $\varphi \in$ Form(aa). By an *assignment s for φ*, we mean a map s such that dom$(s) \in \omega$ and $\{i < \omega \mid v_i$ is a variable occurring in $\varphi\} \subseteq$ dom(s).

Let $\varphi \in$ Form, and s an assignment for φ. If x is a set such that range$(s) \subseteq x$, then the meaning of $(x; \in) \models \varphi[s]$ is self-evident.

The following lemma is the key to define the satisfaction relation for first-order formulas over the universe.

5.2 Lemma. *For every $\varphi \in$ Form and every assignment s for φ,*

$$(\mathit{aa}\, u)((u; \in) \models \varphi[s]) \quad \textit{iff} \quad (\mathit{stat}\, u)((u; \in) \models \varphi[s]).$$

PROOF. The proof will be carried by induction on the complexity of φ. This argument is ensured by the axiom of regularity and the separation scheme ranging over all formulas $\mathcal{L}(\mathit{aa})$. For example, we show that the lemma is true for the case that φ is

$$(\forall v_m)\psi(v_0 \dots v_{m-1} v_m v_{m+1} \dots v_{n-1}),$$

providing that the lemma is true for $\psi(v_0 \dots v_{n-1})$.
Let s be an assignment for φ. Suppose that

$$(\mathit{stat}\, u)((u; \in) \models \varphi[s]).$$

Then, we have

$$(\mathit{stat}\, u)(\forall x \in u)((u; \in) \models \psi[s_{m/x}], \quad \text{where } s_{m/x} = \{s - \{\langle m, s(m)\rangle\}\} \cup \{\langle m, x\rangle\}.$$

By (vii) in Lemma 1.7,

$$(\forall x)((\mathsf{aa}\, u)(x \in u) \to ((\mathsf{stat}\, u)((u; \in) \models \psi[s_{m/x}]).$$

By predicate logic and **UB1**,

$$(\forall x)(\mathsf{stat}\, u)((u; \in) \models \psi[s_{m/x}]).$$

It follows that, by the induction hypothesis,

$$(\forall x)(\mathsf{aa}\, u)((u; \in) \models \psi[s_{m/x}]).$$

By the axiom of diagonal intersection,

$$(\mathsf{aa}\, u)(\forall x \in u)((u; \in) \models \psi[s_{m/x}]).$$

It follows that

$$(\mathsf{aa}\, u)((u; \in) \models (\forall v_m)\varphi). \quad \dashv$$

5.3 Definition. We define a binary predicate St by:

$$\mathrm{St}(\varphi, s) \quad \text{iff} \quad \varphi \in \mathrm{Form},\ s \text{ is an assignment for } \varphi \text{ and } (\mathsf{aa}\, u)((u; \in) \models \varphi[s]).$$

(We shall often write $(\mathbf{V}; \in) \models \varphi[s]$ instead of $\mathrm{St}(\varphi, s)$.)

The following theorem can be clear from the definition of St and Lemma 5.2.

5.4 Theorem. St *constitute a definition of satisfaction for first-order formulas over the universe, that is, it satisfies the following properties:*

(i) $\mathrm{St}(v_i = v_j, s) \leftrightarrow s_i = s_j$,
(ii) $\mathrm{St}(v_i \in v_j, s) \leftrightarrow s_i \in s_j$,
(iii) $\mathrm{St}(\neg\varphi, s) \leftrightarrow \neg \mathrm{St}(\varphi, s)$,
(iv) $\mathrm{St}(\varphi \wedge \psi, s) \leftrightarrow \mathrm{St}(\varphi, s) \wedge \mathrm{St}(\psi, s)$,
(v) $\mathrm{St}((\forall v_i)\varphi, s) \leftrightarrow (\forall x)\, \mathrm{St}(\varphi, s_{i/x})$.

Since the collection and separation scheme ranging over formulas containing predicate St are provable in **ZF**(aa), we can have the following as an immediate corollary of Theorem 5.4.

5.5 Corollary. $(\mathbf{V}; \in) \models \mathbf{ZF}$.

By Corollary 5.5, we have **ZF**(aa) $\vdash$ Cons(**ZF**). Therefore, **ZF**(aa) is an essentially stronger than **ZF**. In fact, we can say much more more things,

5.6 Lemma. $(\text{aa}\, u)((u;\in) \prec (\mathbf{V};\in))$, *where* $(u;\in) \prec (\mathbf{V};\in)$ *means that* $(u;\in)$ *is an elementary substructure of* $(\mathbf{V};\in)$.

PROOF. By the definition of St, we have

$$(\varphi \in \text{Form})(\forall s)(s \text{ is an assignment for } \varphi \\ \rightarrow ((\text{aa}\, u)((u;\in) \models \varphi[s]) \leftrightarrow (\mathbf{V};\in) \models \varphi[s])).$$

By using Lemma 5.2,

$$(\varphi \in \text{Form})(\forall s)(s \text{ is an assignment for } \varphi \rightarrow (\text{aa}\, u)((u;\in) \models \varphi[s] \leftrightarrow (\mathbf{V};\in) \models \varphi[s])).$$

It follows that

$$(\forall\varphi \in \text{Form}(\text{aa}\, u)(\forall s \in u)(s \text{ is an assignment for } \varphi \\ \rightarrow ((u;\in) \models \varphi[s] \leftrightarrow (\mathbf{V};\in) \models \varphi[s]).$$

It is clear that

$$(\text{aa}\, u)(\forall s)(s \text{ is an assignment for } \varphi \rightarrow (\text{range}(s) \subseteq u \rightarrow s \in u)).$$

It follows that

$$(\varphi \in \text{Form})(\text{aa}\, u)(\forall s)(s \text{ is an assignment for } \varphi \wedge \text{range}(s) \subseteq u \\ \rightarrow ((u;\in) \models \varphi[s] \leftrightarrow (\mathbf{V};\in) \models \varphi[s])).$$

That is,

$$(\text{aa}\, u)((u;\in) \prec (\mathbf{V};\in). \dashv$$

5.7 Corollary. $(\text{aa}\, x)(\text{aa}\, y)((x;\in) \prec (y;\in))$. $\dashv$

If we restate Lemma 5.6 and Corollary 5.7 by using the quantifier "almost all ordinals", they will be more impressive.

5.8 Lemma.
(vi) $(\text{aa}\, \alpha)(\text{aa}\, \beta)((R_\alpha;\in) \prec (\mathbf{V};\in))$.
(vii) $(\text{aa}\, \alpha)(\text{aa}\, \beta)((R_\alpha;\in) \prec (R_\beta;\in))$. $\dashv$

In [17], Montague-Vaught discussed the theory **ZF′** defined as follows:

(i) **ZF′** is formulated on the language obtained by adding a new binary predicate symbol St to the language of **ZF**,

(ii) the axiom of **ZF′** are all the axioms of **ZF** (the separation and the collections schemes range over all formulas in the extended language), and the axioms claiming that St constitutes a definition of satisfaction for formulas φ ($\varphi \in$ Form) over the universe.

Theorem 5.4 says that **ZF′** is interpretable in **ZF**(*aa*). Actually, *zfaa* is much stronger than **ZF′**. For as we shall see in the sequel, we can define the satisfaction relation over the structure $(\mathbf{V};\in,\text{St})$ in **ZF**(*aa*).

5.9 Definition. Let α be an ordinal. $\mathcal{L}_\alpha$ is the language obtained by adding a new binary predicate symbol St_β for each $\beta < \alpha$ to the language $\mathcal{L}$ of **ZF**.

We associate each symbol of $\mathcal{L}_\alpha$ with a set as follows: $v_i = \langle 0, i \rangle$, $\mathrm{St}_\beta = \langle 1, \beta \rangle$, $= \,= \langle 2, 0 \rangle$, $\in = \langle 2, 1 \rangle$, $\neg = \langle 2, 2 \rangle$, $\wedge = \langle 2, 3 \rangle$ and $\forall = \langle 2, 4 \rangle$. then, a formula of $\mathcal{L}_\alpha$ can be regarded as a finite sequence of

$$\{ \langle 0, i \rangle \mid i < \omega \} \cup \{ \langle 1, \beta \rangle \mid \beta < \alpha \} \cup \{ \langle 2, 0 \rangle , \langle 2, 1 \rangle , \langle 2, 2 \rangle , \langle 2, 3 \rangle , \langle 2, 4 \rangle \}.$$

Notice that $\varphi \in M$ for any formula φ of $\mathcal{L}_\alpha$ whenever M is a transitive set such that $\alpha \subseteq M$ and $s \in M$ for any finite subset s of M.

5.10 Definition. Let M be a transitive set such that $\alpha \subseteq M$ and $s \in M$ for any finite subset s of M. We define a binary relation $\mathrm{St}_\beta^{(M)}$ on M for each $\beta < \alpha$ by induction on β. by:

$$\mathrm{St}_\beta^{(M)} = \{ \langle \varphi, s \rangle \mid \varphi \text{ is a formula of } \mathcal{L}_\beta,\ s \text{ is an assignment for } \varphi \text{ with range}(s) \subseteq M, \text{ and } (M; \in, \mathrm{St}_\gamma^{(M)})_{\gamma < \beta} \models \varphi[s] \}.$$

5.11 Definition. Let M be a transitive set such that $s \in M$ for any finite subset s of M. We define a ternary relation $\mathrm{Sat}^{(M)}$ on M by:

$$\mathrm{Sat}^{M)} = \{ \langle \alpha, \varphi, s \rangle \mid \alpha \text{ is an ordinal in } M \text{ and } \langle \varphi, s \rangle \in \mathrm{St}_\alpha^{(M)} \}.$$

5.12 Lemma. *For every ordinal α, every formula φ of $\mathcal{L}_\alpha$ and every assignment s for φ,*

$$(\mathit{stat}\, u)(\langle \alpha, \varphi, s \rangle \in \mathrm{Sat}^{(u)}) \quad \text{iff} \quad (\mathit{aa}\, u)(\langle \alpha, \varphi, s \rangle \in \mathrm{Sat}^{(u)}).$$

PROOF. The proof will be carried out by transfinite induction on α.
Case 1. $\alpha = 0$: This is Lemma 5.2 itself.
Case 2. $\alpha > 0$: We shall show that the lemma is true for all formulas α of $\mathcal{L}_\alpha$ by induction on the complexity of φ, assuming that the lemma is true for all β less than α. The case that φ is of the form $\mathrm{St}_\beta(v_i, v_j)$ $(\beta < \alpha)$ is only problematic. The rest can be proved just as in Lemma 5.2.

For any transitive set u closed under finite subset, we can see that the following are equivalent.

(a) $\langle \alpha, \mathrm{St}_\beta(v_i, v_j), s \rangle \in \mathrm{Sat}^{(u)}$.
(b) $\langle \mathrm{St}_\beta(v_i, v_j), s \rangle \in \mathrm{St}_\alpha^{(u)}$.
(c) $(u; \in, \mathrm{St}_\beta^{(u)})_{\beta < \alpha} \models \mathrm{St}_\beta(v_i, v_j)[s]$.
(d) s_i is a formula of $\mathcal{L}_\beta$, s_j is an assignment for s_i such that $\mathrm{range}(s_j) \subseteq u$ and $\langle s_i, s_j \rangle \in \mathrm{St}_\beta^{(U)}$.
(e) s_i is a formula of $\mathcal{L}_\beta$, s_j is an assignment for s_i such that $\mathrm{range}(s_j) \subseteq u$ and $\langle \beta, s_i, s_j \rangle \in \mathrm{Sat}^{(u)}$.

If we have

$$(\mathit{stat}\,u)(\langle\alpha, \mathrm{St}_\beta(v_i, v_j), s\rangle \in \mathrm{Sat}^{(u)}),$$

then

$$(\mathit{stat}\,u)(\langle\beta, s_i, s_j\rangle \in \mathrm{Sat}_{(u)}), \quad s_i \text{ is a formula of } \mathcal{L}_\beta, \text{ and } s_j \text{ is an assignment for } s_i.$$

By the induction hypothesis on β and the above,

$$(\mathit{stat}\,u)(\langle\alpha, \mathrm{St}_\beta(v_i, v_j), s\rangle \in \mathrm{Sat}^{(u)} \rightarrow (\mathit{aa}\,u)(\langle\beta, s_i, s_j\rangle \in \mathrm{Sat}^{(u)}).$$

It follows that

$$(\mathit{stat}\,u)(\langle\alpha, \mathrm{St}_\beta(v_i, v_j), s\rangle \in \mathit{Sat}^{(u)}) \rightarrow (\mathit{aa}\,u)(\langle\alpha, \mathrm{St}_\beta(v_i, v_j), s\rangle \in \mathrm{Sat}^{(u)}). \dashv$$

5.13 Definition. we define a ternary predicate **Sat** in **ZF**(aa) by:

$$\mathbf{Sat}(\alpha, \varphi, s) \leftrightarrow \alpha \text{ is an ordinal, } \varphi \text{ is a formula of } \mathcal{L}_\alpha, s \text{ is an assignment for } \varphi, \text{ and } (\mathit{aa}\,u)(\langle\alpha, \varphi, s\rangle \in \mathrm{Sat}^{(u)}).$$

The next theorem shows that $\{\, \mathbf{Sat}(\alpha, \varphi, s) \mid \alpha \in On \,\}$ constitute the hierarchy of satisfaction predicates. That is, if we put $\mathrm{St}_\beta(\varphi, s) \equiv \mathbf{Sat}(\beta, \varphi, s)$, then St_α is a satisfaction predicate for $\mathcal{L}_\alpha$ over $(\mathbf{V}; \in, \mathrm{St}_\beta)_{\beta<\alpha}$. The proof of the next theorem can be easily seen by Lemma 5.12.

5.14 Theorem. *The predicate* **Sat** *has the following properties:*

(1) $\mathbf{Sat}(\alpha, \mathbf{V}_i = v_j, s) \leftrightarrow s_i = s_j$,
(2) $\mathbf{Sat}(\alpha, v_i \in v_j, s) \leftrightarrow s_i \in s_j$,
(3) $\mathbf{Sat}(\alpha, \mathrm{St}_\beta(v_i, v_j), s) \leftrightarrow \mathbf{Sat}(\beta, s_i, s_j)$ forall$\beta < \alpha$
(4) $\mathbf{Sat}(\alpha, \varphi \wedge \psi, s) \leftrightarrow \mathbf{Sat}(\alpha, \varphi, s) \wedge \mathbf{Sat}(\alpha, \psi, s)$,
(5) $\mathbf{Sat}(\alpha, \neg\varphi, s) \leftrightarrow \neg\,\mathbf{Sat}(\alpha, \varphi, s)$,
(6) $\mathbf{Sat}(\alpha, (\forall v)_i\varphi, s) \leftrightarrow (\forall x)\,\mathbf{Sat}(\alpha, \varphi, s_i(x))$.

5.15 Definition. A theory **ZFS** is formulated on the language $\mathcal{L}_{\mathbf{Sat}}$, which is obtained by adding a ternary predicate symbol **Sat**. The axiom of **ZFS** are:

(i) All axioms of **ZF** with the separation and the collection schemes ranging over all formulas of $\mathcal{L}_{\mathbf{Sat}}$,
(ii) (1) to(6) in Lemma 5.14.

By Theorem 5.14, we can have the following corollary.

5.16 Corollary. **ZFS** *is interpretable in* **ZF**(aa).

In fact, by the same method as in defining the predicate St, we can define the satisfaction predicate for $\mathcal{L}_{\mathbf{Sat}}$ over $(\mathbf{V}; \in, \mathbf{Sat})$. Therefore, we can have a stronger result that $(\mathbf{V}; \in, \mathbf{Sat})$ is a model of **ZFS**.

§6. Kaufmann's hierarchy of formulas in $\mathcal{L}(aa)$

In §5, we saw that the satisfaction relation St for first-order formulas over the universe can be defined in**ZF**(aa). If we examine the definition of St, we can see that St is defined by a formula containing only one quantifier aa. Moreover, all the results in §5 except Cor. 5.7 and Lemma 5.8 are relevant only to formulas obtained by the first-order connectives and quantifiers from the set of formulas containing only one quantifier aa.These suggest us a hierarchy of formulas of $\mathcal{L}(aa)$ just like as Lévy's hierarchy of formulas of the language of **ZF**. Indeed, Kaufmann introduced a hierarchy of formulas of $\mathcal{L}(aa)$ according to the essential number of aa, and obtained the satisfaction relation for formulas in each level of this hierarchy.

6.1 Definition. We define a proper n-formula for each natural number n inductively as follows:

(i) φ or a proper 0-formula iff φ is a formula of $\mathcal{L}$.

(ii) Suppose $n > 0$.

(1) if φ is a proper $(n-1)$-formula, then $(aa\,x)\varphi$ is a proper n-formula,

(2) if φ and ψ are proper n-formulas, then so are $\neg\varphi$, $\varphi \wedge \psi$ and $(\forall x)\varphi$.

(3) There are no proper n-formulas other than constructed in (1) or (2).

6.2 Definition. We say that a formula φ of $\mathcal{L}(aa)$ is an n-formula if φ is a proper m-formula for some $m \leq n$. We denote the set of n-formulas by Form^n.

By the n-separation scheme (resp. the n-collection scheme), we mean the separation scheme (resp. the collection scheme) ranging over all n-formulas.

6.3 Definition. (i) A theory $\mathbf{ST}^n$ $(1 \leq n < \omega)$ is the theory obtained by restricting all the axioms of **ST** to n-formulas.

(ii) A theory $\mathbf{ZF}^n$ is obtained by adding the axiom of extensionality, regularity, power set, the n-separation scheme to $\mathbf{ST}^n$. By $\mathbf{ZF}^0$, we mean the theory **ZF**.

(iii) A theory $\mathbf{ZF}^{+n}$ is obtained by adding the n-collection scheme to $\mathbf{ZF}^n$.

By checking the proof of results in §5, we can have the following theorem.

6.4 Theorem. *Lemma 5.2, Theorem 5.4, Lemma 5.12 and Theorem 5.14 hold in* $\mathbf{ZF}^1$. *Corollary 5.5 and 5.16 hold in* $\mathbf{ZF}^{+1}$. *Lemma 5.6,, Corollary 5.7 and Lemma 5.8 hold in* $\mathbf{ZF}^2$. $\dashv$

6.5 Remark. Theorem 6.4 shows that even $\mathbf{ZF}^{+1}$ is fairy strong. For example, we can have $\mathbf{ZF}^{+1} \vdash \mathrm{Cons}(\mathbf{ZFS})$.

6.6 Remark. Though St can be defined in $\mathbf{ZF}^1$, $\mathbf{ZF}^1$ is a conservative extension of **ZF**. (See §9.) Only we can prove in $\mathbf{ZF}^1$ is Cons(**ZF** − Collection).

Kaufmann showed that the satisfaction predicate St^n for n-formulas over the universe can be defined in **ZF**(aa) (in fact, in $\mathbf{ZF}^{2n+1}$). We shall quote his result as theorem, though our definition of St^n is different form his.

6.7 Theorem. (See [10, Theorem 3.6]) *By induction on n, we can define a binary predicate St^n in $\mathbf{ZF}^{2n+1}$ and a unary operation q^{n+1} in $\mathbf{ZF}^{2n+2}$ as follows:*

$n = 0$: $\mathrm{St}^0(\varphi, s) \leftrightarrow \mathrm{St}(\varphi, s)$ (defined in §5).
$q^1(x) = \{\{y \in x \mid (x; \in) \models \varphi[s_{i/y}]\} \mid \varphi \in \mathrm{Form}^0 \wedge$ s is an assignment for $(\mathsf{aa}\, v_i)\varphi$ such that $\mathrm{range}(s) \subseteq x \wedge (\mathsf{aa}\, v)\, \mathrm{St}^0(\varphi, s_{i/v})\}$.

$n > 0$: $\mathrm{St}^n(\varphi, s) \leftrightarrow \varphi \in \mathrm{Form}^n \wedge$ s is an assignment for φ $\wedge (\mathsf{aa}\, x)(((x; \in), q^n(x)) \models \varphi[s])$.
$q^{n+1}(x) = \{\{y \in x \mid ((x; \in), q^n(x)) \models \varphi[s_{i/y}]\} \mid \varphi \in \mathrm{Form}^n \wedge$ s is an assignment for $(\mathsf{aa}\, v_i)\varphi$ such that $\mathrm{range}(s) \subseteq x \wedge (\mathsf{aa}\, v)\, \mathrm{St}^n(\varphi, si/v)\}$.

Then, St^n constitute a definition of satisfaction for Form^n, that is, the following can be proved in $\mathbf{ZF}^{n+1}$:

(i) $\mathrm{St}^n(v_i = v_j, s) \leftrightarrow s_i = s_j$.
(ii) $\mathrm{St}^n(v_i \in v_j, s) \leftrightarrow s_i \in s_j$.
(iii) $\mathrm{St}^n(\neg\varphi, s) \leftrightarrow \neg\mathrm{St}^n(\varphi, s)$.
(iv) $\mathrm{St}^n(\varphi \wedge \psi, s) \leftrightarrow \mathrm{St}^n(\varphi, s) \wedge \mathrm{St}^n(\psi, s)$.
(v) $\mathrm{St}^n((\forall x_i))\varphi, s) \leftrightarrow (\forall x)\, \mathrm{St}^n(\varphi, s_{i/x})$.
(vi) $\mathrm{St}^n(\mathsf{aa}\, v_i)\varphi, s) \leftrightarrow (\mathsf{aa}\, x)\, \mathrm{St}^{n-1}(\varphi, s_{i/x})$.
(vii) $\mathrm{St}^n(\varphi, s) \leftrightarrow \mathrm{St}^{n-1}(\varphi, s)$ for $\varphi \in \mathrm{Form}^{n-1}$.

The proof of Theorem 6.7 is not so hard but tedious. The only essential point is the following lemma.

6.8 Lemma. *In $\mathbf{ZF}^{2n+1}$, we can prove the following:*

For every $\varphi \in \mathrm{Form}^n$ and every assignment s for φ,

$$(\mathsf{aa}\, u)((u; \in, q^n(u)) \models \varphi[s]) \quad \text{iff} \quad (\mathsf{stat}\, u)(u; \in, q^n(u)) \models \varphi[s]).$$

The proofs of Theorem 6.7 and Lemma 6.8 are left to the reader. (The reader may be referred to Lemma 8.7 for the proof of Theorem 6.7.)

Part II

§7. The axiom of self-duality

So far we often encountered the fact that

$$(\mathsf{aa}\,x)\varphi \leftrightarrow (\mathsf{stat}\,x)\varphi \tag{1}$$

for some particular φ. That is, the quantifier aa is self-dual for some particular formula. For example, we know that

$$(\mathsf{aa}\,x)\varphi^{(x)} \leftrightarrow (\mathsf{stat}\,x)\varphi^{(x)}$$

is a theorem scheme of **ST** (Lemma 2.15).

If (1) holds for all formulas φ of $\mathcal{L}(\mathsf{aa})$, we say that aa is a self-dual quantifier. The aim of Part II is to study theories with the self-dual quantifier aa.

7.1 Definition. (i) By the duality scheme, in symbol, **Dual** we mean the scheme

$$(\mathsf{aa}\,x)\varphi \vee (\mathsf{aa}\,x)\neg\varphi, \quad \text{where } \varphi \text{ is a formula of } \mathcal{L}(\mathsf{aa}). \tag{2}$$

By n-**Dual**, we mean the scheme obtained by weakening **Dual** so that φ ranges over $(n-1)$-formulas in (2).
(ii) By **SD**, we mean the theory obtained by adding **Dual** to **ST**, and by $\mathbf{SD}^n$, the theory obtained by weakening **SD** so that all the axioms are n-formulas.

7.2 Definition. For every formula φ of $\mathcal{L}(\mathsf{aa})$ we define the relativization $\varphi^{(u)}$ of φ to u, which does not occur in φ;
(1) if φ is an atomic formula, then $\varphi^{(u)}$ is φ,
(2) $(\neg\varphi)^{(u)}$ is $\neg\varphi^{(u)}$,
(3) $(\varphi \wedge \psi)^{u)}$ is $\varphi^{(u)} \wedge \psi^{(u)}$,
(4) $((\forall x)\varphi)^{(u)}$ is $(\forall x \in u)\varphi^{(u)}$,
(5) $((\mathsf{aa}\,x)\varphi(x))^{(u)}$ is $\varphi(u)$.

7.3 Remark. (1) We should note that the relativization in the above definition is different from the one given in Def. 2.14. In Part II, by the relativization, we always mean the one given in the above definition.
(2) We should note that $\varphi^{(u)}$ is a Σ_0-formula if φ is a 0-formula, and an $(n-1)$-formula if φ is an n-formula for $n \geq 1$

The following lemma can be proved by induction on the complexity of formulas. The proof is left to the reader. (Cf. (iii) of Lemma 2.12.)

7.4 Lemma. $\mathbf{SD}^n \vdash (\text{aa}\, u)\varphi^{(u)} \leftrightarrow \varphi$, *where φ is an n-formula in which u does not occur.* ⊣

7.5 Definition. By $\mathbf{RF}^n$ (n-Reflection Principle, $n \geq 0$), we mean the following scheme:

$$(\exists u)(u \text{ is transitive} \wedge (\forall x_1 \cdots \forall x_m \in u) \bigwedge_{k=1}^{l} (\varphi^k \leftrightarrow \varphi_k^{(u)})),$$

where each φ_k is an n-formulas with free variables among $x_1, \ldots, x_m$.

7.6 Lemma. $\mathbf{RF}^{n-1}$ *is provable in* $\mathbf{SD}^n$. *In fact, the following stronger scheme provable in* $\mathbf{SD}^n$:

$$(\text{aa}\, u)(u \text{ is transitive} \wedge (\forall x_1 \cdots \forall x_m \in u) \bigwedge_{k=1}^{l} (\varphi_k \leftrightarrow \varphi_k^{(u)})),$$

where each φ_k is an $(n-1)$-formula with free variables among $x_1, \ldots, x_m$.

PROOF. Let $\varphi_l, \ldots, \varphi_l$ be $(n-1)$-formulas with free variables among $x_1, \ldots, x_m$. By Lemma 7.4, we have

$$\varphi_k \leftrightarrow (\text{aa}\, u)\varphi_k^{(u)}$$

for each φ_k. Since each φ_k is an $(n-1)$-formula in which u does not occur, we have

$$(\text{aa}\, u)(\varphi_k \leftrightarrow \varphi_k^{(u)}).$$

It follows that

$$(\text{aa}\, u) \bigwedge_{k=1}^{l} (\varphi_k \leftrightarrow \varphi_k^{(u)}),$$

and so

$$(\forall x_1 \cdots \forall x_m)(\text{aa}\, u) \bigwedge_{k=1}^{l} (\varphi_k \leftrightarrow \varphi_k^{(u)}).$$

By the axiom of diagonal intersection,

$$(\text{aa}\, u)(\forall x_1 \cdots x_m \in u) \bigwedge_{k=1}^{l} (\varphi_k \leftrightarrow \varphi_k^{(u)}).$$

Since $(\text{aa}\, u)($ is transitive), we have the required formula. ⊣

7.7 Definition. Let $\mathcal{L}_c$ be the language obtained by adding a new constant c to $\mathcal{L}$. $\mathbf{SD}_c^{n-1}$ is a theory on the language $\mathcal{L}_c(\text{aa})$ obtained by adding the following axioms to $\mathbf{SD}^{n-1}$:

(1) c is transitive,
(2) $(\forall x_1 \cdots \forall x_m \in c)(\varphi \leftrightarrow \varphi^{(c)})$, where φ is an $(n-1)$-formula of $\mathcal{L}(\text{aa})$ with free variables among $x_1, \ldots, x_m$. (For $n = 1$, $\mathbf{SD}_c^0$ denotes a first-order theory on $\mathcal{L}_c$, of which axioms are (1) and (2).)

By noting that $\mathbf{SD}_c^{n-1} \vdash \mathbf{RF}^{n-1}$, the following lemma is easily proved.

7.8 Lemma. $\mathbf{SD}_c^{n-1}$ *is a conservative extension of* $\mathbf{SD}^{n-1} + \mathbf{RF}^{n-1}$, *that is,*

$$\mathbf{SD}_c^{n-1} \vdash \varphi \quad \textit{iff} \quad \mathbf{SD}^{n-1} + \mathbf{RF}^{n-1} \vdash \varphi$$

for any $(n-1)$*-formula* φ *of* $\mathcal{L}(\text{aa})$. $\dashv$

7.9 Lemma. *Let* $\varphi(x_1, \ldots, x_m)$ *be an* n*-formula of* $\mathcal{L}(\text{aa})$ *with free variables among* $x_1, \ldots, x_m$. *Then,*

$$\mathbf{SD}^n \vdash \varphi \quad \textit{iff} \quad \mathbf{SD}_c^{n-1} \vdash x_1 \in c \rightarrow \cdots \rightarrow x_m \in c \rightarrow \varphi^{(c)}.$$

PROOF. *From the left to the right:* The proof will be carried out by induction on the length of the proof of $\mathbf{SD}^n \vdash \varphi$. It is enough to show that $\mathbf{SD}_c^{n-1} \vdash x_1 \in c \rightarrow \cdots \rightarrow x_m \in c \rightarrow \varphi^{(c)}$ for any axiom φ of $\mathbf{SD}^n$. For sample, we show that φ is the axiom of diagonal intersection. The remainder is left to the reader. So let φ be of the form

$$(\forall x)(\text{aa}\, y)\psi(xyx_1 \ldots x_m) \rightarrow (\text{aa}\, y)(\forall x \in y)\psi(xyx_1 \ldots x_m).$$

Then, $\varphi^{(c)}$ is of the form

$$(\forall x \in c)\psi(xcx_1 \ldots x_m) \rightarrow (\forall x \in c)\psi(xcx_1 \ldots x_m).$$

This is logically valid. Therefore, it is trivial that

$$\mathbf{SD}_c^{n-1} \vdash x_1 \in c \rightarrow \cdots \rightarrow x_m \in c \rightarrow \varphi^{(c)}.$$

From the right to the left: Suppose that

$$\mathbf{SD}_c^{n-1} \vdash x_1 \in c \rightarrow \cdots \rightarrow x_m \in c \rightarrow \varphi^{(c)}.$$

Then,

$$\mathbf{SD}^{n-1} \vdash (c \text{ is transitive}) \wedge (\forall y_1 \cdots \forall y_p \in c) \bigwedge_{k=1}^{l} (\varphi_k \leftrightarrow \varphi_k^{(c)})$$
$$\rightarrow (x_1 \in c \rightarrow \cdots \rightarrow x_m \in c \rightarrow \varphi^{(c)}),$$

where each φ_k is an $(n-1)$-formula with free variables among $y_1, \ldots, y_p$.

Since c does not appear in $\mathbf{SD}^{n-1}$, we can have

$$\mathbf{SD}^{n-1} \vdash (\forall u)[(u \text{ is transitive} \wedge (\forall y_1 \cdots \forall y_p \in u) \bigwedge_{k=1}^{l} (\varphi_k \leftrightarrow \varphi_k^{(u)})$$
$$\rightarrow (x_1 \in u \rightarrow \cdots \rightarrow x_m \in u \rightarrow \varphi^{(u)})].$$

It follows that

$$\mathbf{SD}^n \vdash (\text{aa}\, u)(u \text{ is transitive} \wedge (\forall y_1 \cdots \forall y_p \in u) \bigwedge_{k=1}^{l} (\varphi_k \leftrightarrow \varphi_k^{(u)})$$
$$\rightarrow (\text{aa}\, u)(x_1 \in u \rightarrow \cdots \rightarrow x_m \in u \rightarrow \varphi^{(u)})).$$

By Lemma 7.6, we have

$$\mathbf{SD}^n \vdash (\text{aa}\, u)(x_1 \in u \to \cdots \to x_m \in u \to \varphi^{(u)}).$$

Since $(\text{aa}\, u)(x_1 \in u \wedge \cdots \wedge x_m \in u)$, we have

$$\mathbf{SD}^n \vdash (\text{aa}\, u)\varphi^{(u)}.$$

By Lemma 7.4, we have

$$\mathbf{SD}^n \vdash \varphi. \dashv$$

7.10 Theorem. *$\mathbf{SD}^n$ is a conservative extension of $\mathbf{SD}^{n-1} + \mathbf{RF}^{n-1}$, that is,*

$$\mathbf{SD}^n \vdash \varphi \quad \textit{iff} \quad \mathbf{SD}^{n-1} + \mathbf{RF}^{n-1} \vdash \varphi$$

for any $(n-1)$-formula φ.

PROOF. Let $\varphi(_1, \ldots, x_m)$ be an $(n-1)$-formula with free variables among $x_1, \ldots, x_m$. Suppose that $\mathbf{SD}^n \vdash \varphi$. Then, by Lemma 7.9, we have

$$\mathbf{SD}_c^{n-1} \vdash x_1 \in c \to \cdots \to x_m \in c \to \varphi^{(u)}.$$

By (2) of the axiom of $\mathbf{SD}_c^{n-1}$, we have

$$\mathbf{SD}_c^{n-1} \vdash \varphi.$$

By Lemma 7.8, we have

$$\mathbf{SD}^n \vdash \varphi.$$

For the converse, it is clear from Lemma 7.6. ⊣

For a special case of the above theorem, we have the following corollary.

7.11 Corollary. *$\mathbf{SD}^1$ is a conservative extension of the first-order theory with $\mathbf{RF}^0$ as its axioms.* ⊣

Lemma 7.9 suggest to us that we can eliminate the quantifier aa from n-formulas by adding new n constants $c_0, \ldots, c_{n-1}$ to $\mathcal{L}$. To describe this precisely, we have the following definition.

7.12 Definition. $\mathbf{SD}^{n-k}_{c_0,\ldots,c_{k-1}}$ $(0 < k \le n)$ is a theory on the language $\mathcal{L}_{c_0,\ldots,c_{k-1}}(\text{aa})$ which is obtained by adding the following axioms to $\mathbf{SD}^{n-k}$:

(1) c_0 is transitive.

(2) $c_0 \in c_1$ (for the case that $1 < k$).

(3) $(\forall x_1 \cdots \forall x_m \in c_0)(\varphi \leftrightarrow \varphi^{(c_0)})$, where φ is an $(n-1)$-formula of $\mathcal{L}(\text{aa})$ with free variables among $x_1, \ldots, x_m$.

(4) $(\forall x_1 \cdots \forall x_m \in c_0)(\varphi(c_0 \ldots c_{k-2} x_1 \ldots x_m) \leftrightarrow \varphi(c_{i_0} \ldots c_{i_{k-2}} x_1 \ldots x_m)$, where $\varphi(y_0 \ldots y_{k-2} x_1 \ldots x_m)$ is an $(n-k)$-formula of $\mathcal{L}(\text{aa})$ with free variables among $y_0, \ldots, y_k, x_1, \ldots, x_k$, and $i_0 < \cdots < i_{k-2} < k$.

(For the notational convenience, $\mathbf{SD}^{n-k}_{c_0,\ldots,c_{k-1}}$ means $\mathbf{SD}^n$ for $k = 0$.)

Our nest task is to prove the following theorem.

7.13 Theorem. *Let $\varphi(x_1 \dots x_m)$ be an n-formula of $\mathcal{L}(aa)$ with free variables among $x_1, \dots, x_m$. Then,*

$$\mathbf{SD}^n \vdash \varphi \quad \textit{iff} \quad \mathbf{SD}^0_{c_0,\dots,c_{n-1}} \vdash x_1 \in c_0 \to \cdots \to x_m \in c_0 \to \varphi^{(c_0,\dots,c_{n-1})},$$

where $\varphi^{(c_0,\dots,c_{n-1})}$ means $(\cdots(\varphi^{(c_0)})^{(c_1)}\dots)^{(c_{n-1})}$.

To prove the theorem, we need some lemmas.

7.14 Lemma. *The following are provable in $\mathbf{SD}^{n-k}_{c_0,\dots,c_{k-1}}$.*

(i) $(\forall x_1 \cdots \forall x_m \in c_{i_0})(\varphi(c_{i_0} \dots c_{i_{k-2}} x_1 \dots x_m) \leftrightarrow \varphi(c_{j_0} \dots c_{j_{k-2}} x_1 \dots x_m))$, *where $\varphi(y_0 \dots y_{k-2} x_1 \dots x_m)$ is an $(n-k)$-formula of $\mathcal{L}(aa)$ with free variables among $y_0, \dots, y_{k-2}, x_1, \dots, x_m$, and $i_0 < \dots < i_{k-2} < k$, $j_0 < \dots < i_{k-2} < k$, $j_0 < \dots < j_{k-2} < k$, $i_0 \le j_0$.*

(ii) *$c_i (0 \le i < k)$ is transitive.*

(iii) *$c_i \in c_j$ for $0 \le i < j < k$.*

(iv) $(\forall x_1 \cdots \forall x_m \in c_i)(\varphi(c_0 \dots c_{i-1} x_1 \dots x_m) \leftrightarrow \varphi^{(c_i)}(c_0 \dots c_{i-1} x_1 \dots x_m)$, *where $0 < i < k$, and $\varphi(y_0 \dots y_{i-1} x_1 \dots x_m)$ is an $(n-k)$-formula of $\mathcal{L}(aa)$ with free variables among $y_0, \dots, y_{i-1}, x_1, \dots, x_m$.*

PROOF. (i), (ii) and (iii) are clear. For (iv), let $\varphi(y_0 \dots y_{i-1} x_1, \dots, x_m)$ be an $(n-k)$-formula of $\mathcal{L}(aa)$ with free variables among $y_0, \dots, y_{i-1}, x_1, \dots, x_m$. Then, we have

$$(\forall y_0 \cdots y_{i-1} \in c_0)(\forall x_1 \cdots \forall x_m \in c_0)(\varphi(y_0 \dots y_{i-1} x_1 \dots x_m) \leftrightarrow \varphi^{(c_0)}(y_0 \dots y_{i-1} x_1 \dots x_m)).$$

It follows that

$$(\forall y_0 \cdots \forall y_{i-1} \in c_i)(\forall x_1 \cdots \forall x_m \in c_i)(\varphi(y_0 \dots y_{i-1} x_1 \dots x_m) \to \varphi^{(c_i)}(c_0 \dots c_{i-1} x_1 \dots x_m)).$$

Since $c_k \in c_i$ $(k = 0, \dots, i-1)$ by (iii), we have

$$(\forall x_1 \cdots x_m \in c_i)(\varphi(c_0 \dots c_{i-1} x_1 \dots x_m) \leftrightarrow \varphi^{(c_i)}(c_0 \dots c_{i-1} x_1 \dots x_m)). \dashv$$

7.15 Lemma. *For any $(n-k)$-formula $\varphi(x_1 \dots x_m)$ of $\mathcal{L}_{c_0,\dots,c_{k-1}}(aa)$ with free variables among $x_1, \dots, x_m$,*

$$\mathbf{SD}^{n-k}_{c_0,\dots,c_{k-1}} \vdash \varphi(x_1 \dots x_m)$$
$$\textit{implies } \mathbf{SD}^{n-(k+1)}_{c_0,\dots,c_k} \vdash x_1 \in c_k \to \cdots \to x_m \in c_k \to \varphi^{(c_k)}(x_1 \dots x_m).$$

PROOF. For $k = 0$, we have already proved the lemma. (Lemma 7.9.) So suppose $0 < k$. It is enough to show that the following (i) and (ii) are provable in $\mathbf{SD}^{n-(k+1)}_{c_0,\dots,c_k}$.

(i) $(\forall x_1 \cdots \forall x_m \in c_0)(\varphi^{(c_k)}(x_1 \dots x_m) \leftrightarrow \varphi^{(c_0,c_k)}(x_1 \dots x_m))$, where $\varphi(x_1 \dots x_m)$ is an $(n-k)$-formula of $\mathcal{L}(aa)$ with free variables among $x_1, \dots, x_m$.

(ii) $(\forall x_1 \cdots \forall x_m \in c_0)(\varphi^{(c_k)}(c_0 \dots c_{k-2} x_1 \dots x_m)$, where $\varphi(y_0 \dots y_{k-2} x_1 \dots x_m)$ is an $(n-k)$-formula of $\mathcal{L}(aa)$ with free variables among $y_0, \dots, y_{k-2}, x_1, \dots, x_m$, and $i_0 < \dots < i_{k-2} < k$.

To prove (i), let $\varphi(x_1 \ldots x_m)$ be an $(n-k)$-formula of $\mathcal{L}(aa)$ with free variables among $x_1, \ldots, x_m$. By (i) of Lemma 7.14, we have

$$(\forall x_1 \cdots x_m \in c_0)(\varphi^{c_0)}(x_1 \ldots x_m) \leftrightarrow \varphi^{(c_k)}(x_1 \ldots x_m)). \tag{1}$$

By (iv) of Lemma 7.14, we have

$$(\forall x_1 \cdots x_m \in c_0)(\varphi^{(c_0)}(x_1 \ldots x_m) \leftrightarrow \varphi^{(c_0, c_k)}(x_1 \ldots x_m)). \tag{2}$$

Clearly, by (1) and (2), we have (i). (ii) is clear from (i) of Lemma 7.14. ⊣

The following lemma can be proved by using Lemma 7.4 iteratively.

7.16 Lemma. *The following scheme is provable in* $\mathbf{SD}^n$:

$$\varphi \leftrightarrow (aa\, u_0) \cdots (aa\, u_{n-1})\varphi^{(u_0, \ldots, u_{n-1})},$$

where φ is an n-formula in which $u_0, \ldots, u_{n-1}$ do not occur. ⊣

7.17 *Proof of Theorem 7.13.* For the part of "*from the left to the right*", we can prove it by using Lemma 7.15 iteratively. To prove the other half part, suppose that

$$\mathbf{SD}^0_{c_0, \ldots, c_{n-1}} \vdash x_1 \in c_0 \to \cdots \to x_m \in c_0 \to \varphi^{(c_0, \ldots, c_{n-1})},$$

where $\varphi(x_1 \ldots x_m)$ be an n-formula with free variables among $x_1, \ldots, x_m$.

Then, we can find a sentence $\sigma(c_0 \ldots c_{n-1})$ of $\mathcal{L}_{c_0, \ldots, c_{n-1}}$ satisfying the following (1) and (2):

(1) $\vdash \sigma(c_0 \ldots c_{n-1}) \to (x_1 \in c_0 \to \cdots \to x_m \in c_0 \to \varphi^{(c_0, \ldots, c_{n-1})})$.

(2) $\sigma(c_0 \ldots c_{n-1})$ is a finite conjunction of the sentences with the following form:

- (i) $(\forall z_1 \cdots z_l \in c_0)(\theta(z_1 \ldots z_l) \leftrightarrow \varphi^{(c_0)}(z_1 \ldots z_l))$, where $\theta(z_1 \ldots z_l)$ is a 0-formula with free variables among $z_1, \ldots, z_l$,
- (ii) c_0 is transitive,
- (iii) $c_0 \in c_1$,
- (iv) $(\forall y_1 \cdots y_k \in c_0)\psi(c_0 \ldots c_{n-2} y_1 \ldots y_k) \leftrightarrow \psi(c_{i_0} \ldots c_{i_{n-2}} y_1 \cdots y_k))$, where $\psi(w_0 \ldots w_{n-2} y_1 \ldots y_k)$ is a 0-formula with free variables among $w_0, \ldots, w_{n-2}, y_1, \ldots, y_k$, and $i_0 < \cdots < i_{n-2} < n$.

From (1), we have

$$\vdash (\forall u_0 \cdots u_{n-1})(\sigma(u_0 \ldots u_{n-1}) \to (x_1 \in u_0 \to \cdots \to x_m \in u_0 \to \varphi^{(u_0, \ldots, u_{n-1})}).$$

It follows that

$$\mathbf{SD}^n \vdash (aa\, u_0) \cdots (aa\, u_{n-1})\sigma(u_0, \ldots, u_{n-1} \to (aa\, u_0) \cdots (aa\, u_{-1})\varphi^{(u_0, \ldots, u_{n-1})}).$$

By Lemma 7.16,if we have

$$\mathbf{SD}^n \vdash (aa\, u_0) \cdots (aa\, u_{n-1})\sigma(u_0, \ldots, u_{n-1}), \tag{*}$$

then

$$\mathbf{SD}^n \vdash \varphi.$$

To prove $(*)$, we have to show that the following are provable in $\mathbf{SD}^n$.

(i′) $(\text{aa}\, u_0)(\theta(z_1 \ldots z_l) \leftrightarrow \theta^{(u_0)}(z_1 \ldots z_l))$,

(i′) $(\text{aa}\, u_0)(u_0$ is transitive),

(iii′) $(\text{aa}_0)(\text{aa}\, u_1)(u_0 \in u_1)$,

(iv′) $(\text{aa}\, u_0) \cdots (\text{aa}\, u_{n-1})\psi(u_0 \ldots u_{n-2} y_1 \ldots y_k) \leftrightarrow \psi(u_{i_0} \ldots u_{i_{n-2}} y_1 \ldots y_k))$.

(i′) is form Lemma 7.4, and (ii′) and (ii′) are clear. For (iv′), let $v_0, \ldots, v_{n-2}$ be new variables. Then,

$$(\text{aa}\, u_0) \cdots (\text{aa}\, u_{n-2})\psi(u_0 \ldots u_{n-2} y_1 \ldots y_k) \leftrightarrow (\text{aa}\, v_0) \cdots (\text{aa}\, v_{n-2})\psi(v_0 \ldots v_{n-2} y_1 \ldots y_k),$$

and

$$(\text{aa}\, u_{i_0}) \cdots (\text{aa}\, u_{i_{n-2}})\psi(u_{i_0} \ldots u_{i_{n-2}} y_1 \ldots y_k) \leftrightarrow (\text{aa}\, v_0) \cdots (\text{aa}\, v_{n-2})\psi(v_0 \ldots v_{n-2} y_1 \ldots y_k).$$

It follows that

$$(\text{aa}\, u_0) \cdots (\text{aa}\, u_{n-2})((\text{aa}\, u_{n-1})\psi(u_0 \ldots u_{n-2} y_1 \ldots y_k) \\ \leftrightarrow (\text{aa}\, v_0) \cdots (\text{aa}\, v_{n-2})\psi(v_0 \ldots v_{n-2} y_1 \ldots y_k),$$

and

$$(\text{aa}\, u_0) \cdots (\text{aa}\, u_{n-1})\psi(u_{i_0} \ldots u_{i_{n-2}} y_1 \ldots y_k) \leftrightarrow (\text{aa}\, v_0) \cdots (\text{aa}\, v_{n-1})\psi(v_0 \ldots v_{n-1} y_1 \ldots y_k).$$

Thus,

$$(\text{aa}\, u_0) \cdots (\text{aa}\, u_{n-1})\psi(u_0 \ldots u_{n-2} y_1 \ldots y_k) \leftrightarrow (\text{aa}\, u_0) \cdots (\text{aa}\, u_{n-1})\psi(u_{i_0} \ldots u_{i_{n-2}} y_1 \ldots y_k).$$

By using n-**Dual**, we have (iv′). ⊣

7.18 Corollary.

$\mathbf{SD}^n \vdash (\text{aa}\, x_1) \cdots (\text{aa}\, x_n)\varphi(x_1 \ldots x_n)$ *iff* $\mathbf{SD}^0_{c_0,\ldots,c_{n-1}} \vdash \varphi(c_0 \ldots c_{n-1})$,
where $\varphi(x_1 \ldots x_n)$ *is a formula of* $\mathcal{L}$ *with free variables among* $x_1, \ldots, x_n$. ⊣

7.19 Corollary. *Let* Γ *be the set of sentences* $\varphi(c_0 \ldots c_{n-1})$ *of* $\mathcal{L}_{c_0,\ldots,c_{n-1}}$ *such that* $\mathbf{SD}^n \vdash (\text{aa}\, x_1) \cdots (\text{aa}\, x_n)\varphi(x_1 \ldots x_n)$. *Then,* Γ *coincides with the set of sentences provable in* $\mathbf{SD}^0_{c_0,\ldots,c_{n-1}}$. ⊣

7.20 Definition. Let $\mathcal{L}_\infty$ be the language obtained by adding new constant symbols c_n ($n < \omega$). $\mathbf{SD}^0_\infty$ is a theory formulated on the language $\mathcal{L}_\infty$, of which axioms are:

(i) c_0 is transitive,

(ii) $c_0 \in c_1$,

(iii) $(\forall x_1 \cdots x_m \in c_0)(\varphi \leftrightarrow \varphi^{(c_0)})$, where φ is a formula of $\mathcal{L}$ with free variables among $x_1, \ldots, x_m$,

(iv) $(\forall x_1 \cdots x_m \in c_0)(\varphi(c_0 \ldots c_{n-1} x_1 \ldots x_m) \leftrightarrow \varphi(c_{i_0} \ldots c_{i_{n-1}} x_1 \ldots x_m))$, where $\varphi(y_1 \ldots y_{n-1} x_1 \ldots x_m)$ is a formula of $\mathcal{L}$ with free free variables among $y_1, \ldots, y_{n-1}, x_1, \ldots, x_m$, and $i_0 < \cdots < i_{n-1} < \omega$.

Noting that $\mathbf{SD} = \bigcup_{n<\omega} \mathbf{SD}^n$ and $\mathbf{SD}^0_\infty = \bigcup_{n<\omega} \mathbf{SD}^0_{c_0,\ldots,c_{n-1}}$, we can easily see the following theorem by Theorem 7.13. (Since it is not necessary that $\mathbf{SD} \vdash \varphi$ implies $\mathbf{SD}^n \vdash \varphi$ for all n-formulas φ, we need a little care to apply Theorem 7.13 to proving the following theorem.)

7.21 Theorem. (i) $\mathbf{SD} \vdash \varphi(x_1 \ldots x_m)$ *iff* $\mathbf{SD}^0_\infty \vdash x_1 \in c_0 \to \cdots \to x_m \in c_0 \to \varphi^{(c_0,\ldots,c_{n-1})}(x_1 \ldots x_m)$, *where* $\varphi(x_1 \ldots x_m)$ *is an n-formula with free variables among* $x_1 \ldots x_m$.
(ii) $\Gamma = \{\varphi(c_0 \ldots c_{n-1}) \mid \varphi(x_1 \ldots x_n)$ is a formula of $\mathcal{L}$ with free variables among $x_1, \ldots, x_n$ and $\mathbf{SD} \vdash (\text{aa}\, x_i) \cdots (\text{aa}\, x_n)\varphi\}$ *coincides with the set of sentences provable in* $\mathbf{SD}^0_\infty$. ⊣

We conclude this section to give a lemma which will be needed in §10.

7.22 Lemma. *Let* **Dual′** *denote the following scheme:*

$$(\text{aa}\, x_1) \cdots (\text{aa}\, x_n)\varphi \vee (\text{aa}\, x_1) \cdots (\text{aa}\, x_n)\neg\varphi,$$

where φ *is a formula of* $\mathcal{L}$.

Then, we have

$$\mathbf{ST} + \mathbf{Dual}' \vdash \mathbf{Dual}. \dashv$$

§8. Set theory with the self-dual quantifier

In this section, we shall see how the axiom **Dual** behaves in **ZF**(aa). Throughout this section, to simplify descriptions, we assume that $\mathcal{L}$ is the language of **ZF**, that is, $\mathcal{L}$ is a first-order language with equality and one binary predicate symbol $\in$. However, with a little modification, all arguments in this section are applicable to any language with non-logical symbol other than $\in$.

8.1 Definition. For each natural number n, we define the theory $\mathbf{ZD}^n$ as follows:

(i) $\mathbf{ZD}^0$ is a theory on $\mathcal{L}$, of which axioms are:

(1) Extensionality,
(2) Regularity,
(3) Σ_0-separation,
(4) Power Set.

(ii) $\mathbf{ZD}^n = \mathbf{SD}^n + \mathbf{ZD}^0$ $(n \geq 1)$.

8.2 Lemma. *n-Separation Scheme is probable in* $\mathbf{ZD}^n$ $(n \geq 1)$.

PROOF. It suffices to show that

$$\mathbf{ZD}^n + (m-1)\text{--}\mathbf{Sep} \vdash m\text{--}\mathbf{Sep} \quad \text{for } m \leq n,$$

where $(m-1)$-**Sep** means Σ_0-**Sep** for $m = 0$.

So let φ be an m-formula. We want to show that the following is provable in $\mathbf{ZD}^n +$ $(m-1)$-**Sep**:

$$(\exists z)(\forall x)(x \in z \leftrightarrow x \in y \wedge \varphi), \quad \text{where } z \text{ does not occur free in } \varphi.$$

Let u be a new variable. It is clear that

$$(\exists z)(\forall x)(x \in z \leftrightarrow x \in y \wedge \varphi^{(u)}) \wedge (\forall z)(z \subseteq y \rightarrow z \in u) \rightarrow (\exists z \in u)(\forall x)(x \in z \leftrightarrow x \in y \wedge \varphi^{(u)}).$$

By using $(m-1)$-**Sep**, we have

$$(\forall z)(z \subseteq y \rightarrow z \in u) \rightarrow (\exists z \in u)(\forall x)(x \in z \leftrightarrow x \in y \wedge \varphi^{(u)}).$$

It follows that

$$(\text{aa}\, u)(\forall z)(z \subseteq y \rightarrow z \in u) \rightarrow (\text{aa}\, u)(\exists z \in u)(\forall x)(x \in z \leftrightarrow x \in y \wedge \varphi^{(u)}).$$

By Power Set, we have

$$(\exists u)(\forall z)(z \subseteq y \rightarrow z \in u),$$

so by (iii) in Lemma 2.2,

$$(\text{aa}\, u)(\forall z)(z \subseteq y \rightarrow z \in u).$$

Therefore, we have

$$(\text{aa}\, u)(\exists z \in u)(\forall x)(x \in z \leftrightarrow x \in y \wedge \varphi^{(u)}).$$

It follows that

$$(\exists z)(\forall x)(x \in z \leftrightarrow x \in y \wedge (\text{aa}\, u)\varphi^{(u)}).$$

By Lemma 7.4, we have

$$(\exists z)(\forall x)(x \in z \leftrightarrow x \in y \wedge \varphi). \dashv$$

8.3 Remark. (i) Indeed, Power Set Axiom is essentially needed in the above proof. (ii) Notice that $(n-1)$-Collection Scheme is provable in $\mathbf{ZD}^n$. However, n-Collection Scheme is, in general, not provable in $\mathbf{ZD}^n$.

8.4 Definition. By $\mathbf{RF}'^n$, we mean the following reflection scheme:

$$(\forall x)(\exists u)(x \in u \wedge u \text{ is transitive} \wedge (\forall x_1 \cdots x_n \in u)(\varphi \leftrightarrow \varphi^{(u)})),$$

where φ is an n-formula with free variables $x_1 \dots x_n$.

The proof of the following lemma is due to A. Lévy ([15]).

8.5 Lemma. *$\mathbf{RF}^n$ and $\mathbf{RF}'^n$ are equivalent under $\mathbf{ZD}^n$.*

PROOF. First, we show that $\mathbf{ZD}^n + \mathbf{RF}'^n \vdash \mathbf{RF}^n$. By using Σ_0-Separation, we can define the empty set 0. Applying Σ_0–Separation and $\mathbf{RF}'^0$ iteratively, we can define the set $\boldsymbol{n}$ for every meta-natural number n, where $\mathbf{0} = 0$, and $(\forall x)(x \in \boldsymbol{n+1} \leftrightarrow x \in \boldsymbol{n} \vee x = \boldsymbol{n})$. Now, let $\varphi_0, \dots, \varphi_l$ be n-formulas with free variables among $x_1, \dots, x_m$. Let ψ be an n-formula defined by

$$\psi \leftrightarrow \bigvee_{k=0}^{l} (y = \boldsymbol{k} \wedge \varphi_k).$$

Then, by $\mathbf{RF}'^n$,

$$(\exists u)(\boldsymbol{l} \in u \wedge u \text{ is transitive} \wedge (\forall x_1 \cdots x_m y \in u)(\psi \leftrightarrow \psi^{(u)})).$$

By this, it can be easily seen that

$$(\exists u)(u \text{ is transitive} \wedge (\forall x_1 \cdots \forall x_m \in u) \bigwedge_{k=0}^{l} (\varphi_k \leftrightarrow \varphi_k^{(u)})).$$

To prove that $\mathbf{ZD}^n + \mathbf{RF}^n \vdash \mathbf{RF}'^n$, recall that $\mathbf{SD}^{n+1}$ is a consevative extension of $\mathbf{SD}^n + \mathbf{RF}^n$. (Theorem 7.10.) By Lemma 7.6, it is clear that $\mathbf{RF}'^n$ is provable in $\mathbf{SD}^{n+1}$. Therefore, $\mathbf{RF}'^n$ is provable in $\mathbf{SD}^n + \mathbf{RF}^n$, it a fortiori, in $\mathbf{ZD}^n + \mathbf{RF}^n$. $\dashv$

8.6 Lemma. *$\mathbf{ZD}^1$ is a conservative extension of $\mathbf{ZF}$, that is, $\mathbf{ZD}^1 \vdash \varphi$ iff $\mathbf{ZF} \vdash \varphi$ for any formula φ of $\mathcal{L}$.*

PROOF. By Theorem 7.10, $\mathbf{ZD}^1$ is a consevative extension of $\mathbf{ZD}^0 + \mathbf{RF}^0$. By Lemma 8.2 and Remark 8.3, Collection and Separation are provable in $\mathbf{ZD}^1$. It follows that $\mathbf{ZD}^1$ is an extension of $\mathbf{ZF}$. Since $\mathbf{RF}^0$ is provable in $\mathbf{ZF}$, $\mathbf{ZF}$ is an extension of $\mathbf{ZD}^0 + \mathbf{RF}^0$.

It follows that **ZF** and $\mathbf{ZD}^0 + \mathbf{RF}^0$ are equivalent theories. Thus, $\mathbf{ZD}^1$ is a conservative extension of **ZF**.

By Lemma 8.6 and Lemma 8.2, we can have $\mathbf{ZD}^n = \mathbf{ZF}^n + \mathbf{Dual}^n$ $(n \geq 1)$, and so $\bigcup_{n<\omega} \mathbf{ZD}^n = \mathbf{ZF}(\text{aa}) + \mathbf{Dual}$.

Now, let us discuss the definability of satisfaction predicates of n-formulas. In the case that the quantifier aa is self-dual, circumstances are much neater than the general case of $\mathbf{ZF}(\text{aa})$. (See §6.)

By induction on n $(n \geq 1)$, we defined, in $\mathbf{ZD}^n$, a binary predicate $\mathbf{ST}^{n-1}$ and a unary operation q^n as follows:

$n = 1$: $\mathrm{St}^0(\varphi, s) \leftrightarrow \varphi \in \mathrm{Form}^0 \wedge s$ is an assignment for $\varphi \wedge (\text{aa}\, x)(\langle x; \in\rangle \models \varphi[s])$;

$q^1(x) = \{\{y \in x \mid \langle x; \in\rangle \models \varphi[s_{i/y}]\} \mid \varphi \in \mathrm{Form}^0 \wedge s$ is an assignment for φ such that $\mathrm{range}(s) \subseteq x \wedge \mathrm{St}^0(\varphi, s_{i/x}) \wedge i \in \mathrm{dom}(s)\}$.

$n > 1$: $\mathrm{St}^{n-1}(\varphi, s) \leftrightarrow \varphi \in \mathrm{Form}^0 \wedge s$ is an assignment for φ $\wedge\, (\text{aa}\, x)((\langle x; \in\rangle, q^{n-1}(x)) \models \varphi[s])$;

$q^n(x) = \{\{y \in x \mid (\langle x; \in\rangle, q^{n-1}(x) \models \varphi[s_{i/y}]\} \mid \varphi \in \mathrm{Form}^{n-1} \wedge s$ is an assignment for $(\text{aa}\, v_i)\varphi$ such that $\mathrm{range}(s) \subseteq x \wedge \mathrm{St}^{n-1}(\varphi, s_{i/x})\}$.

8.7 Lemma. *The following are provable in* $\mathbf{ZD}^n$:

$(\mathrm{i})_n$ $(n > 1)$: $(\text{aa}\, x)(\forall \varphi \in \mathrm{Form}_{n-2})(\forall s)(s$ is an assignment for φ such that $\mathrm{range}(s) \subseteq x \rightarrow ((\langle x; \in\rangle, q^{n-2}(x)) \models \varphi[s] \leftrightarrow \mathrm{St}^{n-2}(\varphi, s)))$.

(ii) : $\mathrm{St}^0(v_i = v_j, s) \leftrightarrow s_i = s_j$.

(iii) : $\mathrm{St}^0(v_i \in v_j, s) \leftrightarrow s_i \in s_j$.

$(\mathrm{iv})_n$: $\mathrm{St}^{n-1}(\neg\varphi, s) \leftrightarrow \neg\mathrm{St}^{n-1}(\varphi, s)$.

$(\mathrm{v})_n$: $\mathrm{St}^{n-1}(\varphi \wedge \psi, s) \leftrightarrow \mathrm{St}^{n-1}(\varphi, s) \wedge \mathrm{St}^{n-1}(\psi, s)$.

$(\mathrm{vi})_n$: $\mathrm{St}^{n-1}((\forall v)_i\varphi, s) \leftrightarrow (\forall x)\, \mathrm{St}^{n-1}(\varphi, s_{i/x})$.

$(\mathrm{vii})_n$ $(n > 1)$: $\mathrm{St}^{n-1}((\text{aa}\, v_i)\varphi, s) \leftrightarrow (\text{aa}\, x)\, \mathrm{St}^{n-2}(\varphi, s_{i/x})$.

$(\mathrm{viii})_n$ $(n > 1)$: $\mathrm{St}^{n-1}(\varphi, s) \leftrightarrow \mathrm{St}^{n-2}(\varphi, s)$ for all $\varphi \in \mathrm{Form}^{n-2}$.

Proof. The proof will be carried out by induction on n. To simplify the notation, (1) $x \models_k \varphi[s]$, (2) $\models_k \varphi[s]$ and (3) $x \prec_k \mathbf{V}$ mean

(1) $(\langle x; \in\rangle, q^k(x)) \models \varphi[s]$,

(2) $\mathrm{St}^k(\varphi, s)$,

and

(3) $(\forall \varphi \in \mathrm{Form}^k)(\forall s)(s$ is an assignment for φ such that $\mathrm{range}(s) \subseteq x \rightarrow (x \models_k \varphi[s] \leftrightarrow \models_k \varphi[s])$

respectively.

For the case that $n = 1$, (ii), (iii), $(\text{iv})_1$, $(\text{v})_1$ and $(\text{vi})_1$ are provable in $\mathbf{ZF}^1$, *a fortiori*, in $\mathbf{ZD}^1$.

Suppose that $n > 1$. $(\text{i})_n$ is clear from the definition of St^{n-2}. It is easily seen that $(\text{iv})_n$, $(\text{v})_n$ and $(\text{vi})_n$ are provable in $\mathbf{ZD}^n$. To show the remainder, we need the following claim:

Claim. Let $\theta_k(x)$ denote the following formula:

$$(\forall\varphi \in \mathrm{Form}^{k-2})(\forall s)(s \text{ is an assignment for } \varphi \text{ such that} \\ \rightarrow \mathrm{range}(s) \subseteq x)(x \models_{k-1} \varphi[s] \leftrightarrow x \models_{k-2} \varphi[s])).$$

Then, $(\mathrm{aa}\, x)\theta_n(x)$ is provable in $\mathbf{ZD}^n$.

Proof of Claim: Let x be such that $x \prec_{n-2} \mathbf{V}$ and $\theta_{n-2}(x)$. We show, by induction on the complexity of φ, that $\theta_n(x)$. The only non-trivial induction step is that φ is of the form $(\mathrm{aa}\, v_i)\psi$. Suppose that

$$x \models_{n-1} (\mathrm{aa}\, v_i)\psi[s], \quad \text{were } s \text{ is an assignment for } (\mathrm{aa}\, v_i)\psi \text{ such that } \mathrm{range}(s) \subseteq x.$$

Then, we have

$$\{\, y \in x \mid x \models_{n-1} \psi[s_{s/i}] \,\} \in q^{n-1}(x).$$

By the induction hypothesis,

$$\{\, y \in x \mid x \models_{n-2} \psi[s_{i/y}] \,\} = \{\, y \in x \mid x \models_{n-1} \psi[s_{i/y}] \,\} \in q^{n-1}(x).$$

By the definition of $q^{n-1}(x)$, there exists a $\theta \in \mathrm{Form}^{n-2}$ and an assignment t for $(\mathrm{aa}\, v_j)\theta$ such that

$$\{\, y \in x \mid x \models_{n-2} \psi[s_{i/y}] \,\} = \{\, y \in x \mid x \models_{n-2} \theta[t_{j/y}] \,\} \quad \text{and} \quad \models_{n-2} \theta[t_{j/x}].$$

From $x \prec_{n-2} \mathbf{V}$, it follows that

$$\models_{n-2} \psi[s_{i/x}],$$

so by $(\text{viii})_{n-1}$,

$$\models_{n-3} \psi[s_{i/x}]. \tag{1}$$

Since $\theta_{n-1}(x)$, we have

$$\{\, y \in x \mid x \models_{n-3} \psi[s_{i/y}] \,\} = \{\, y \in x \mid x \models_{n-2} \psi[s_{i/y}]. \tag{2}$$

So, by the definition of $q^{n-2}(x)$ and (1), we have

$$\{\, y \in x \mid x \models_{n-3} \psi[s_{i/y}] \,\} \in q^{n-2}(x).$$

so by (2),

$$\{\, y \in x \mid x \models_{n-2} \psi[s_{i/y}] \,\} \in q^{n-2}(x).$$

Therefore,

$$x \models_{n-2} (\mathrm{aa}\, v_i)\psi[s].$$

Conversely, suppose that

$$x \models_{n-2} (\mathrm{aa}\, v_i)\psi[s].$$

Then,

$$\{\, y \in x \mid x \models_{n-2} \psi[s_{i/y}] \,\} \in q^{n-2}(x).$$

By the definition of $q^{n-2}(x)$, there exists a $\theta \in \mathrm{Form}^{n-3}$ and an assignment t for $(\mathit{aa}\, v_j)\theta$ such that

$$\{\, y \in x \mid x \models_{n-2} \psi[s_{i/y}] \,\} = \{\, y \in x \mid x \models_{n-3} \theta[t_{j/y}] \,\} \quad \text{and} \quad \models_{n-3} \theta[t_{j/x}]. \tag{3}$$

Since $\theta_{n-1}(x)$, we have

$$\{\, y \in x \mid x \models_{n-2} \theta[t_{j/y}] \,\} = \{\, y \in x \mid x \models_{n-3} \theta[t_{j/y}] \,\}. \tag{4}$$

By (3) and (4),

$$\{\, y \in x \mid x \models_{n-2} \psi[s_{i/y}] \,\} = \{\, y \in x \mid x \models_{n-2} \theta[t_{j/y}] \,\}. \tag{5}$$

On the other hand, by $(\mathrm{viii})_{n-1}$,

$$\models_{n-2} \theta[t_{j/x}]. \tag{6}$$

By $x \prec_{n-2} \mathbf{V}$, (5) and (6),

$$\models_{n-2} \psi[s_{i/x}].$$

Then, by the definition of $q^{n-1}(x)$,

$$\{\, y \in x \mid x \models_{n-2} \psi[s_{i/y}] \,\} \in q^{n-1}(x).$$

By using the induction hypothesis, we have

$$\{\, y \in x \mid x \models_{n-1} \psi[s_{i/y}] \,\} \in q^{n-1}(x).$$

It follows that

$$x \models_{n-1} (\mathit{aa}\, v_i)\psi[s].$$

Thus, we have shown that

$$x \prec_{n-2} \mathbf{V} \wedge \theta_{n-1}(x) \rightarrow \theta_n(x). \tag{7}$$

By $(\mathrm{viii})_{n-1}$, it is clear that $(\mathit{aa}\, x)\theta_{n-1}(x)$. By $(\mathrm{i})_n$, $(\mathit{aa}\, x)(x \prec_{n-2} \mathbf{V})$. Therefore, it follows from (7) that $(\mathit{aa}\, x)\theta_n(x)$. Thus, our claim is complete.

Now, $(\mathrm{viii})_n$ is clear from our claim. To show that $(\mathrm{vii})_n$, let x be such that $x \prec_{n-2} \mathbf{V}$ and $\theta_n(x)$. Suppose that $x \models_{n-1} (\mathit{aa}\, v_i)\varphi[s]$, where $\varphi \in \mathrm{Form}^{n-2}$ and s is an assignment for $(\mathit{aa}\, v_i)\varphi$ such that $\mathrm{range}(s) \subseteq x$. then, we have

$$\{\, y \in x \mid x \models_{n-1} \varphi[s_{i/y}] \,\} \in q^{n-1}(x).$$

By using $\theta_n(x)$, we have

$$\{\, y \in x \mid x \models_{n-2} \varphi[s_{i/y}] \,\} \in q^{n-1}(x).$$

By using $x \prec_{n-2} \mathbf{V}$, we can see that

$$\models_{n-2} \varphi[s_{i/x}].$$

Conversely, suppose that

$$\models_{n-2} \varphi[s_{i/x}].$$

Then, by the definition of $q^{n-1}(x)$,

$$\{ y \in x \mid x \models_{n-2} \varphi[s_{i/y}] \} \in q^{n-1}(x).$$

Since $\theta_n(x)$, we have

$$\{ y \in x \mid x \models_{n-1} \varphi[s_{i/y}] \} \in q^{n-1}(x).$$

That is,

$$x \models_{n-1} (\text{aa}\, v_i)\varphi[s].$$

Thus, we have shown that

$$x \prec_{n-2} \mathbf{V} \wedge \theta_n(x) \rightarrow (\forall \varphi \in \mathrm{Form}^{n-2})(\forall s)(s \text{ is an assignment for } (\text{aa}\, v_i)\varphi$$
$$\text{such that } \mathrm{range}(s) \subseteq x \rightarrow (x \models_{n-1} (\text{aa}\, v_i)\varphi[s] \leftrightarrow \models_{n-2} \varphi[s_{i/x}])).$$

Since $(\text{aa}\, x)(x \prec_{n-2} \mathbf{V})$ and $(\text{aa}\, x)\theta_n(x)$, we have

$$(\text{aa}\, x)\left(x \models_{n-1} (\text{aa}\, v_i)\varphi[s]\right) \leftrightarrow (\text{aa}\, x) \models_{n-2} \varphi[s_{i/x}] \quad \text{for all } \varphi \in \mathrm{Form}^{n-2}.$$

That is,

$$\models_{n-1} (\text{aa}\, v_i)\varphi[s] \leftrightarrow (\text{aa}\, x) \models_{n-2} \varphi[s_{i/x}].$$

8.8 Definition. By $\mathbf{ZD}^{+n}$, we mean the theory obtained by adding n-Collection Scheme to $\mathbf{ZD}^n$. By Lemma 8. 7, we have the following theorem.

8.9 Theorem. *In* $\mathbf{ZD}^n$ $(n \geq 1)$, *we can define a binary predicate* St^{n-1} *which constitutes a definition of satisfaction for n-formulas over the universe, that is, which satisfies* (ii) *to* (viii) *in Lemma 8.7. Moreover, in* $\mathbf{ZD}^{+n}$ $(n \geq 1)$, *the collection scheme ranging over all formulas of the language* $\mathcal{L} \cup \{\mathrm{St}^{n-1}\}$ *is provable.* ⊣

8.10 Corollary. *Let n be a natural number* ≥ 1. *Then,*

(i) $\mathbf{ZD}^n \vdash \mathrm{Cons}(\mathbf{ZD}^{n-1})$.
(ii) $\mathbf{ZD}^{+n} \vdash (\forall \alpha)(\exists \beta > \alpha)(\exists q \subseteq R_{\beta+1})\left((\langle R_\beta ; \in \rangle, q) \prec_{n-1} \mathbf{V}\right)$.
(iii) $\mathbf{ZD}^{n+1} \vdash (\text{aa}\, x)(x \prec_{n-1} \mathbf{V})$. ⊣

§9. The rôle of Global Choice Axiom

It is an interesting problem to axiomatize the set of sentences φ of $\mathcal{L}$ such that $\mathbf{ZF}(aa) + \mathbf{Dual} \vdash \varphi$. Though the present author does not know any answer of this problem, we can give an answer for this problem if the axiom of global choice is added to $\mathbf{ZF}(aa) + \mathbf{Dual}$.

In this section, we assume that $\mathcal{L}_\sigma$ is the language obtained by adding a unary function symbol σ to the language of **ZF**.

9.1 Definition. (i) We call the following sentence $\mathbf{AC}_\sigma$ the axiom of global choice:

$$(\forall x)(x \neq 0 \rightarrow \sigma(x) \in x).$$

(ii) By $\mathbf{ZF}_\sigma$, we mean the theory obtained by adding $\mathbf{AC}_\sigma$ to **ZF** (relative to $\mathcal{L}_\sigma$). The meanings of $\mathbf{ZF}_\sigma(aa)$, $\mathbf{ZD}_\sigma^n$ are self-evident.

9.2 Definition. (i) Let A be a set of ordinals. A sequence

$$\overline{S} = \langle S_{\beta_1 \ldots \beta_n} : \beta_1 < \cdots < \beta_n, \beta_1, \ldots, \beta_n \in A \rangle$$

is an (n, A)-sequence if $S_{\beta_1 \ldots \beta_n} \subseteq \beta_1$ for all $\beta_1, \ldots, \beta_n$.

(ii) A set X is homogeneous for an (n, A)-sequence $\overline{S}$ iff $X \subseteq A$ and for any two sequences $\beta_1 < \cdots < \beta_n$, $\beta_1' < \cdots < \beta_n'$ from X, if $\beta_1 \leq \beta_1'$ then $S_{\beta_1 \ldots \beta_n} = \beta_1 \cap S_{\beta_1' \ldots \beta_n'}$.

9.3 Definition. (i) A cardinal κ is said to be n-subtle if for every (n, κ)-sequence $\overline{S}$ and every closed unbounded subset $C \subseteq \kappa$, there is an $x \subseteq C$ such that $\overline{\overline{x}} = n + 1$ and x is homogeneous for $\overline{S}$.

(ii) A cardinal κ is said to be n-ineffable if every (n, κ)-sequence has a homogeneous set which is stationary on κ.

For more detail about n-subtle and n-ineffable cardinals, the reader should refer to [2].

9.4 Definition. Let $\mathbf{RF}_\infty$ denote the following reflection scheme:

$$(\exists \alpha)(\alpha \text{ is } n\text{-subtle} \wedge (\forall x_1 \cdots x_m \in R_\alpha)(\varphi \leftrightarrow \varphi^{(R_\alpha)}))$$

where n is a (meta)natural number and φ is a formula of the language of **ZF** with free variables among $x_1, \ldots, x_m$.

The following theorem is the main result of this section.

9.5 Theorem. $\mathbf{ZF}_\sigma(aa) + \mathbf{Dual}$ *is a conservative extension of* $\mathbf{ZFC} + \mathbf{RF}_\infty$.

Theorem 9.5 can be considered as a refinement of the following theorem, which was earlier obtained by Kaufmann ([**10**,Th. 5.2]).

9.6 Theorem. $\mathbf{ZF}(aa) + \mathbf{Dual}$ *and* $\mathbf{ZFC} + \{(\exists \kappa)(\kappa \text{ is } n\text{-ineffable} \mid n < \omega\}$ *are equiconsistent.*

We shall commence to see the strength of $\mathbf{ZD}_\sigma^n$.

9.7 Lemma. $\mathbf{ZD}^1_\sigma + 1\text{-}Collection \vdash (\mathsf{aa}\,\alpha)(\alpha \text{ is a strongly inaccessible cardinal})$.

PROOF. Since $(\mathsf{aa}\,x)(\exists\alpha)(x = R_\alpha)$, it suffices to show that

$$(\mathsf{aa}\,x)(\forall y \in x)(\forall f)(f : y \to x \text{ implies } (\exists u \in x)(f\text{''}y \subseteq u)).$$

Suppose not. Then, by **Dual**[1], we have

$$(\mathsf{aa}\,x)(\exists y \in x)(\exists f)(f : y \to x \wedge (\forall u \in x)(f\text{''}y \not\subseteq u)).$$

It follows that

$$(\exists y)(\mathsf{aa}\,x)(\exists f)(f : y \to x \wedge (\forall u \in x)(f\text{''}y \not\subseteq u)).$$

Fix a set y such that

$$(\mathsf{aa}\,x)(\exists f)(f : y \to x \wedge (\forall u \in x)(f\text{''}y \not\subseteq u)).$$

Set $f_x = \sigma(\{ f \mid f : y \to x \wedge (\forall u \in x)(f\text{''}y \not\subseteq u) \})$. Then,

$$(\mathsf{aa}\,x)(f_x : y \to x \wedge (\forall u \in x)(f\text{''}_x y \not\subseteq u)),$$

and so,

$$(\mathsf{aa}\,x)(\forall z \in y)(\exists w \in x)(f_x : y \to x \wedge f_x(z) = w \wedge (\forall u \in x)(f_x\text{''}y \not\subseteq u)).$$

It follows that

$$(\forall z \in y)(\exists w)(\mathsf{aa}\,x)(f_x : y \to x \wedge f_x(z) = w \wedge (\forall u \in x)(f_x\text{''}y \not\subseteq u)).$$

By 1-Collection, we have

$$(\exists u)(\forall z \in y)(\exists w \in u)(\mathsf{aa}\,x)(f_x : y \to x \wedge f_x(z) = w \wedge (\forall u \in x)(f_x\text{''}y \not\subseteq u)).$$

It follows that

$$(\exists u)(\mathsf{aa}\,x)(\forall z \in y)(\exists w \in u)(f_x : y \to x \wedge f_x(z) = w \wedge (\forall u \in x)(f_x\text{''}y \not\subseteq u)).$$

That is,

$$(\exists u)(\mathsf{aa}\,x)(f_x : y \to x \wedge f_x\text{''}y \subseteq u \wedge (\forall u \in x)(f_x\text{''}y \not\subseteq u)).$$

It follows that

$$(\exists u)(\mathsf{aa}\,x)(u \notin x).$$

This is a contradiction. ⊣

9.8 Definition. Let $\mathbf{IRF}_\sigma$ denote the following reflection scheme:

$$(\exists\alpha)(\alpha \text{ is strongly inaccessible} \wedge (\forall x_1 \cdots x_m \in R_\alpha)(\varphi \leftrightarrow \varphi^{(R_\alpha)})),$$

where φ is a formula of the language $\mathcal{L}_\sigma$ with free variables $x_1, \ldots, x_m$.

9.9 Lemma. $\mathbf{ZD}^1_\sigma + 1$-Collection *is a conservative extension of* $\mathbf{ZF}_\sigma + \mathbf{IRF}_\sigma$.

PROOF. By Lemma 7.6,

$$(\mathsf{aa}\,\alpha)(\forall x_1 \cdots x_m \in R_\alpha)(\varphi \leftrightarrow \varphi^{(R_\alpha)})$$

(where φ is a formula of $\mathcal{L}_\sigma$ with free variables $x_1, \ldots, x_m$) is provable in $\mathbf{ZD}^1_\sigma$. By Lemma 9.7 and this, $\mathbf{IRF}_\sigma$ is provable in $\mathbf{ZD}^1_\sigma$+1-Collection. Thus, $\mathbf{ZD}^1_\sigma$+1-Collection is an extension of $\mathbf{ZF}_\sigma + \mathbf{IRF}_\sigma$. By the method of Lemma 8.5, we can easily see that

$$\mathbf{SD}^0_c\,(\text{relative to } \mathcal{L}_\sigma) + \mathbf{ZF}_\sigma + (\exists\alpha)(\alpha \text{ is strongly inaccessible} \wedge c = R_\alpha)$$

is a conservative extension of $\mathbf{ZF}_\sigma + \mathbf{IRF}_\sigma$. Let φ be a sentence of $\mathcal{L}_\sigma$. Suppose that $\mathbf{ZD}^1_\sigma$ + 1-Collection $\vdash \varphi$. Then, by Lemma 7.9, we have

$$\mathbf{SD}^0_c\,(\text{relative to } \mathcal{L}_\sigma) + \mathbf{ZF}_\sigma \vdash \theta_1^{(c)} \wedge \cdots \wedge \theta_n^{(c)} \rightarrow \varphi^{(c)},$$

where each θ_i had the scheme of 1-Collection. Therefore, it is enough to show that

$$\mathbf{SD}^0_c\,(\text{relative to } \mathcal{L}_\sigma) + \mathbf{ZF}_\sigma + (\exists\alpha)(\alpha \text{ is strongly inaccessible} \wedge c = R_\alpha) \vdash \theta^{(c)},$$

where θ has the scheme of 1-Collection.

We may suppose that θ is the universal closure of the following formula:

$$(\forall z)[(\forall x \in z)(\exists y)(\text{aa}\, w)\psi(x,y,z,w) \rightarrow (\exists u)(\forall x \in z)(\exists y \in u)((\text{aa}\, w)\psi(x,y,z,w)]$$

where ψ is a formula of $\mathcal{L}_\sigma$ in which u does not occur free.

Thus, it is enough to show that

$$(\forall z \in c)[(\forall x \in z)(\exists y \in c)\psi(x,y,z,c) \rightarrow (\exists u \in c)(\forall x \in z)(\exists y \in u)\psi(x,y,z,c)]$$

is provable in

$$\mathbf{SD}^0_c\,(\text{relative to } \mathcal{L}_\sigma) + \mathbf{ZF}_\sigma + (\exists\alpha)(\alpha \text{ is strongly inaccessible} \wedge c = R_\alpha).$$

However, it is clear from the fact that

$$(\forall z \in R_\alpha)(\forall f)(f : z \rightarrow R_\alpha \quad \text{implies} \quad f\text{''} z \in R_\alpha)$$

for any strongly inaccessible cardinal α. ⊣

9.10 Lemma. $\mathbf{ZD}^n_\sigma \vdash n$-Collection $(n \geq 2)$.

PROOF. We would like to show that, on the condition that φ is an n-formula in which u does not occur free,

$$(\forall x \in z)(\exists y)\varphi \rightarrow (\exists u)(\forall x \in z)(\exists y \in u)\varphi$$

is provable in $\mathbf{ZD}^n_\sigma$. By Lemma 7.4, it is enough to show that

$$(\text{aa}\, w)(\forall z \in w)[(\forall x \in z)(\exists y \in w)\varphi^{(w)} \rightarrow (\exists u \in w)(\forall x \in z)(\exists y \in u)\varphi^{(w)}]. \quad (1)$$

But, we have, for any inaccessible cardinal α,

$$(\forall z \in R_\alpha)[(\forall x \in z)(\exists y \in w)\varphi^{(R_\alpha)} \rightarrow (\exists u \in R_\alpha)(\forall x \in z)(\exists y \in u)\varphi^{(R_\alpha)}]. \quad (2)$$

By Lemma 9.7, we have

$$(\text{aa}\, \alpha)(\alpha \text{ is strongly inaccessible}) \quad (3)$$

is provable in $\mathbf{ZD}^n_\sigma$ ($n \geq 2$), because

$$\mathbf{ZD}^n_\sigma \vdash \text{1-Collection}$$

for $n \geq 2$. Therefore, by (2) and (3), (1) is provable in $\mathbf{ZD}^n_\sigma$. ⊣

The proof of the following lemma is a syntactical version of the proof that $\{\, \alpha < \kappa \mid \alpha \text{ is an } n\text{-subtle cardinal} \,\}$ is in any normal ultrafilter on a measurable cardinal κ. The proof of this type first appeared in [**10**, Th. 5.6]. We can see a syntactical version of flipping type arguments in this proof.

9.11 Lemma. $\mathbf{ZD}^{n+2}_\sigma \vdash (\mathrm{aa}\,\alpha)(\alpha \text{ is } n\text{-subtle})$, *where* $n \geq 1$

PROOF. Suppose not. By using $\mathbf{AC}_\sigma$, for almost all ordinal α, we can choose a (n,α)-sequence $\langle S^\alpha_{\zeta_0\ldots\zeta_{n-1}} : \zeta_0 < \cdots < \zeta_{n-1} < \alpha \rangle$ and a cub subset K_α of α so that $S^\alpha_{\zeta_0\ldots\zeta_{n-1}} \neq \zeta_0 \cap S^\alpha_{\zeta_1\ldots\zeta_n}$ for any $\zeta_0 < \cdots < \zeta_n$ from K_α. We define an (n, On)-sequence $\langle S_{\zeta_0\ldots\zeta_{n-1}} : \zeta_0 < \cdots < \zeta_{n-1} \rangle$ as follows:

$$S_{\zeta_0\ldots\zeta_{n-1}} = \{\, \beta \in \zeta_0 \mid (\mathrm{aa}\,\alpha)(\beta \in S^\alpha_{\zeta_0\ldots\zeta_{n-1}}) \,\}$$

We define $\theta_0(\zeta), \ldots, \theta_{n-1}(\zeta)$ as follows:

$$\begin{aligned} \theta_0(\zeta_0) \leftrightarrow (\forall \beta < \zeta_0)((\mathrm{aa}\,\zeta_1)\cdots(\mathrm{aa}\,\zeta_{n-1})\,(\beta \in S_{\zeta_0\ldots\zeta_{n-1}}) \\ \leftrightarrow (\mathrm{aa}\,\zeta_0)\cdots(\mathrm{aa}\,\zeta_{n-1})\,(\beta \in S_{\zeta_0\ldots\zeta_{n-1}})), \end{aligned}$$

$$\vdots$$

$$\begin{aligned} \theta_i(\zeta_i) \leftrightarrow (\forall \zeta_{i-1} < \zeta_i)(\forall \zeta_{i-2} < \zeta_{i-1})\cdots(\forall \beta < \zeta_0)((\mathrm{aa}\,\zeta_{i+1})\cdots(\mathrm{aa}\,\zeta_{n-1})(\beta \in S_{\zeta_0\ldots\zeta_{n-1}}) \\ \leftrightarrow (\mathrm{aa}\,\zeta_i)\cdots(\mathrm{aa}\,\zeta_{n-1})(\beta \in S_{\zeta_0\ldots\zeta_{n-1}})), \end{aligned}$$

$$\vdots$$

$$\begin{aligned} \theta_{n-1}(\zeta_{n-1}) \leftrightarrow (\forall \zeta_{n-2} < \zeta_{n-1})(\forall \zeta_{n-3} < \zeta_{n-2})\cdots(\forall \beta < \zeta_0)(\beta \in S_{\zeta_0\ldots\zeta_{n-1}} \\ \leftrightarrow (\mathrm{aa}\,\zeta_{n-1})(\beta \in S_{\zeta_0\ldots\zeta_{n-1}})). \end{aligned}$$

It is not hard to see that $(\mathrm{aa}\,\zeta)\theta_i(\zeta)$ for each $i < n$. Put $\theta(\zeta) \equiv \bigwedge_{i<n} \theta_i(\zeta)$. Then, $(\mathrm{aa}\,\zeta)\theta(\zeta)$. We claim that

$$\begin{aligned} (\forall \zeta_0 \cdots \forall \zeta_{n-1})(\forall \eta_0 \cdots \forall \eta_{n-1})(\zeta_0 < \cdots < \zeta_{n-1} \to \eta_0 < \cdots < \eta_{n-1} \to \zeta_0 \leq \eta_0 \\ \to \bigwedge_{i<n} \theta(\zeta_i) \to \bigwedge_{i<n} \theta(\eta_i) \to S_{\zeta_0\ldots\zeta_{n-1}} = \zeta_0 \cap S_{\eta_0\ldots\eta_{n-1}}). \end{aligned}$$

To prove this claim, let $\zeta_0 < \cdots < \zeta_{n-1}$ and $\eta_0 < \cdots < \eta_{n-1}$ be such that $\zeta_0 \leq \eta_0$, $\bigwedge_{i<n} \theta(\zeta_i)$ and $\bigwedge_{i<n} \theta(\eta_i)$. Since $\bigwedge_{i<n} \theta(\zeta_i)$ and $\bigwedge_{i<n} \theta(\eta_i)$, we can easily see that

$$(\forall \beta < \zeta_0)(\beta \in S_{\zeta_0\ldots\zeta_{n-1}} \leftrightarrow (\mathrm{aa}\,\zeta_0)\cdots(\mathrm{aa}\,\zeta_{n-1})(\beta \in S_{\zeta_0\ldots\zeta_{n-1}})),$$

and

$$(\forall\beta < \eta_0)(\beta \in S_{\eta_0\ldots\eta_{n-1}} \leftrightarrow (\text{aa}\,\eta_0)\cdots(\text{aa}\,\eta_{n-1})(\beta \in S_{\eta_0\ldots\eta_{n-1}})).$$

Since $\zeta_0 \leq \eta_0$, it follows that

$$(\forall\beta < \zeta_0)(\beta \in S_{\zeta_0\ldots\zeta_{n-1}} \leftrightarrow \beta \in S_{\eta_0\ldots\eta_{n-1}}).$$

Thus, our claim has been proved.
Now, we define

$$\psi(\zeta) \equiv (\text{aa}\,\alpha)(\zeta \in K_\alpha).$$

By the same argument of Lemma 2.18, we have

$$(\text{aa}\,\zeta)\psi(\zeta).$$

Since $(\text{aa}\,\zeta)(\theta(\zeta) \wedge \psi(\zeta))$, we can have an increasing sequence $\zeta_0 < \cdots < \zeta_n$ of ordinals so that

$$\bigwedge_{i\leq n} (\theta(\zeta_i) \wedge \psi(\zeta_i)).$$

Then,

$$(\text{aa}\,\alpha)(\zeta_0 < \cdots < \zeta_n < \alpha \wedge \zeta_0, \ldots, \zeta_n \in K_\alpha)$$

By the choice of $\langle S^\alpha_{\zeta_0\ldots\zeta_{n-1}} : \zeta_0 < \cdots < \zeta_{n-1} < \alpha \rangle$ and K_α, we have

$$(\text{aa}\,\alpha)(\exists\beta < \zeta_0)(\neg(\beta \in S^\alpha_{\zeta_0\ldots\zeta_{n-1}} \leftrightarrow \beta \in S^\alpha_{\zeta_1\ldots\zeta_n}))$$

It follows that

$$(\exists\beta < \zeta_0)\neg(\beta \in S_{\zeta_0\ldots\zeta_{n-1}} \leftrightarrow \beta \in S_{\zeta_1\ldots\zeta_n}) \tag{1}$$

On the other hand, since $\bigwedge_{i\leq n} \theta(\zeta_i)$, we have

$$S_{\zeta_0\ldots\zeta_{n-1}} = \zeta_0 \cap S_{\zeta_1\ldots\zeta_{n-1}}. \tag{2}$$

Thus, (1) and (2) contradict each other. ⊣

9.12 Lemma (ZFC). *Let κ be an n-subtle cardinal, and F be a choice function on R_κ, that is, $F : R_\kappa \to R_\kappa$ and $(\forall x \in R_\kappa)(x \neq 0 \to F(x) \in x)$. Then, there exists an increasing sequence $\alpha_0 < \cdots < \alpha_n < \kappa$ such that $\langle R_\kappa; \in, F, R_{\alpha_0}, \ldots, R_{\alpha_n} \rangle$ is a mode of* $\mathbf{SD}^0_{c_0\ldots c_n}$ (relative to $\mathcal{L}_\sigma$) $+ \mathbf{ZF}_\sigma$, *where c_k is interpreted by R_{α_k} and σ is interpreted by F.*

PROOF. Let Form_α $(\alpha \leq \kappa)$ be the set of all formulas of $\mathcal{L}_\sigma$ with parameters from R_α. Let $\varphi \mapsto \ulcorner\varphi\urcorner$ be an ono-one correspondence between Form_κ and κ. We identify φ with $\ulcorner\varphi\urcorner$. Put $K = \{\alpha < \kappa \mid \text{Form}_\alpha \subseteq \alpha\}$. It is clear that K is *cub* in κ. We define an (n, K)-sequence $\langle S_{\alpha_0\ldots\alpha_{n-1}} : \alpha_0 < \cdots < \alpha_{n-1}, \alpha_0, \ldots, \alpha_{n-1} \in K \rangle$ as follows:

$$S_{\alpha_0\ldots\alpha_{n-1}} = \{\varphi(x_0\ldots x_{n-1}) \in \text{Form}_{\alpha_0} \mid \langle R_\kappa; \in, F \rangle \models \varphi[R_{\alpha_0}, \ldots, R_{\alpha_{n-1}}]\}$$

Let $C = \{\alpha < \kappa \mid \langle R_\alpha; \in, F \mid R_\alpha \rangle \prec \langle R_\kappa; \in, F \rangle\}$. then, C is *cub* in κ. Since κ is n-subtle, there exists an increasing sequence $\alpha_0 < \cdots < \alpha_n < \kappa$ such that $\{\alpha_0, \ldots, \alpha_n\}$

is homogeneous for $\langle S_{\alpha_0 \dots \alpha_{n-1}} : \alpha_0 < \cdots < \alpha_n \rangle$ and $\{\alpha_0, \dots, \alpha_n\} \subseteq C \cap K$. Clearly, $(R_\kappa; \in, F, R_{\alpha_0}, \dots, R_{\alpha_n})$ is a model of $\mathbf{SD}^0_{c_0 \dots c_n}$ (relative to $\mathcal{L}_\sigma$) + $\mathbf{ZF}_\sigma$.

9.13 *Proof of Theorem 9.5.* By Lemma 9.11, it is clear that $\mathbf{ZF}_\sigma(aa)$ + **Dual** is an extension of **ZFC** + $\mathbf{RF}_\infty$. Suppose that $\mathbf{ZF}_\sigma(aa)$ + **Dual** $\vdash \varphi$, where φ is a sentence of the language of **ZF**. By the compactness, $\mathbf{ZD}^n_\sigma \vdash \varphi$ for some $n \geq 2$. By Theorem 7.13, $\mathbf{SD}^0_{c_0, \dots, c_{n-1}}$ (relative to $\mathcal{L}_\sigma$) + $\mathbf{ZF}_\sigma \vdash \varphi$. By $\mathbf{RF}_\infty$, **ZFC** + $\mathbf{RF}_\infty \vdash$ $(\exists\kappa)(\kappa$ is $(n-1)$-subtle $\wedge (\varphi \leftrightarrow \varphi^{(R_\kappa)}))$. We work in the theory **ZFC** + $\mathbf{RF}_\infty$. Let κ be such that κ is $(n-1)$-subtle and $\varphi \leftrightarrow \varphi^{(R_\kappa)}$. By Lemma 9.12, we can choose a choice function F and an increasing sequence $\alpha_0 < \cdots < \alpha_{n-1} < \kappa$ so that $\langle R_\kappa : \in, F, R_{\alpha_0}, \dots, R_{\alpha_{n-1}} \rangle$ is a model of $\mathbf{SD}^0_{c_0 \dots c_{n-1}}$ (relative to $\mathcal{L}_\sigma$) + $\mathbf{ZF}_\sigma$. Since $\mathbf{SD}^0_{c_0 \dots c_{n-1}}$ (relative to $\mathcal{L}_\sigma$) + $\mathbf{ZF}_\sigma \vdash \varphi$, $\langle R_\kappa; \in \rangle \models \varphi$. Since $\varphi \leftrightarrow \varphi^{(R_\kappa)}$, we have φ. That is, φ is provable in **ZFC** + $\mathbf{RF}_\infty$. The proof is complete. ⊣

By the similar method of the proof of Theorem 9.5, we can give the connection between $\mathbf{ZF}_\sigma(aa)$ + **Dual** and Kelly-Morse class theory **KM**. To precisely express this, we need the following definition.

9.14 Definition (KM). *By the global choice axiom* **E**, *we mean the following sentences:*

$$(\exists F)(F : \mathbf{V} \to \mathbf{V} \wedge (\forall x)(\neq 0 \to F(x) \in x)).$$

We denote theory **KM** + **E** *by* **KME**.

9.15 Definition (KM). *Let n be a natural number. The phrase "On is n-subtle" means the following: For every (n, On)-sequence $\langle S_{\alpha_0 \dots \alpha_{n-1}} \mid \alpha_0 < \cdots < \alpha_{n-1} \rangle$ and every closed unbounded subclass K of On there exists a homogeneous set $x\,(\subseteq K)$ such that $\overline{\overline{x}} = n+1$.*

9.16 Theorem. *For every formula φ of the language of* **ZF**,

$$\mathbf{ZF}_\sigma(aa) + \mathbf{Dual} \vdash \varphi \text{ iff } \mathbf{KME} + \{ On \text{ is } n\text{-subtle} \mid n < \omega \} \vdash \varphi. \dashv$$

By Theorem 9.5 and the above theorem, we have the following corollary.

9.17 Corollary. **KME** + $\{ On$ is n-subtle $\mid n < \omega \}$ *is a conservative extension of* **ZFC** + $\mathbf{RF}_\infty$. ⊣

9.18 Remark. (i) As we have seen so far, the global axiom of choice plays the essential rôle in the dual quantifier. In fact, Kaufmann-Shelah showed that $\mathbf{ZF}(aa)$ + **AC** + **Dual** + $\neg(\exists\kappa)(\kappa$ is Mahlo) is consistent relative to $\mathbf{ZF}(aa)$ + **Dual**. (See [11].)

(ii) It is much plausible that $\mathbf{ZD}^{n+2}_\sigma$ is a conservative extension of **ZFC** + $\mathbf{RF}_n$. Here by $\mathbf{RF}'_n$ we mean the following scheme:

$$(\forall\alpha)(\exists\beta > \alpha)(\beta \text{ is } n\text{-subtle } \wedge (\forall x_1 \cdots x_m \in R_\beta)(\varphi \leftrightarrow \varphi^{R_\beta})),$$

where φ is a formula with free variables among $x_1 \dots x_m$.

However, the author does not know the proof at present.

We close this section for considering inner models of **ZF**(aa) + **Dual**. Let $M(x)$ be a predicate definable in **ZF**(aa) + **Dual**. Suppose that $M(x)$ is an inner model of **ZF**, that is, $M(x)$ satisfies the following conditions:

(i) M is transitive, *i.e.*, $(\forall x)(\forall y \in x)(M(x) \rightarrow M(y))$;
(ii) M contains On, *i.e.*, $(\forall x)(x$ is an ordinal $\rightarrow M(x)))$;
(iii) $\theta^{(M)}$ for every axiom θ of **ZF**.

Let $M_\alpha = R_\alpha^{(M)} = R_\alpha \cap M$.

9.19 Definition. Let φ be a formula of $\mathcal{L}(\mathsf{aa})$. The relativization of φ to M, $\varphi^{(M)}$, is defined as usual by induction on the complexity of φ except the following case: if φ is of the form $(\mathsf{aa}\, x)\psi(x)$, then $\varphi^{(M)}$ is $(\mathsf{aa}\, \alpha)(\exists x)(x = M_\alpha \wedge \psi^{(M)}(x))$.

It is easily seen that the relativization of every axiom of **SD** to M is provable in **ZF**(aa) + **Dual**. Therefore, by Lemma 8.2 and 8.6, we have the following lemma.

9.20 Lemma. *M becomes as inner model of* **ZF**(aa) + **Dual** *under the interpretation of aa as in Definition 9.19.* ⊣

If we take the constructible universe **L** as a special case, **L** becomes an inner model of **ZF**(aa) + **Dual** + **V** = **L**. Since **L** has a definable well-ordering, **L** is an inner model of $\mathbf{ZF}_\sigma$(aa) + **Dual**. Thus, we have the following corollary.

9.21 Corollary. *Cons*(**ZF**(aa) + **Dual**) *implies Cons*($\mathbf{ZF}_\sigma$(aa) + **Dual**). ⊣

9.22 Remark. Though we saw as in the above that **ZF**(aa) + **Dual** behaves naturally with respect to inner models, it is not known whether **ZF**(aa) behaves similarly with respect to inner models. For example, it is not known whether *Cons*(**ZF**(aa)) implies *Cons*(**ZF**(aa)) + **V** = **L**).

§10. Skolem ultrapowers.

Let $(\mathcal{A}, q)$ be a model of **SD**. Then, q is an ultrafilter on the Boolean algebra of definable subsets of the universe of $(\mathcal{A}, q)$. Therefore, we can construct the *Skolem ultrapower* of $(\mathcal{A}, q)$, provided that $(\mathcal{A}, q)$ has built-in Skolem functions.

In this section, we shall present an application of Skolem ultrapowers to study the connection between the self-dual quantifier aa and the method of Ehrenfeucht-Mostowski. Throughout this section, we assume that $\mathcal{L}$ is a language with two binary predicate symbols $\in$ and W (possibly with other non-logical symbols).

10.1 Definition. **SK** is the theory on the language $\mathcal{L}$, of which axioms are the following:

(i) $(\exists x)\varphi(x) \to (\exists x)(\varphi(x) \wedge (\forall y)(\varphi(y) \to xWy))$, where $\varphi(x)$ is a formula of $\mathcal{L}$,
(ii) $(\forall x \forall y)(xWy \wedge yWx \to x = y)$.

For any formula $\psi \equiv (\exists y)\varphi$ of $\mathcal{L}$, we define a function symbol F_ψ in **SK** as follows: Suppose that ψ has exactly the free variables $x_1, \ldots, x_n$. Then, F_ψ is an n-ary function symbol, and its defining axiom is:

$$y = F_\psi(x_1 \ldots x_n) \leftrightarrow ((\exists y)\varphi(yx_1 \ldots x_n) \wedge (\forall z)(\varphi(zx_1 \ldots x_n) \to yWz)) \\ \vee (\neg(\exists y)\varphi(yx_1 \ldots x_n) \wedge (\forall z)(yWz)).$$

It is clear that the defining axiom in the above satisfies the uniqueness and existence condition.

By a *Skolem term* of $\mathcal{L}$, we mean a term of the expanded language by adding a new function symbol F_ψ for each formula $\psi \equiv (\exists x)\varphi$ of $\mathcal{L}$ as in the above. If $\mathcal{A}$ is a model of **SK** and $\tau(x_1 \ldots x_n)$ is a Skolem term of $\mathcal{L}$, the meaning of $\tau^{\mathcal{A}}[x_1 \ldots x_n]$ is clear for $a_1, \ldots, a_n$ in the universe of $\mathcal{A}$.

Let $\mathcal{A}$ be a model of **SK**, and X be a subset of the universe of $\mathcal{A}$. We put $H_{\mathcal{A}}(X) = \{ \tau^{\mathcal{A}}[a_1 \ldots a_n] \mid \tau(x_1 \ldots x_n)$ is a Skolem term of $\mathcal{L}$, and $a_1, \ldots, a_n \in X \}$, and call it the *Skolem hull* of X in $\mathcal{A}$. We let $\mathcal{H}_{\mathcal{A}}(X)$ denote the corresponding elementary submodel of $\mathcal{A}$ determined by the set $H_{\mathcal{A}}(X)$.

10.2 Definition. $\mathbf{SD}_W$ is the theory on the language $\mathcal{L}(\mathit{aa})$ obtained by adding to **SD** (relative to $\mathcal{L}$) the axioms of **SK**.

10.3 Lemma. *The scheme obtained by strengthening the scheme* (i) *in Definition 10.1 so that φ ranges over all formulas of $\mathcal{L}(\mathit{aa})$ is provable in $\mathbf{SD}_W$.*

Proof. Let (W_n) denote the scheme obtained by strengthening the scheme (i) in Definition 10.1 so that $\varphi(x)$ ranges over all n-formulas. Then, it suffices to show that

$$\mathbf{SD} + (W_{n-1}) \vdash (W_n) \quad (n \geq 1).$$

So suppose that $\varphi(x)$ is an n-formula We have to show that

$$(\exists x)\varphi(x) \to (\exists x)(\varphi(x) \wedge (\forall y)(\varphi(y) \to xWy))$$

is provable in $\mathbf{SD}+(W_{n-1})$. For this, by Lemma 7.4, it suffices to show that

$$(\text{aa}\,u)((\exists x\in u)\varphi^{(u)}(x)\to(\exists x\in u(\varphi^{(u)}(x)\wedge(\forall y\in u)(\varphi^{(u)}(y)\to xWy))) \tag{1}$$

is provable in $\mathbf{SD}+(W_{n-1})$. However, by (W_{n-1}),

$$(\exists x\in u)\varphi^{(u)}(x)\to(\exists\in u)(\varphi^{(u)}(x)\wedge(\forall\in u)(\varphi^{(u)}(y)\to xWy)). \tag{2}$$

It is clear that (2) implies (1). ⊣

By Lemma 10.3, we can define a function symbol F_ψ for each formula $\psi\equiv(\exists y)\varphi$ of $\mathcal{L}(\text{aa})$ just as in **SK**. By a Skolem term of $\mathcal{L}(\text{aa})$, we mean a term of the language obtained by adding a function symbol F_ψ for each formula $\psi\equiv(\exists x)\varphi$ of $\mathcal{L}(\text{aa})$ to $\mathcal{L}(\text{aa})$.

Let $(\mathcal{A},q)$ be a model of $\mathbf{SD}_W$, and X be a subset of the universe of $\mathcal{A}$. We can define the Skolem hull $H_{(\mathcal{A},q)}$ of X in $(\mathcal{A},q)$ as in the case of $\mathcal{L}$. Set $q_X=\{B\cap H_{(\mathcal{A},q)}\mid B$ is definable in $(\mathcal{A},q)$ with some parameters from $H_{(\mathcal{A},q)}(X)$ and $B\in q\}$. Then, it is clear that $(\mathcal{H}_{(\mathcal{A},q)}(X),q_X)\prec(\mathcal{A},q)$, where $(\mathcal{H}_{(\mathcal{A},q)}(X),q_X)$ is the corresponding elementary submodel of $\mathcal{A}$.

10.4 Lemma. *Let $(\mathcal{A},q)$ be a model of $\mathbf{SD}_W$. Then, there exists a model $(\mathcal{B},r)$ having the following properties.*

(i) $(\mathcal{A},q)\prec(\mathcal{B},r)$.
(ii) *There exists an element e of B such that*
  (1) $B=H_{(\mathcal{B},r)}(A\cup\{e\})$,
  (2) *for any formula $\varphi(x)$ of $\mathcal{L}(\text{aa})$ with at most one free variable x and some constants from A, $(\mathcal{A},q)\models(\text{aa}\,x)\varphi(x)$ iff $(\mathcal{B},r)\models\varphi[e]$.*
(iii) *If e is an element of B satisfying the conditions* (1) *and* (2) *in* (ii), *then $e_F=A$, where $\mathcal{A}=(A;E\dots)$ and $\mathcal{B}=(B;\dots)$.*

Moreover, if $(\mathcal{B},r)$ and $(\mathcal{B}'r')$ are models having properties (i) *and* (ii), *then there exists an isomorphism over $(\mathcal{A},q)$ between $(\mathcal{B},r)$ and $(\mathcal{B}',r')$.*

PROOF. We shall first prove the latter part of the lemma. So let $(\mathcal{B},r)$ and $(\mathcal{B}',r')$ be models with properties (i) and (ii). Let e and e' be elements of B and B' satisfying the conditions (1) and (2) in (ii) respectively. Define a function f so that

$$f=\{\langle\tau^{(\mathcal{B},r)}[ea_1\dots a_n],\tau^{(\mathcal{B}',r')}[e'a_1\dots a_n]\rangle\mid\tau(xy_1\dots y_n)\text{ is a Skolem term, and }a_1,\dots,a_n\in A\}.$$

To make sure that f is an isomorphism over $(\mathcal{A},q)$ between $(\mathcal{B},r)$ and $(\mathcal{B}',r')$, it is enough to see the following claim:

Claim. Let $\varphi(yx_1,\dots,x_n)$ be a formula of $\mathcal{L}(\text{aa})$, and $a_1\dots a_n\in A$. Then,

$$(\mathcal{B},r)\models\varphi[ea_1\dots a_n]\text{ iff }(\mathcal{B}',r')\models\varphi[e'a_1\dots a_n].$$

However, the claim is clear from (2) in (ii).

Secondly, we shall show that the model $(\mathcal{B},r)$ with properties (i) and (ii) has the property (iii). So let e be an element of B satisfying the conditions (1) and (2) in (ii). Now, let b be an element of B such that $b\,F\,e$. By (1) in (ii), b can be written in a form

$\tau^{(\mathcal{B},r)}[ea_1 \dots a_n]$, where $\tau(xy_1 \dots y_n)$ is a Skolem term and $a_1, \dots, a_n \in A$. By (2) in (ii), we have

$$(\mathcal{A}, q) \models (aa\, x)(\tau(xa_1 \dots a_n) \in x).$$

Thus, we have

$$(\mathcal{A}, q) \models (aa\, x)(\exists z \in x)(z = \tau(xa_1 \dots a_n).$$

Since $(\mathcal{A}, q)$ is a model of $\mathbf{SD}_W$, we have

$$(\mathcal{A}, q) \models (\exists z)(aa\, x)(z = \tau(xa_1 \dots a_n)).$$

Hence, there is an element $a \in A$ such that

$$(\mathcal{A}, q) \models (aa\, x)(a = \tau(xa_1 \dots a_n).$$

Again, by (2) in (ii), we have

$$(\mathcal{B}, r) \models a = \tau(ea_1 \dots a_n).$$

It follows that

$$b = \tau^{(\mathcal{B},r)}[ea_1 \dots a_n] = a \in A.$$

Conversely, suppose that $b \in A$. Then, we have

$$(\mathcal{A}, q) \models (aa\, x)(b \in x).$$

By (2) in (ii), we have $b\, F\, e$. Thus, we have $e_F = A$.

Finally, we shall show the existence of $(\mathcal{B}, r)$ having the properties (i) and (ii). Let c be a new constant symbol. Let Γ be the set all $\varphi(c)$ such that $\varphi(x)$ is a formula of $\mathcal{L}(aa)$ with at most one free variable and some constants from A, and $(\mathcal{A}, q) \models (aa\, x)\varphi(x)$. It is not hard to see that Γ is a maximal consistent set of sentences. Let $(\mathcal{B}', r')$ be a model of Γ. We can identify the interpretation of a (the constant corresponding the element a of A) with a. Let $(\mathcal{B}'', r')$ be the reduct of $(\mathcal{B}', r')$ to $\mathcal{L}(aa)$. Then, $(\mathcal{A}, q) \prec (\mathcal{B}'', r')$. Let e be the element which interprets c. Set $\mathcal{B} = \mathcal{H}^{(\mathcal{B}'',r')}(A \cup \{e\})$ and $r = r'_{(A \cup \{e\})}$. Then, it is clear that $(\mathcal{B}, r)$ is a required one. ⊣

In fact, we can construct a ,model$(\mathcal{B}, r)$ having the properties (i) and (ii) via the method of Skolem ultrapowers. Let D be the of all functions $f : A \to A$ definable in $(\mathcal{A}, q)$ with some parameters, in other words, the set of all functions determined by 1-placed Skolem term with some constants from A. Define an equivalence relation $\sim_q$ on D as follows:

Let $f, g \in D$. Then,

$$f \sim_q g \text{ iff } \{\, a \in A \mid f(a) = g(a) \,\} \in g.$$

Set $B = D/\sim_q$. As in the case of usual ultrapowers, we can define the interpretations of $\mathcal{L}$. Thus, we have the structure $\mathcal{B} = (B; \cdots)$ of $\mathcal{L}$. Let $\varphi(yx)$ be a formula of $\mathcal{L}(aa)$ with at most two free variables x nd y, and some constants from A. Let S_φ be the set of $\tilde{f} \in B$ such that $\{a \in A \mid (\mathcal{A}, q) \models \varphi[f(a)a]\} \in q$, where $\tilde{f}$ denotes the equivalence class of f by $\sim_q$, and r be the set of S_φ such that $\varphi(xy)$ is a formula of $\mathcal{L}(aa)$ with some constants from A and at most two free variables, say, y and x, and

$(\mathcal{A}, q) \models (\text{aa}\, x)(\text{aa}\, y)\varphi(yx)\}$.Thus, we have the structure $(\mathcal{B}, r)$ of $\mathcal{L}(\text{aa})$. It is not hard to see that the following Łos Theorem holds:

$$(\mathcal{B}, r) \models \varphi[f_0 \dots f_{n-1}] \text{ iff } \{\, a \in A \mid (\mathcal{A}, q) \models \varphi[f_0(a) \dots f_{n-1}(a)] \,\} \in q,$$
$$\text{where } \varphi(x_1 \dots x_n) \text{ is a formula of } \mathcal{L}(\text{aa}) \text{ and } f_0, \dots, f_{n-1} \in D.$$

If we identify an element a of$A$ with $\tilde{c}_a$ and set $e = \widetilde{id}$, then it is clear that $(\mathcal{B}, r)$ is a required one, where c_a is the constant function $c_a(x) \equiv a$ and id is the identity function on A.

10.5 Definition. The model $(\mathcal{B}, r)$ constructed above is said to be the Skolem ultrapower of $(\mathcal{A}, q)$, and $e = \widetilde{id}$ is said to be the canonical representation of the universe A of $\mathcal{A}$.

We should note that the cardinality of the universe of the Skolem ultrapower of $(\mathcal{A}, q)$ does not exceed $\min(2^{\overline{\overline{A}}}, \max(\overline{\overline{\mathcal{L}}}, \overline{\overline{A}}))$.

We can now construct iterated Skolem ultrapowers. Let $(\mathcal{A}, q)$ be a model of **SD**$_W$. We define an elementary power $(\mathcal{A}_\alpha, q_\alpha)$, $\alpha \in On$ as follows:

(i) $(\mathcal{A}_0, q_0) = (\mathcal{A}, q)$;
(ii) $(\mathcal{A}_{\alpha+1}, q_{\alpha+1}) =$ the Skolem ultrapower of $(\mathcal{A}_\alpha, q_\alpha)$;
(iii) $(\mathcal{A}_\lambda, q) = \bigcup_{\alpha<\lambda}(\mathcal{A}_\alpha, q_\alpha)$ if λ is a limit ordinal.

For each $\alpha \in On$, we let $e(\alpha)$ denote the canonical representation of the universe of $\mathcal{A}_\alpha$.

As in the case of usual iterated ultrapowers ([**14**, Lemma 4.5.]), we have the following lemma.

10.6 Lemma. *Let λ be a limit ordinal. Let B be a subset of the universe A_λ of $\mathcal{A}_\lambda$ which is definable in $(\mathcal{A}_\lambda, q_\lambda)$ with parameters. Then,*

$$B \in q_\lambda \text{ iff } \{\, e(\alpha) \mid \beta \le \alpha < \lambda \,\} \subseteq B \quad \text{for some } \beta < \lambda.$$

PROOF. Let $\varphi(yx_1 \dots x_n)$ be a formula of $\mathcal{L}(\text{aa})$ and $a_1 \dots a_n \in A_\lambda$ such that

$$B = \{\, a \in A \mid (\mathcal{A}_\lambda, q_\lambda) \models \varphi[aa_1 \dots a_n] \,\}.$$

Let $\beta < \lambda$ be such that $a_1, \dots, a_n \in A_\beta$. Suppose that $B \in q_\lambda$. If we put $\psi(x_1 \dots x_n) \equiv (\text{aa}\, y)\varphi(yx_1 \dots x_n)$, then we have,

$$(\mathcal{A}_\lambda, q_\lambda) \models \psi[a_1 \dots a_n].$$

Thus, for each α, $\beta \le \alpha < \lambda$

$$(\mathcal{A}_\alpha, q_\alpha) \models \psi[a_1 \dots a_n],$$

and so,

$$(\mathcal{A}_{\alpha+1}, q_{\alpha+1}) \models \varphi[e(\alpha)a_{1n} \dots a],$$

and so,

$$(\mathcal{A}_\lambda, q_\lambda) \models \varphi[e(\alpha)a_1 \dots a_n].$$

Thus, we have

$$\{e(\alpha) \mid \beta \le \alpha < \lambda\} \subseteq B \quad \text{for some } \beta < \lambda.$$

The converse is clear from the fact that $B \in q_\lambda$ or $A_\lambda - B \in q_\lambda$. ⊣

By Lemma 10.6, we can have the following theorem, which is relevant to the results in §3.

10.7 Theorem. *Let Σ be a set of sentences of $\mathcal{L}(aa)$. If Σ is consistent with* **SD**$_W$*, then for any cardinal κ such that $\overline{\overline{\mathcal{L}}} < \kappa$ there exists a model $(\mathcal{B}, r)$ of Σ and κ-cumulation e for $\mathcal{B}$ such that, for any $X \subseteq B$, $X \in r$ iff X is definable with parameters and $\{e(\alpha) \mid \beta \le\!< \kappa\} \subseteq X$ for some $\beta < \kappa$. In particular, if κ is regular,*

$(\mathcal{B}, r) \models \varphi[a_1 \dots a_n]$ iff $(\mathcal{B}, \mathcal{F}_e) \models \varphi[a_1 \dots a_n]$

for any formula $\varphi(x_1 \dots x_n)$ *of $\mathcal{L}(aa)$ and $a_1, \dots, a_n \in B$.* ⊣

Now, let Σ be a maximal consistent set of sentences of $\mathcal{L}(aa)$ extending **SD**$_W$. For each ordinal α, we define $\mathcal{K}(\Sigma, \alpha) = (\mathcal{A}_\alpha, q_\alpha)$ as follows:
(i) $\alpha = 0$: Let T be the set of variable-free Skolem terms of $\mathcal{L}(aa)$. Define an equivalence relation $\sim_\Sigma$ on T as follows: Let τ, $\sigma \in T$. Then,

$$\tau \sim_\Sigma \sigma \text{ iff } \tau = \sigma \in \Sigma.$$

Let $A_0 = T/\!\sim_\Sigma$. Let P be an n-ary predicate symbol of $\mathcal{L}$. We define the interpretation R, on A_0, of P by:

$R(\tilde{\tau}_0 \dots \tilde{\tau}_{n-1})$ iff $P(\tau_0 \dots \tau_{n-1}) \in \Sigma$,

where $\tilde{\tau}$ denotes the equivalence class of τ by $\sim_\Sigma$.

We can define the interpretation of function symbols and constant symbols similarly. Thus, we have a structure $\mathcal{A}_0$ with its universe A_0. Let $\varphi(x)$ be a formula of $\mathcal{L}(aa)$ with at most one free variable x. Set $S_\varphi = \{\tilde{\tau} \in A_0 \mid \varphi(\tau) \in \Sigma\}$. Now, let q_0 be the set of S_φ such that φ is a formula of $\mathcal{L}(aa)$ with at most one free variable, say x, and $(aa\,x)\varphi(x) \in \Sigma$. It is not hard to see that $(\mathcal{A}_0, q_0)$ is a model of Σ and $\tau^{(\mathcal{A}_0, q_0)} = \tilde{\tau}$ for any variable-free Skolem term τ.
(iii) $\alpha = \beta + 1$: $(\mathcal{A}_\alpha, q_\alpha) =$ the Skolem ultrapower of $(\mathcal{A}_\beta, q_\beta)$.
(iii) α is a limit ordinal: Set $(\mathcal{A}_\alpha, q'_\alpha) = \bigcup_{\beta<\alpha}(\mathcal{A}_\beta, q_\beta)$, and q_α be the set of all subsets X of A_α such that X is definable in $(\mathcal{A}_\alpha, q'_\alpha)$ with some parameters from A_α and $X \subseteq q'_\alpha$. (It is clear that $(\mathcal{A}_\alpha, q_\alpha) \models \varphi[a_1 \dots a_n]$ iff $(\mathcal{A}_\alpha, q'_\alpha) \models \varphi[a_1 \dots a_n]$.)

10.8 Lemma. (i) *Every element of the universe of $\mathcal{A}_\alpha$ can be written in a form $\tau^{(\mathcal{A}_\alpha)}[e(\beta_1) \dots (\beta_n)]$, where $\tau(x_1 \dots x_n)$ is a Skolem term of $\mathcal{L}(aa)$ and $\beta_1 < \dots < \beta_n < \alpha$, that is , $\mathcal{H}_{(\mathcal{A}_\alpha, q_\alpha)}(\{e(\beta) \mid \beta < \alpha\}) = A_\alpha$.*
(ii) *For each limit ordinal α, every element of the universe of $\mathcal{A}_\alpha$ can be written in a form $\tau^{(\mathcal{A}_\alpha)}[e(\beta_1) \dots (\beta_n)]$, where $\tau(x_1 \dots x_n)$ is a Skolem term of $\mathcal{L}$ and $\beta_1 < \dots < \beta_n < \alpha$, that is, $\mathcal{H}_{\mathcal{A}_\alpha}(\{e(\beta) \mid \beta < \alpha\}) = A_\alpha$.*

Proof. (i) can be easily proved by induction on α. For (ii), let a be an element of the universe A_α of $\mathcal{A}_\alpha$. By (i), a has a form $\tau^{(\mathcal{A}_\alpha, q_\alpha)}[e(\beta_1 \dots e(\beta_1)]$, where $\tau(x_1 \dots x_n)$

s a Skolem term of $\mathcal{L}(aa)$, and $\beta_1 < \cdots < \beta_n < \alpha$. We can choose a formula $\varphi(yx_1 \ldots x_n y_1 \ldots y_m)$ of $\mathcal{L}$ so that

$$y = \tau(x_1 \ldots x_n) \leftrightarrow (aa\, y_1) \cdots (aa\, y_m)\varphi(yx_1 \ldots x_n y_1 \ldots y_m)$$

s provable in $\mathbf{SD}_W$. (See Lemma 7.16.) Set

$$\theta(yx_1 \ldots x_n y_1 \ldots y_m) \equiv y \in y_1 \wedge \varphi(yx_1 \ldots x_n y_1 \ldots y_m).$$

Choose $\gamma_1 < \cdots < \gamma_n$ so that $\beta_n < \gamma_1$. Then, $a\ E_\alpha\ e(\gamma_1)$. Clearly, we have

$$\mathcal{A}_\alpha \models \theta[a\, e(\beta_1) \ldots e(\beta_n)\, e(\gamma_1) \ldots e(\gamma_m)].$$

Thus, a is the unique element b of satisfying

$$\mathcal{A}_\alpha \models \theta[b\, e(\beta_1) \ldots e(b_n)\, e(\gamma_1) \ldots e(\gamma_m)]. \dashv$$

To suit the method of Ehrenfeuht-Mostowski to our work, we shall generalize the concept of sets of indiscernibles a little bit.

10.9 Definition. Let $\mathcal{A}\langle A; \ldots\rangle$ be a model of **SK**. Let X be an infinite set linearly ordered by $<$. A map $e : X \to A$ is said to be an *indiscernible indexization of X for $\mathcal{A}$* f e satisfies the following condition:
For any formula $\varphi(v_0 \ldots v_{n-1})$ of $\mathcal{L}$ and any increasing sequences $x_1 < \cdots < x_n$ and $y_1 < \cdots < y_n$ from X, we have

$$\mathcal{A} \models \varphi[e(x_1) \ldots e(x_n)] \quad \text{iff} \quad \mathcal{A} \models \varphi[e(y_1) \ldots e(y_n)].$$

Let $\mathcal{L}_\infty$ be the language obtained by adding a new constant c_n for each $n < \omega$. Let $\mathcal{A} = \langle A; \ldots\rangle$ be a model of **SK**, and e be an indiscernible indexization of X for $\mathcal{A}$. Let Γ be the set of $\varphi(c_0 \ldots c_{n-1})$ such that $\varphi(v_0 \ldots v_{n-1})$ is a formula of $\mathcal{L}$ and $\mathcal{A} \models \varphi[e(x_1) \ldots e(x_n)]$ for some increasing sequence $x_1 < \cdots < x_n$ from X. Then, it is clear that Γ satisfies the following conditions (a) and (b).

(a) Γ is a maximal consistent set of sentences of $\mathcal{L}_\infty$ which extends **SK**.
(b) Let $\varphi(v_0 \ldots v_{n-1})$ be a formula of $\mathcal{L}$. Then,

$$\varphi(c_{i_1} \ldots c_{i_n}) \leftrightarrow \varphi(c_{j_1} \ldots c_{j_n})$$

belongs to Γ, where $i_1 < \cdots < i_n < \omega$ and $j_1 < \cdots < j_n < \omega$.

10.10 Definition. A set Γ of sentences of $\mathcal{L}_\infty$ is said to be an *Ehrenfeucht-Mostowski set* (or, simply, *E-M set*) if Γ satisfies the above conditions (a) and (b).

10.11 Definition. Let Γ be an *E-M* set, and X be a linearly ordered infinite set. A pair $(\mathcal{A}, e)$ of a model $\mathcal{A}$ of **SK** and an indiscernible indexization e of X for $\mathcal{A}$ is said to be a *(Γ, X)-pair* if $(\mathcal{A}, e)$ satisfies the following conditions:
(i) $\mathcal{A} = \mathcal{H}_\mathcal{A}(\{\, e(x) \mid x \in X \,\})$,
(ii) $\Gamma = \{\, \varphi(c_0 \ldots c_{n-1}) \mid \varphi(v_0 \ldots v_{n-1})$ is formula of $\mathcal{L}$ and $\mathcal{A} \models \varphi[e(x_1) \ldots e(x_n)]$ for some increasing sequence $x_1 < \cdots < x_n$ from $X \,\}$

The following two lemmas can be proved as in the usual way.

10.12 Lemma. *Let Γ be an E-M set, and X be a linearly ordered infinite set. Then, there exists a (Γ, X)-pair, (A, e). Moreover, (Γ, X)-pair is uniquely determined in the following sense: Let (A, e) and (A′, e′) be (Γ, X)-pairs. Then, there is a unique isomorphism* $f : \mathcal{A} \to \mathcal{A}'$ *be such that the following diagram is commutative.*

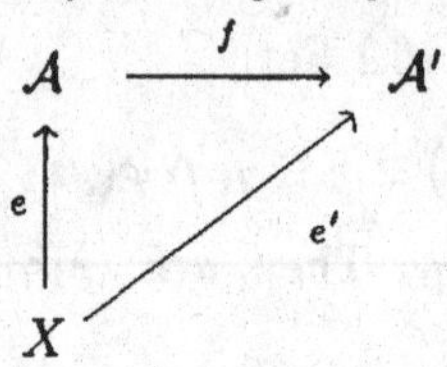

⊣

10.13 Lemma. *Let Γ be an E-M set, and X and Y be linearly ordered infinite sets. Let* $g : X \to Y$ *be an order-preserving map. Let* $(\mathcal{A}_X, e_X)$ *and* $(\mathcal{A}_Y, e_Y)$ *be the (Γ, X)-pair and (Γ, Y)-pair respectively. Then, there is a unique elementary embedding* $f : \mathcal{A}_X \to \mathcal{A}_Y$ *be such that the following diagram is commutative.*

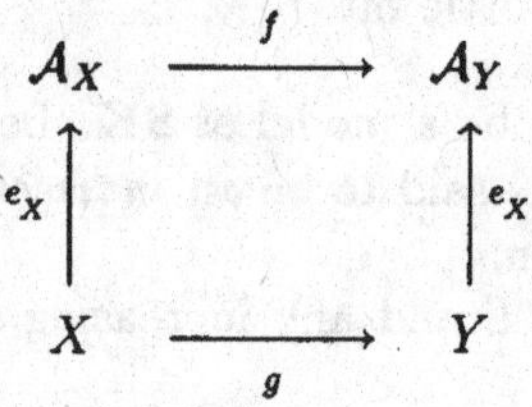

⊣

Let Σ be a maximal consistent set of sentences of $\mathcal{L}(\mathit{aa})$ which extends $\mathbf{SD}_W$. Set $\Sigma_\infty = \{\, \varphi(c_0 \dots c_{n-1}) \mid (\mathit{aa}\, v_0) \cdots (\mathit{aa}\, v_{n-1}) \varphi(v_0 \dots v_{n-1}) \in \Sigma \,\}$. Let λ be a limit ordinal. Let $(\mathcal{A}_\lambda, q_\lambda) = \mathcal{K}(\Sigma, \lambda)$ and $e_\lambda = e|\lambda$. Then, clearly, we have

$$\varphi(c_0 \dots c_{n-1}) \in \Sigma_\infty \quad \text{iff} \quad \mathcal{A}_\lambda \models \varphi[e(\alpha_1) \dots e(\alpha_n)]$$
$$\text{for any } \alpha_1 < \cdots < \alpha_n < \lambda.$$

Therefore, Σ_∞ is an *E-M* set. By Lemma 10.8, $(\mathcal{A}_\lambda, e_\lambda)$ is a (Σ_∞, λ)-pair. This (Σ_∞, λ)-pair $(\mathcal{A}_\lambda, e_\lambda)$ has the following important property: For every limit ordinal $\gamma < \lambda$,

$$e_\lambda(\gamma)_{E_\lambda} = H_{\mathcal{A}_\lambda}(\{\, e_\lambda(\alpha) : \alpha < \lambda \,\}).$$

Concerning this property, we can have the following lemma. The lemma can be proved by the standard argument.

10.14 Lemma. *Let Γ be an E-M set. We let (*) denote the following condition about the (Γ, λ)-pair (A, e) :*

(*) *For every limit ordinal* $\gamma < \lambda$, $e(\gamma)_E = H_{\mathcal{A}}(\{\, e(\alpha) : \alpha < \gamma \,\})$.

Then, the following conditions about Γ are equivalent.

(i) *For all limit ordinal λ, the (Γ, λ)-pair satisfies the condition* $(*)$.
(ii) *For some limit ordinal $\lambda > \omega$, the (Γ, λ)-pair satisfies the condition* $(*)$.
(iii) Γ *has the following properties:*
(1) *Let $\tau(v_0 \dots v_{n-1})$ is a Skolem term of $\mathcal{L}$. Then, the sentence $\tau(c_0 \dots c_{n-1}) \in c_n$ belongs to Γ.*
(2) *Let $\tau(v_0 \dots v_{n-1} v_n \dots v_{n+m-1})$ be a Skolem term of $\mathcal{L}$. Then, the sentences*

$$\tau(c_0 \dots c_{n-1} c_n \dots c_{n+m-1}) \in c_n,$$
$$\tau(c_0 \dots c_{n-1} c_{i_1} \dots c_{i_m}) = \tau(c_0 \dots c_{n-1} c_{j_1} \dots c_{j_m})$$

belong to Γ, where $n-1 < i_1 < \dots < i_m < \omega$ and $n-1 < j_1 < \dots < j_m < \omega$. ⊣

10.15 Definition. An E-M set Γ is said to be remarkable if it satisfies the condition (iii) in Lemma 10.14 and contains the sentences

"c_0 is transitive"

Then, we can have the following lemma.

10.16 Lemma. *Let Σ be a maximal consistent set of sentences of $\mathcal{L}(\mathit{aa})$ which extends* $\mathbf{SD}_W$. *Then, Σ_∞ is a remarkable E-M set.* ⊣

Our next task is to find a maximal consistent set Σ of sentences of $\mathcal{L}(\mathit{aa})$ extending $\mathbf{SD}_W$ so that $\Sigma_\infty = \Gamma$ for each remarkable E-M set Γ.

The following lemma can be easily proved by Definition 10.15 and (Lemma 10.14).

10.17 Lemma. *Let Γ be a remarkable E-M set and κ be a regular cardinal with $\overline{\overline{\mathcal{L}}} < \kappa$. Let $(\mathcal{A}, e)$ be the (Γ, κ)-pair. Then, e is a κ-cumulation for $\mathcal{A}$.* ⊣

10.18 Lemma. *Let κ be a regular uncountable cardinal such that $\overline{\overline{\mathcal{L}}} < \kappa$. Let $\mathcal{A}$ be a model of* **SK**, *and e be an indiscernible indexization of κ for $\mathcal{A}$ such that $A = H_{\mathcal{A}}(\{e(\alpha) \mid \alpha < \kappa\})$. If e is a κ-cumulation for $\mathcal{A}$, then $(\mathcal{A}, \mathcal{F}_e)$ is a model of* $\mathbf{SD}_W$.

PROOF. By Lemma 3.5, it suffices to show that $(\mathcal{A}, \mathcal{F}_e) \models$ **Dual**. By the indiscernibility and the definition of $\mathcal{F}_e$, we can see the following claim.

Claim 1. Let $\psi(x_1 \dots x_n y_1 \dots y_n)$ be a formula of $\mathcal{L}$, and $\alpha_1 < \dots < \alpha_n < \kappa$. Then, we have

$$(\mathcal{A}, \mathcal{F}_e) \models \varphi[e(\alpha_1) \dots e(\alpha_n)] \quad \text{iff} \quad \mathcal{A} \models \varphi[e(\alpha_1) \dots e(\alpha_n)\, e(\beta_1) \dots e(\beta_m)]$$
$$\text{for all} \quad \beta_1 < \dots < \beta_m \quad \text{with} \quad \beta_m < \kappa \quad \text{and} \quad \alpha_n < \beta_1,$$

where $\varphi(x_1 \dots x_n) \equiv (\mathit{aa}\, y_1) \cdots (\mathit{aa}\, y_m) \psi(x_1 \dots x_n y_1 \dots y_m)$.

By Claim 1 and the assumption for e that $A = H_{\mathcal{A}}(\{ e(\alpha) : \alpha < \kappa \})$, we have the following claim.

Claim 2. Let $\varphi(x_1 \dots x_n)$ be a formula of $\mathcal{L}$ with some constants from A. Then, there is some ordinal $\beta < \kappa$ such that the following (1), (2) and (3) are equivalent:

(1) $(\mathcal{A},\mathcal{F}_e) \models (\mathrm{aa}\, x_1)\cdots(\mathrm{aa}\, x_n)\varphi(x_1 \ldots x_n)$,
(2) $\mathcal{A} \models \varphi[e(\alpha_1)\ldots e(\alpha_n)]$ for some $\alpha_1 < \cdots < \alpha_n$ such that $\beta < \alpha_1$ and $\alpha_n < \kappa$,
(3) $\mathcal{A} \models \varphi[e(\alpha_1)\ldots e(\alpha_n)]$ for all $\alpha_1 < \cdots < \alpha_n$ such that $\beta < \alpha_1$ and $\alpha_n < \kappa$.

By Claim 2, we can easily see the following claim.

Claim 3. Let $\varphi(x_1 \ldots x_n)$ be a formula of $\mathcal{L}$ with some constants from A. Then,

$$(\mathrm{aa}\, x_1)\cdots(\mathrm{aa}\, x_n)\varphi \vee (\mathrm{aa}\, x_1)\cdots(\mathrm{aa}\, x_n)\neg\varphi$$

holds in $(\mathcal{A},\mathcal{F}_e)$. By Claim 3 and Lemma 7.22, we can see that the **Dual** holds in $(\mathcal{A},\mathcal{F}_e)$. ⊣

Let Γ be a remarkable E-M set, and κ be regular cardinal with $\overline{\overline{\mathcal{L}}} < \kappa$. Let $(\mathcal{A}, e)$ be the (Γ,κ)-pair. Then, by Lemmas 10.17 and 10.18, we see that $(\mathcal{A},\mathcal{F}_e)$ is a model of $\mathbf{SD}_W$. Set $\Sigma = \{\sigma : \sigma$ is a sentence of $\mathcal{L}(\mathrm{aa})$ and $(\mathcal{A},\mathcal{F}_e) \models \sigma\}$. Then, it is clear that $\Gamma = \Sigma_\infty$.

10.19 Theorem. *The map $\Sigma \mapsto \Sigma_\infty$ gives a one-one correspondence between the collection of maximal consistent sets of sentences of $\mathcal{L}(\mathrm{aa})$ extending* $\mathbf{SD}_W$ *and the collection of remarkable E-M sets*

PROOF. It is enough to show that the map $\Sigma \mapsto \Sigma_\infty$ is injective. So let Σ nd Σ' are maximal consistent sets of sentences of $\mathcal{L}(\mathrm{aa})$ extending SD_W such that $\Sigma_\infty = \Sigma'_\infty$. Let $\sigma \in \Sigma$. By Lemma 7.16, there exists a formula $\varphi(x_1 \ldots x_n)$ with free variables among $x_1, \ldots, x_m$ of $\mathcal{L}$ such that

$$\mathbf{SD}_W \vdash \sigma \leftrightarrow (\mathrm{aa}\, x_1)\cdots(\mathrm{aa}\, x_n)\varphi(x_1 \ldots x_n).$$

It follows that $(\mathrm{aa}\, x_1)\cdots(\mathrm{aa}\, x_n)\varphi(x_1 \ldots x_n) \in \Sigma$. Therefor, $\varphi(c_0 \ldots c_{n-1})$ belongs to $\Sigma_\infty = \Sigma'_\infty$. It follows that $(\mathrm{aa}\, x_1)\cdots(\mathrm{aa}\, x_n)\varphi(x_1 \ldots x_n) \in \Sigma'$, and so $\sigma \in \Sigma'$. Thus, we have $\Sigma \subseteq \Sigma'$. By the maximality of Σ, we have $\Sigma = \Sigma'$.

Let Γ be a remarkable E-M set, and Σ be a maximal consistent set of sentences of $\mathcal{L}(\mathrm{aa})$ such that $\Gamma = \Sigma_\infty$. Then, by Theorem 10.19 and (v) in Lemma 2.12, we can see that the following are equivalent:

(i) Γ contains $\neg c_0 \in c_0$,
(ii) Σ contains $\neg(\mathrm{aa}\, x)(x \in x)$,
(iii) Σ contains $\neg(\exists y)(\forall x)(x \in y)$,
(iv) Γ contains $\neg(\exists y)\forall x(x \in y)$.

Let e be an indiscernible indexization of X for $\mathcal{A}$ such that

$$\Gamma = \{\varphi(c_0 \ldots c_{n-1}) \mid \varphi(v_0 \ldots v_{n-1}) \text{ is a formula of } \mathcal{L} \text{ and } \mathcal{A} \models \varphi[e(x_1)\ldots e(x_n)] \text{ for some } x_1 < \cdots < x_n \text{ from X}\}$$

Clearly, the sufficient and necessary condition that e is an isomorphism between $\langle X; <\rangle$ and $\langle \{e(x) \mid x \in X\}; E\rangle$ is that Γ contains the sentence $\neg c_0 \in c_0$. Thus, we have the following lemma.

10.20 Lemma. *Let e be an indiscernible indexization of X for $\mathcal{A}$ such that the set $\{\varphi(c_0 \ldots c_{n-1})) \mid \varphi(v_0 \ldots v_{n-1})$ is a formula of $\mathcal{L}$ and $\mathcal{A} \models \varphi[e(x_1) \ldots e(x_n)]$ for some $x_1 < \cdots < x_n$ from $X\}$ is a remarkable E-M set. Then, the sufficient and necessary condition that e is an isomorphism between $\langle X, < \rangle$ and $\langle \{e(x) \mid x \in X\}; E \rangle$ is that $\neg(\exists y)(\forall x)(x \in y)$ holds in $\mathcal{A}$.* ⊣

§11. $\mathcal{M}$-ultrafilter and $0^\#$

In this section, we shall apply our observations in §10 to study $\mathcal{M}$-ultrafilters and $0^\#$. First, we shall that the axiom of power set can be proved in $\mathbf{SD}_W$+Extensionality.

11.1 Lemma. $\mathbf{SD}_W$ + Extensionality $\vdash (\forall x)(\exists y)(\forall z)(z \subseteq x \to z \in y)$.

PROOF. Suppose $(\exists x)(\forall y)(\exists z)(z \subseteq x \wedge z \notin y)$. Fix an x so that $(\forall y)(\exists z)(z \subseteq x \wedge z \notin y)$. Set $\theta(xyz) \equiv (z \subseteq x \wedge z \notin y)$. Set $z_{x,y} = F_{(\exists z)\theta(zxy)}(xy)$. Then, $(\forall y)(z_{x,y} \subseteq x \wedge z_{x,y} \notin y)$. Let $\varphi(xy)$ denote the formula $(\forall w \in y)(w \in z_{x,y} \leftrightarrow (aa\, y)(w \in z_{x,y}))$. Then, clearly, we have the following[2]:

$$(aa\, y)\varphi(xy) \tag{1}$$

$$\varphi(xy_1) \wedge \varphi(xy_2) \wedge x \subseteq y_1 \wedge x \subseteq y_2 \to z_{x,y_1} = z_{x,y_2} \tag{2}$$

Therefore, we have the unique z_x such that

$$\varphi(xy) \wedge x \subseteq y \to z_{x,y} = z_x \tag{3}$$

Since $(aa\, y)\varphi(xy)$, $(aa\, y)(x \subseteq y)$ and $(aa\, y)(z_x \in y)$, we have a y such that $\varphi(xy) \wedge x \subseteq y \wedge z_x \in y$. By (3), we have $z_{x,y} = z_x \in y$. this is a contradiction because $z_{x,y} \notin y$. ⊣

By Lemma 11.1, we see that $\mathbf{SD}_W$ + Extensionality + Σ_0 − Separation + Regularity is an equivalent theory to $\mathbf{ZF}_\sigma(aa)$ + **Dual**. (Notice that, in $\mathbf{ZF}_\sigma(aa)$ + **Dual**, we can define a binary predicate W which well-orders the universe.)

In §9, we saw that $\mathbf{ZF}_\sigma(aa)$ + **Dual** $\vdash \varphi$ iff **KME** + {"On is n-subtle" $\mid n < \omega$} $\vdash \varphi$ for any formula φ of the language of **ZF**. Now, we can see that, for any model $(\mathcal{A}, q)$ of $\mathbf{ZF}_\sigma(aa)$ + **Dual**, $\mathcal{A}$ is extendable to **KME** + {On is n-subtle $\mid n < \omega$}. To see this, we first see the following lemma.

11.2 Lemma. *Let $(\mathcal{A}, q)$ be a model of $\mathbf{ZF}_\sigma(aa)$ + **Dual**. Let $(\mathcal{B}, r)$ be the Skolem ultrapower of $(\mathcal{A}, q)$ and e be the element of the universe B of $\mathcal{B}$ which represents the universe A of $\mathcal{A}$, that is, $e_F = A$, where $\mathcal{A} = \langle A; E \ldots \rangle$ and $\mathcal{B} = \langle B; F \ldots \rangle$. Then, $D(\mathcal{A}, q) = \{x_F \mid x \in B$ and $x_F \subseteq e_F\}$, where $D(\mathcal{A}, q) = \{X \subseteq A \mid X$ is definable in $(\mathcal{A}, q)$ with some parameters from $A\}$.*

PROOF. Suppose that $X \in D(\mathcal{A}, q)$. Let $\varphi(x)$ be a formula of $\mathcal{L}(aa)$ with some constants form A such that $X = \{a \in A \mid (\mathcal{A}, q) \models \varphi[a]\}$. By applying Separation inside $(\mathcal{B}, r)$, there exists an element $x \in B$ such that $(\mathcal{B}, r) \models (\forall y)(y \in x \leftrightarrow y \in e \wedge \varphi(x))$. Thus, $x_F \subseteq e_F$ and $x_F = X$. Conversely, suppose that $x \in B$ and $x_F \subseteq e_F$. Then, there

---

[2] Note that we need the axiom of extensionality to prove (2).

exists a Skolem term $\tau(zx_1 \ldots x_{n-1})$ of $\mathcal{L}(aa)$ such that $x = \tau^{(\mathcal{B},r)}[ea_1 \ldots a_{n-1}]$, where $a_1, \ldots, a_{n-1} \in A$. Let $\psi(zx_1 \ldots x_n y)$ be formula of $\mathcal{L}(aa)$ such that $\psi(zx_1 \ldots x_n y) \leftrightarrow y \in \tau(zx_1 \ldots x_{n-1})$ is provable in $\mathbf{ZF}_\sigma(aa) + \mathbf{Dual}$. Then,

$$(\mathcal{B}, r) \models \psi[ea_1 \ldots a_{n-1}a] \leftrightarrow a \in x_F \quad \text{for all } a \in A.$$

If we set $\varphi(x_1 \ldots x_{n-1}y) \equiv (aa\, z)\psi(zx_1 \ldots x_{n-1}y)$, then

$$(\mathcal{B}, r) \models \psi[ea_1 \ldots a_{n-1}a] \leftrightarrow (\mathcal{A}, q) \models \varphi[a_1 \ldots a_{n-1}a] \quad \text{for all } a \in A.$$

Therefore, we have

$$x_F = \{\, a \in A \mid (\mathcal{A}, q) \models \varphi[a_1 \ldots a_{n-1}a] \,\} \in D(\mathcal{A}, q). \dashv$$

11.3 Theorem. *Let $(\mathcal{A}, q)$ be a model of* $\mathbf{ZF}_\sigma(aa) + \mathbf{Dual}$. *Then, $(\mathcal{A}, D(\mathcal{A}, q))$ is a model of* **KME** + { "*On* is a *n*-subtle" | $n < \omega$ }.

Proof. Let $(\mathcal{B}, r)$ be the Skolem ultrapower of $(\mathcal{A}, r)$ and e be the representation of the universe of $\mathcal{A}$. Since, for every natural number n, $(\mathcal{A}, q) \models (aa\, x)(\exists \alpha)(\alpha$ is n-subtle), then $(\mathcal{B}, r) \models e = R_\theta \wedge$ "θ is n-subtle" for every natural number n. By Lemma 11.2, we have $(R_{\theta+1})_F^{(\mathcal{A})} = D(\mathcal{A}, q)$. From this, it is clear that $(\mathcal{A}, D(\mathcal{A}, q))$ is a model of **KME** + { "*On* is n-subtle" | $n < \omega$ }. $\dashv$

As a by-product of Lemma 11.2, we can have the connection between the self-dual quantifier aa and the notion of normal $\mathcal{M}$-ultrafilters in the sense of [**14**].

11.4 Definition. Let $\mathcal{M} = \langle M; E \rangle$ be a model of **ZFC**, and θ be an ordinal in $\mathcal{M}$. A subset U of $\{\, x_E \mid x \in M$ and $x_E \subseteq \theta_E \,\}$ is said to be an $\mathcal{M}$-ultrafilter on θ if:

(i) $\{\zeta\} \notin U$ for every ordinal ζ in $\mathcal{M}$ less than θ;
(ii) $(\forall x \in M)(x_E \subseteq \theta_E \rightarrow x_E \in U \vee \theta_E - x_E \in U)$;
(iii) $(\forall x \in M)(\forall y \in M)(x_E \subseteq y_E \subseteq \theta \wedge x_E \in U \rightarrow y_E \in U)$;
(iv) Let $f \in M$ be such that $\mathcal{M} \models$ "$f : \zeta \rightarrow P(\theta)$", and $f(\xi)_E \in U$ for all ordinal ξ in $\mathcal{M}$ less than ζ, where ζ is an ordinal in $\mathcal{M}$ less than θ. Then, $\bigcap_{\xi E \zeta}(f(\zeta))_E \in U$;
(v) Let $f \in M$ be such that $\mathcal{M} \models$ "$f \rightarrow P(\theta)$". Then, there is an element x of M such that $x_E = \{\, \zeta \mid \zeta$is an ordinal in $\mathcal{M}$ less than θ and $(f(\zeta))_E \in U \,\}$.

We say that an $\mathcal{M}$-ultrafilter on θ is normal if, for every $f \in M$ such that $\mathcal{M} \models$ "$f : \theta \rightarrow P(\theta)$" and $f(\zeta)_E \in U$ for every ordinal ζ in $\mathcal{M}$ less than θ, $(\Delta_{\zeta<\theta} f(\zeta))_E \in U$ where $\Delta_{\zeta<\theta} f(\zeta)$ denotes the diagonal intersection of f inside M.

Let $(\mathcal{A}, q)$ be a model of $\mathbf{ZF}_\sigma(aa) + \mathbf{Dual}$. Let $q \mid On = \{\, \{\, \alpha \in A \mid \alpha$ is an ordinal in $\mathcal{A}$ and $(\mathcal{A}, q) \models \varphi[\alpha] \,\} \mid \varphi(y)$ is a formula of $\mathcal{L}(aa)$ with some constants from A and $(\mathcal{A}, q) \models (aa\, x)(\exists \alpha)(x = R_\alpha \wedge \varphi(\alpha)) \,\}$.

11.5 Lemma. *Let $(\mathcal{A}, q)$ be a model of $\mathbf{ZF}_\sigma(aa) + \mathbf{Dual}$. Let $(\mathcal{B}, r)$ be the Skolem ultrapower of $(\mathcal{A}, q)$ and θ be an ordinal such that $R_\theta^{(\mathcal{B})}$ is the representation of the universe of $\mathcal{A}$. Then, $q \mid On$ is a normal $\mathcal{B}$-ultrafilter on θ.*

PROOF. We shall show that $q \mid On$ satisfies (v) in Definition 11.4. The remainders will be left to the reader. So let $f \in B$ be such that $\mathcal{B} \models$ "$f : \theta \to P(\theta)$". We should note that $\theta_F = \{ a \in A \mid \mathcal{A} \models$ "a is an ordinal" $\}$. By Lemma 11.2, there is a formula $\varphi(xy)$ of $\mathcal{L}(aa)$ with some constants from A such that

$$(\mathcal{A}, q) \models \varphi[\xi\zeta] \text{ iff } \xi \in f(\zeta) \text{ for all ordinals } \zeta, \xi \in A.$$

Set $\psi(y) \equiv (aa\, x)(\exists \xi)(x = R_\xi \wedge \varphi(\xi y) \wedge y$ is an ordinal. Then,

$$(\mathcal{A}, q) \models \psi[\zeta] \text{ iff } (f(\zeta))_F \in q \mid On.$$

Choose $x \in B$ so that $x_F = \{ a \in A \mid (\mathcal{A}, q) \models \psi[a] \}$. Then, $x_F = \{ \zeta \mid \zeta$ is an ordinal in $\mathcal{A}$ and $(f(\zeta))_F \in q \mid On \}$. ⊣

We can have the converse of Lemma 11.5. Let $\mathcal{M} = \langle M; E \rangle$ be a model of **ZFC** and U be a normal $\mathcal{M}$-ultrafilter on an ordinal θ in $\mathcal{M}$. Set $A = \{ a \in M \mid \mathcal{M} \models$ "$a \in R_\theta$" $\}$. Define $q_U = \{ x_E \mid x_E \subseteq A$ and $\{ \alpha \mid \mathcal{M} \models$ "α is an ordinal in $\mathcal{M}$ and $R_\alpha \in x$" $\} \in U \}$. Set $\mathcal{A} = \langle A; E \mid A, x_E \rangle_{x_E \subseteq A}$.

11.6 Lemma ([9]). $(\mathcal{A}, q_U) \models \mathbf{ZF}_\sigma(aa) + \mathbf{Dual}$.

PROOF. Since θ is an inaccessible cardinal in M, we can clearly have the following claim by the property of (v) in Definition 11.4.

Claim 1. Let $f \in M$ be such that $\mathcal{M} \models$ "$f : R_\theta \times \cdots \times R_\theta \to R_\theta$". Then, there exists an element $x \in M$ such that $x_E = \{ \langle x_1, \ldots, x_n \rangle^{(M;E)} \mid f(x_1, \ldots, x_n)_E \in q_U \}$.

By the following claim, we can easily see the lemma.

Claim 2. Let $\varphi(x_1 \ldots x_n)$ be a formula of $\mathcal{L}(aa)$. Then, there exists an element $x \in M$ such that

$$x_E = \{ \langle a_1, \ldots, a_n \rangle^{(\mathcal{A})} \mid (\mathcal{A}, q_U) \models \varphi[a_1 \ldots a_n] \}.$$

The proof of Claim 2 can be carried out by induction on the complexity of $\varphi(x_1 \ldots x_n)$. The induction step that $\varphi(x_1 \ldots x_n)$ is $(aa\, y)\varphi(y x_1 \ldots x_n)$ is only non-trivial case. Let $z \in M$ be such that

$$z_E = \{ \langle b, a_1, \ldots, a_n \rangle^{(\mathcal{M})} \mid (\mathcal{A}, q_U) \models \psi[b a_1 \ldots a_n] \}.$$

Define $f \in M$ so that

$$\mathcal{M} \models \text{"} f(a_1, \ldots, a_n) = \{ b \mid \langle b a_1 \ldots a_n \rangle \in z \} \text{"}.$$

By Claim 1, there is an element $x \in M$ such that

$$x_E = \{ \langle a_1, \ldots, a_n \rangle^{(\mathcal{M})} \mid (f(a_1, \ldots, a_n)_E \in q_U \}.$$

It follows that

$$x_E = \{ \langle a_1, \ldots, a_n \rangle^{(\mathcal{M})} \mid (\mathcal{A}, q_U) \models \varphi[a_1 \ldots a_n] \}. \dashv$$

Powell formulated a set theory with an extra binary predicate $\ni$, called *predication* other than the membership relation $\in$. In fact, his system and $\mathbf{ZF}_\sigma(aa) + \mathbf{Dual}$ are essentially the same. More precisely, we have the following theorem.

11.7 Theorem. *If φ is a formula of* **ZF**, *then* $\mathbf{ZF}_\sigma(aa) + \mathbf{Dual} \vdash \varphi$ *iff* $\mathbf{P} \vdash \varphi$. *Here* **P** *is the set theory of Powell's.*

Before proving the theorem, we shall give the precise formulation of **P**.

11.8 Definition. **P** is a theory formulated on the first-order language with equality, two binary predicates $\in$ and $\ni$ and a constant symbol V. Sets are denied to be elements of V and classes to be collections of sets. The variables P, Q, R range over classes. The axioms of **P** are as follows:

(A) $x \in y \in V \rightarrow x \in V$,

(B) $\forall x \in V(P \ni x \leftrightarrow x \in P)$,

(C) $\forall x \in V(P \ni x \leftrightarrow Q \ni x) \rightarrow P = Q$,

(D) $\forall x_1, \ldots, x_n \in V \exists Q \forall y(Q \ni y \leftrightarrow \Phi(P_1, \ldots, P_m; x_1, \ldots, x_n, y))$, where Φ is a formula such that

- (i) all its free variables are displayed,
- (ii) the P's are the only variables occurring on the left in predication,
- (iii) the P's occur only on the left in predication, and
- (iv) V does not occur.

We also add axioms that

(E) V satisfies the axiom of choice and

(F) V satisfies the axiom of regularity.

Now, let $(\mathcal{A}, q)$ be a model of $\mathbf{ZF}_\sigma(aa) + \mathbf{Dual}$, and $(\mathcal{B}, r)$, e and $D(\mathcal{A}, q)$ as in Lemma 11.2. We define a binary relation $\exists^*$ on B as follows:

Let b be an element of B such that $x \subseteq e$ in $\mathcal{B}$. By Lemma 11.2, there is a formula $\varphi(x)$ with parameters from A such that

$$b_F = \{ a \in A \mid (\mathcal{A}, q) \models \varphi[a] \}.$$

For every element a in B, we define

$$b \exists^* a \text{ iff } (\mathcal{B}, r) \models \varphi[a].$$

It is clear that the definition of $\exists^*$ is independent of the choice of $\varphi(x)$.

Now we show that $(B; F, \exists^*, e)$ is a model of **P** under the interpretation of $\in$ as F, $\ni$ as $\exists^*$, and V as e. The only non-trivial case is whether the axiom (D) of **P** holds in $(B; F, \exists^*, e)$. So let $\Phi(P_1, \ldots, P_m; x_1, \ldots, x_n, y)$ be a formula satisfying the conditions in (i)–(iv) in (D), and $b_1, \ldots, b_m$ and $a_1, \ldots, a_n$ classes and sets in $(B; F, \exists^*, e)$ respectively. For each $i < m$, let $\psi_i(z_1^{(i)} \ldots z_{l_i}^{(i)} y_i)$ be a formula of $\mathcal{L}(aa)$ such that ${b_i}_F = \{ a \in A \mid$

$(\mathcal{A}, q) \models \psi_i[a_1^{(i)} \dots a_{l_i}^{(i)} a]\}$ for some $a_1^{(i)}, \dots, a_{n_i}^{(i)}$ in A. We define a formula $\varphi(y)$ of $\mathcal{L}(aa)$ with constants from A by replacing each $P_i \ni y_i$ by $\psi_i(a_i^{(i)} \dots a_{l_i}^{(i)} y_i)$ and then x_j by a_j. Then, clearly

$$(\mathcal{B}, r) \models \varphi[b] \text{ iff } (B; F, \exists^*, e) \models \Phi[b_1, \dots, b_m; a_1, \dots, a_n, b] \quad \text{for any } b \in B.$$

Let $c \in B$ be such that $c_F = \{a \in A \mid (\mathcal{A}, q) \models \varphi[a]\}$. Then, clearly

$$(B; F, \exists^*, e) \models \forall y(c \ni y \leftrightarrow \Phi(b_1, \dots, b_m; a_1, \dots, a_n, y)).$$

Thus, (D) holds in $(B; F, \exists^*, e)$. By the above argument it is clear that $\mathbf{P} \vdash$ implies $\mathbf{ZF}_\sigma(aa) + \mathbf{Dual} \vdash \varphi$ for every formula φ of **ZF**. For the converse, we can easily see it by the *necessity* part of Th.3 in [6] [3]and Theorem 11.6.

By the corollary of Theorem 11.7, we have the following corollary.

11.9 Corollary. *Powell's set theory is a conservative extension of* $\mathbf{ZFC} + \mathbf{RF}_\infty$. *Recall that* $\mathbf{RF}_\infty$ *is the following scheme:*

$$\{(\exists\alpha)(\alpha \text{ is } n\text{-subtle} \wedge (\forall x_1 \cdots x_m \in R_\alpha)(\varphi \leftrightarrow \varphi^{(R_\alpha)})) \mid n \textit{ is a natural number}\}.$$

Kleinberg proved the following result about normal M-ultrafilters.

11.10 Theorem ([13]). *Let M be a transitive model of* **ZFC**, *and θ be an ordinal in M.*

(i) *If θ has a normal M-ultrafilter, then θ is completely ineffable in M.*
(ii) *If M is countable and θ is completely ineffable in M, then θ has an normal M-ultrafilter.*

Lemma 11.5 shows that the hypothesis of transitivity of M in (i) of Theorem 11.7 cannot be eliminated. That is, we can prove the following lemma by using Lemma 11.5.

11.11 Lemma. *There are a model $\langle B; F\rangle$ of* **ZFC** *and an ordinal θ in $\langle B; F\rangle$ with the following properties, providing that* $\mathbf{ZF}(aa) + \mathbf{Dual}$ *is consistent.*

(i) $\langle B; F\rangle \models$ "θ is not completely ineffable."
(ii) *θ has a normal B-ultrafilter.*

PROOF. First, we claim the following:

Claim. $(\exists\kappa)(\kappa$ is ω-subtle) is not provable in $\mathbf{ZF}_\sigma(aa) + \mathbf{Dual}$, providing that $\mathbf{ZF}(aa) + \mathbf{Dual}$ is consistent. Here, κ is said to be ω-subtle if κ is n-subtle for all $n < \omega$.
For suppose that $\mathbf{ZF}_\sigma(aa) + \mathbf{Dual} \vdash (\exists\kappa)(\kappa$ is ω-subtle). Then,

$$(\exists\kappa)(\langle R_\kappa; \in, R_{\kappa+1}\rangle \text{ is a model of } \mathbf{KME} + \{\text{“}On \text{ is } n\text{-subtle”} \mid n < \omega\})$$

[3] In the proof they assume that M is well-founded. However, the proof of the existence of M-ultrafilter is not affected without the assumption. Also the normality of the M-ultrafilter can be proved by the same fashion.

is provable in $\mathbf{ZF}_\sigma(aa)+\mathbf{Dual}$. Therefor, by Theorem 9.16, the consistency of $\mathbf{ZF}_\sigma(aa)+\mathbf{Dual}$ is provable in $\mathbf{ZF}_\sigma(aa)+\mathbf{Dual}$. By Gödel's incompleteness theorem, $\mathbf{ZF}_\sigma(aa)+\mathbf{Dual}$ is inconsistent, and so by Corollary 9.20, $\mathbf{ZF}(aa)+\mathbf{Dual}$ is inconsistent.

By the claim, there is a model $(\mathcal{A}, q)$ of $\mathbf{ZF}_\sigma(aa)+\mathbf{Dual}$ such that $(\mathcal{A}, q) \models \neg(\exists\kappa)(\kappa$ is ω-subtle). Let $(\mathcal{B}, r)$ be the Skolem expansion of $(\mathcal{A}, q)$, and e be the representation of the universe of $\mathcal{A}$. Let $\mathcal{B} \models$ "$e = R_\theta$". By Lemma 11.5, $q \mid On$ is a normal B-ultrafilter on θ. Suppose that $\mathcal{B} \models$ "θ is completely ineffable". By Kleinberg, $\mathcal{B} \models$ "$(\forall n < \omega)(\theta$ is n-ineffable), and so $\mathcal{B} \models$ "θ is ω-subtle". Therefore, $\mathcal{A} \models$ "θ is ω-subtle". Therefor, $\mathcal{A} \models$ "$(\exists\kappa)(\kappa$ is ω-subtle)". This is a contradiction.

Let us now investigate the connection between the self-dual quantifier and $0^\#$.

11.12 Definition. Let Σ be a maximal consistent set of sentences of $\mathcal{L}(aa)$ which extends $\mathbf{ZF}_\sigma(aa)+\mathbf{Dual}$. Σ is said to have the property of well-foundedness if $\mathcal{K}(\Sigma.\alpha)$ is well-founded for all α.

By using Lemma 10.8 and 10.13, we can easily have the following lemma.

11.13 Lemma. *Let Σ be a maximal consistent set of sentences of $\mathcal{L}(aa)$ extending* $\mathbf{ZF}(aa)+\mathbf{Dual}$. *Then, the following are equivalent.*

(i) *Σ has the property of well-foundedness.*
(ii) *For some $\alpha \geq \omega_1$ $\mathcal{K}(\Sigma, \alpha)$ is well-founded.*
(iii) *For all $\alpha < \omega_1$ $\mathcal{K}(\Sigma, \alpha)$ is well-founded.* $\dashv$

Now, let Σ be a maximal consistent set of sentences which extends $\mathbf{ZF}_\sigma(aa)+\mathbf{Dual}$ and has the property of well-foundedness. We identify $\mathcal{K}(\Sigma, \alpha)$ with its transitive collapse. Let $(\mathcal{A}_\alpha, q_\alpha) = \mathcal{K}(\Sigma, \alpha)$. Set $\mathcal{A} = \bigcup_{\alpha \in On} \mathcal{A}_\alpha$. Let $\varphi(x_1 \ldots x_n)$ be a formula of $\mathcal{L}(aa)$, and $a_1, \ldots, a_n$ elements of the universe of $\mathcal{A}$. Since $(\mathcal{A}_\alpha, q_\alpha) \prec (\mathcal{A}_\beta, q_\beta)$ for all $\alpha < \beta$, we can define $\mathcal{A} \models \varphi[a_1 \ldots a_n]$ by:

$(\mathcal{A}_\alpha, q_\alpha) \models \varphi[a_1 \ldots a_n]$ for some α such that $a_1, \ldots, a_n$ are in the universe A_α of $\mathcal{A}_\alpha$.

Clearly, we have the following lemmas.

11.14 Lemma. (i) $\mathcal{A} = \mathcal{H}_\mathcal{A}(\{A_\alpha : \alpha \in On\})$.
(ii) *Let $\varphi(x_1 \ldots x_n) \equiv (aa\, y)\psi(x_1 \ldots x_n y)$ be a formula of $\mathcal{L}(aa)$ and $a_1, \ldots, a_n \in A_\beta$. Then,*

$$\mathcal{A} \models \varphi[a_1 \ldots a_n] \quad \textit{iff} \quad \mathcal{A} \models \psi[a_1 \ldots a_n A_\beta] \quad \textit{for all } \alpha \geq \beta. \dashv$$

Noting that $\mathbf{ZF}(aa) \vdash (aa\, x)(\exists\alpha)(x = R_\alpha)$, we have the following lemma.

11.15 Lemma. *Let $\delta = \delta_\Sigma$ be the function defined by $A_\alpha = R^{(\mathcal{A})}_{\delta(\alpha)} = R_{\delta(\alpha)} \cap A$ for all $\alpha \in On$. Then,*

(i) *δ is increasing nd continuous.*
(ii) *$\delta(\kappa) = \kappa$ for any uncountable cardinal κ.*
(iii) *$\{\delta(\alpha) \mid \alpha \in On\}$ is closed unbounded class of indiscernible ordinals for $\mathcal{A}$ such that $\mathcal{A} = \mathcal{H}_\mathcal{A}(\{\delta(\alpha) \mid \alpha \in On\})$.* ⊣

11.16 Definition. Let κ be a limit cardinal. The cardinal filter $\mathcal{F}$ on κ is defined by

$$\mathcal{F} = \{X \subseteq \kappa \mid \{\gamma < \kappa \mid \gamma \text{ is a cardinal and } \alpha \leq \gamma\} \subseteq X \quad \text{for some } \alpha < \kappa\}.$$

For every uncountable cardinal κ, we let denote $\mathcal{F}_\kappa$ the closed unbounded filter on κ or the cardinal filter on κ according as κ is regular or singular respectively.

Noting that $\mathcal{A}_\alpha \prec \mathcal{A}$, we have the following lemma by Lemma 10.6 and 11.2.

11.17 Lemma. *$q_\kappa \mid On = \mathcal{F}_\kappa \cap A$ for any cardinal κ. Henceforth, we have $(\mathcal{A}_\kappa, \mathcal{F}_\kappa) \prec (\mathcal{A}_\lambda, \mathcal{F}_\lambda)$ for any cardinal κ, λ such that $\omega < \kappa < \lambda$.* ⊣

Now, let Σ be a maximal consistent set of sentences of $\mathcal{L}(aa)$ which extends **SD** + Extensionality + Regularity + Σ_0 − Separation + **V** = **L**. By Lemma 11.1, Σ extends **ZF**(aa) + **Dual** + **V** = **L**. By our observations so far, we can conclude that:

11.18 Theorem. *The existence of $0^\#$ is equivalent to the existence of the maximal consistent set Σ of sentences of $\mathcal{L}(aa)$ which extends* **SD** + Extensionality + Regularity + Σ_0 − Separation + **V** = **L** *and has the property of well-foundedness. In this case, we have:*

(i) $0^\# = \{\varphi(c_0 \ldots c_{n-1})$ *is a formula of the language of* **ZF** *and* $(aax_1)\cdots(aax_n)\varphi(x_n \ldots x_n) \in \Sigma\}$.
(ii) $\{\delta_\Sigma(\alpha) \mid \alpha \in On\}$ *is the class of Silver indiscernibles for* **L**.
(iii) $\Sigma = \{\sigma \mid \sigma$ is a sentence of $\mathcal{L}(aa)$ such that $(L_\kappa, \mathcal{F}_\kappa) \models \sigma\}$, *where κ is an uncountable cardinal.*
(iv) $(L_\kappa, \mathcal{F}_\kappa) \prec (L_\lambda, \mathcal{F}_\lambda)$ *for any uncountable cardinals $\kappa < \lambda$.* ⊣

11.19 Theorem. *Let κ be regular uncountable cardinal. If $(L_\kappa, \mathcal{F}_\kappa) \models$* **Dual**, *then $0^\#$ exists.* ⊣

Proof. We define an increasing sequence $\langle \alpha_\zeta : \zeta < \kappa \rangle$ of ordinals less that κ as follows: Choose $\alpha_0 < \kappa$ se that $(L_\kappa, \mathcal{F}_\kappa) \models \varphi[\alpha_0]$ for any formula $\varphi(x)$ of $\mathcal{L}(aa)$ with at most one free variable such that $(L_\kappa, \mathcal{F}_\kappa) \models (aaz)\varphi(x)$. Assume that α_ξ is defined for any $\xi < \zeta$, $\zeta < \kappa$. Choose $\alpha_\xi < \kappa$ so that $\alpha_\xi < \alpha_\zeta$ for any $\xi < \zeta$ and $(L_\kappa, \mathcal{F}_\kappa) \models \varphi[\alpha_\zeta]$ for any formula $\varphi(x)$ with $(L_\kappa, \mathcal{F}_\kappa) \models (aax)\varphi(x)$. Then, it is clear that $\{\alpha_\zeta \mid \zeta < \kappa\}$ is a set of indiscernibles for L_κ. Therefore, $0^\#$ exits. ⊣

Part III

So far we have discussed a filter quantifier "for almost all *sets*". However, the original intentional meaning of aa in Barwise-Kaufmann-Makkai stationary logic is "for almost all *countable sets*". In Part III, we shall discuss a filter quantifier aa to be interpreted as "for almost all *countable sets*" in set theory.

§12. General Framework

Throughout this section, we assume that $\mathcal{L}_U$ is a first-order language with equality having a binary predicate symbol $\in$ and a unary predicate symbol U (possibly with other constants, function or predicate symbols). We refer to $U(x)$ as "x is a U-set", and reserve the letters $s, t, s', t', \ldots$ as variables ranging over U-sets. Also "x is U-transitive" refers to the formula "x is a U-set$\wedge(\forall s \in x)(s \subseteq x)$". As usual, $(\forall s)\varphi(s)$ (resp. $(\exists s)\varphi(s)$) is an abbreviation for $(\forall x)(x$ is a U-set $\rightarrow \varphi(x))$ (resp. $(\exists x)(x$ is a U-set $\wedge\, \varphi(x))$.

12.1 Definition. (1) The theory $\mathbf{UB}_U$ (relative to $\mathcal{L}_U$) is a theory obtained from **UB** by replacing **UB2** with
$\mathbf{UB}_U\mathbf{2}$ $(\forall x \in s)(aa\, y)\varphi \rightarrow (aa\, y)(\forall x \in s)\varphi$,
and adding
$\mathbf{UB}_U\mathbf{0}$ $(aa\, x)(x$ is a U-set$)$.

(2) The theory $\mathbf{ST}_U$ (relative to $\mathcal{L}_U$) is a theory obtained from **ST** by replacing **ST1** with
$\mathbf{ST}_U\mathbf{1}$ $(\forall s)(aa\, x)(s \subseteq x)$.

As in the case of $(\forall s)\varphi(s)$ (resp. $(\exists s)\varphi(s)$), by $(aa\, s)\varphi(s)$ (resp. $(stat\, s)\varphi(s)$) we shall mean the abbreviation of $(aa\, x)(x$ is a U-set $\rightarrow \varphi(x)$ (resp. $(stat\, x)(x$ is a U-set $\wedge\, \varphi(x)))$.

12.2 Definition. (a) By *U-Pair, U-Union*, and *U-Collection* we mean the following:

U-Pair; $(\forall x)(\forall y)(\exists s)(x \in s \wedge y \in s)$,

U-Union; $(\forall s)(\exists t)(\forall s' \in s)(s' \subseteq t)$,

U-Collection; $(\forall x \in s)(\exists y)\varphi \rightarrow (\exists t)(\forall x \in s)(\exists y \in t)\varphi$, where φ is a formula in which t does not occur free.

(b) By $\mathbf{RF}_U$, we mean the following scheme:

$$(\exists s)(s \text{ is } U\text{-transitive} \wedge (\forall x_0 \cdots x_{n-1} \in s) \bigwedge_{k=1}^{l} (\varphi_k \leftrightarrow \varphi_k^{(s)}),$$

where each φ_k is a formula of $\mathcal{L}_U$ with free variables among $x_0, \ldots, x_{n-1}$.

We shall list some results about $\mathbf{UB}_U$ and $\mathbf{ST}_U$ without proofs. The proof can be carried out by the same fashion as we did for the case **UB** and **ST** in §2.

12.3 Theorem. (a) *U-Pair, U-Union and U-Collection (ranging over all formulas in $\mathcal{L}_U(aa)$) are provable in UB_U.*
(b) *The first-order part of UB_U coincides with the first-order theory on $\mathcal{L}_U$ with U-Pair, U-Union and U-Collection as its own axioms.*
(c) $\mathbf{ST}_U$ *is an extension of* UB_U.
(d) *The first-order part of* $\mathbf{ST}_U$ *coincides with the first-order theory on* $\mathcal{L}_U$ *with* $\mathbf{RF}_U$ *as its own axioms.* ⊣

12.4 Definition. (a) A theory $\mathbf{ZF}_U^W(aa)$ is a theory obtained by adding all axioms of **ZF** relative to $\mathcal{L}_U$ to $\mathbf{ST}_U$.
(b) A theory $\mathbf{ZF}_U(aa)$ is obtained by adding the following Separation Scheme for U-sets to $\mathbf{ZF}_U^W(aa)$.

Separation Scheme for U-sets: $(\forall s)(\exists t)(\forall x)(x \in t \leftrightarrow x \in s \wedge \varphi)$, where φ is a formula of $\mathcal{L}_U(aa)$ in which t does not occur free.

By (d) in Theorem 12.3, we have the following corollary.

12.5 Corollary. $\mathbf{ZF}_U^W(aa)$ *is a conservative extension of* **ZF** (relative to $\mathcal{L}_U$)+$\mathbf{RF}_U$. ⊣

§13. Set theory with a quantifier "for almost all countable sets"

In this section, we assume that $\mathcal{L}_U$ is a first-order language with equality having a binary predicate symbol $\in$ and a unary predicate symbol U. All the theorems in this section are theorems of **ZF** relative to $\mathcal{L}_U$ (that is, **ZF** with the separation and collection schemes ranging over formulas of $\mathcal{L}_U$). We also assume tat U is hereditrarily with respect to the inclusion $\subseteq$, that is,

$$(\forall x)(\forall y)(x \text{ is a } U\text{-set} \land y \subseteq x \to y \text{ is a } U\text{-set}).$$

13.1 Definition. Let A be a set. By $P_U(A)$, we always mean the set $\{p \mid p \subseteq A \text{ and } p \text{ is a } U\text{-set}\}$. A sub set X of $P_U(A)$ is ***unbounded*** if $(\forall p \in P_U(A))(\exists q \in X)(p \subseteq q)$, and ***closed*** if for every non-empty subset D of X, which is directed by $\subseteq$ and $q = \bigcup D \in P_U(A)$, we have $q \in X$.

13.2 Definition. By the Diagonal Intersection Principle for U-sets we mean the following:

For every set A, the collection of closed unbounded subsets of $P_U(A)$ is closed under diagonal intersections, that is, for every sequence $\langle K_x : x \in A\rangle$ of closed unbounded sets of $P_U(A)$ the diagonal intersections of $\langle K_x : x \in A\rangle$, $\Delta_{x\in A} K_x = \{p \in P_U(A) \mid (\forall x \in p)(p \in K_x)\}$, is also closed unbounded.

13.3 Definition. (a) By the U-Closure Principle, we mean the following:

Let S be a non-empty set and let R be a binary relation such that for any finite sequence $\langle x_i : i < n\rangle$ from S there is a n element x of S such that $\langle x_i : i < n\rangle\ R\ x$. Then, for every U-set t with $t \subseteq S$, there is a U-set s such that $t \subseteq s \subseteq U$ and for every sequence $\langle x_i : i < n\rangle$ from s there is an element x of s with $\langle x_i : i < n\rangle\ R\ x$.

(b) By the weak form of U-Closure Principle for ternary relations, we mean the following:

Let S be a non-empty set and let R be a ternary relation on S such that $(\forall x \in s)(\forall y \in s)(\exists z \in s)R(x, y, z)$. Then, there is a non-empty U-set s such that $s \subseteq S$ and $(\forall x \in s)(\forall y \in s)(\exists z \in s)R(x, y, z)$.

(c) By the weak form of U-Closure Principle for binary relations, we mean the following:

Let S be a non-empty set and let R be a binary relation on S such that $(\forall x \in s)(\exists y \in s)R(x, y)$. Then, there is a non-empty U-set s such that $s \subseteq S$ and $(\forall x \in s)(\exists y \in s)R(x, y, z)$.

13.4 Definition. By the Reflection Principle for U-sets, we mean the following scheme:

$$(\forall t)(\exists s)(t \subseteq s \land (\forall x_0 \cdots x_{n-1} \in s) \bigwedge_{k=1}^{l} (\varphi_k \leftrightarrow \varphi_k^{(s)})),$$

where each φ_k is a formula of $\mathcal{L}_U$ with free variables among $x_0, \ldots, x_{n-1}$

13.5 Lemma. *Assume U-Pair and U-Union. Then, the following are equivalent:*

(i) *U-Collection and the weak form of U-Closure Principle for ternary relations;*
(ii) *Diagonal Intersection Principle for U-sets;*
(iii) *The U-Closure Principle;*
(iv) *Let $\mathcal{A}$ be a structure of a first-order language such that the set of non-logical symbols is a U-set. Then, for every U-set C which is a subset of the universe of $\mathcal{A}$, there is an elementary submodel $\mathcal{B}$ of $\mathcal{A}$ such that the universe of $\mathcal{B}$ is a U-set and contains C. Moreover, if the universe of $\mathcal{A}$ is transitive, we can take $\mathcal{B}$ so that the universe of $\mathcal{B}$ is U-transitive;*
(v) *The Reflection Principle for U-sets.*

PROOF. (i) ⇒ (ii): First, we show the following claim.

Claim. Let a be a non-empty U-set and $\langle K_x : x \in a \rangle$ be a family of closed unbounded subsets of $P_U(A)$. Then, $K = \bigcap_{x\in a} K_x$ is closed unbounded in $P_U(A)$.

It is clear that $K = \bigcap_{x\in a} K_x$ is closed. To prove that K is unbounded, let t_0 be an arbitrary element of $P_U(A)$. Set $S = \{p \in P_U(A) \mid t_0 \subseteq p\}$. We define a ternary relation R on S as follows:

$$R_1(p_0, p_1, q) \text{ iff } (\forall x \in a)(\exists r)(r \in K_x \wedge p_0 \subseteq r \subseteq q \wedge p_1 \subseteq r \subseteq q).$$

We show that

$$(\forall p_0 \in S)(\forall p_1 \in S)(\exists q \in S) R_1(p_0, p_1, q).$$

So let p_0 and p_1 be arbitrary elements of S. By U-Pair and U-union, $p = p_0 \cup p_1$ is a U-set. Since each K_x is unbounded, we have

$$(\forall x \in a)(\exists r)(r \in K_x \wedge p \subseteq r).$$

By U-Collection, there is a U-set t such that

$$(\forall x \in a)(\exists r \in t)(r \in K_x \wedge p \subseteq r).$$

By U-Union,

$$q = \bigcup \{ r \in P_U(A) \mid r \in t \}$$

is a U-set. Then, clearly, $R_1(p_0, p_1, q)$. Now, by the weak form of U-Closure Principle for ternary relations, there is a non-empty U-set s such that $s \subseteq S$ and

$$(\forall p_0 \in s)(\forall p_1 \in s)(\exists q \in s) R_1(p_0, p_1, q).$$

Set $t_1 = \bigcup s$. Then, $t_1 \in S$. We want to show that $t_1 \in K_x$ for every $x \in a$. So let $x \in a$. Set $D_x = \{r \in K_x \mid (\exists q \in s)(r \subseteq q)\}$. In order to get $t_1 \in K_x$, we have to show that D_x is directed by $\subseteq$ and $t_1 = \bigcup D_x$. To show that D_x is directed let $r_1, r_2 \in D_x$. Then, $r_0 \subseteq p_0$ and $r_1 \subseteq p_1$ for some $p_0, p_1 \in s$. Let $q \in s$ be such that $R_1(p_0, p_1, q)$. Choose $r \in K_x$ so that $p_0 \subseteq r \subseteq q$ and $p_1 \subseteq r \subseteq q$. Then, $r_0, r_1 \subseteq r$ and $r \in D_x$. To show that $t_1 = \bigcup D_x$, it is enough to show that $t_1 \subseteq \bigcup D_x$. So let $p \in s$. Then, for some $q \in s$, $R_1(p, p, q)$. Choose $r \in K_x$ so that $p \subseteq r \subseteq q$. Then, $r \in D_x$ and $p \subseteq r$. Thus, $t_1 \subseteq \bigcup D_x$. We have shown that K is unbounded.

Now, $\langle K_x : x \in A \rangle$ be a family of closed unbounded subsets of $P_U(A)$. We must show that $K = \Delta_{x\in A} K_x = \{p \in P_U(A) \mid (\forall x \in p)(p \in K_x)\}$ is closed unbounded in

$P_U(A)$. To show that K is closed, let $D \subseteq K$ be a non-empty directed subset of K such that $q = \bigcup D \in P_U(A)$. Let $x \in q$. Set $D_x = \{ p \in D \mid x \in p \}$. Clearly, $D_x \subseteq K_x$, D_x is a non-empty directed set and $q = \bigcup D_x$. Since K_x is closed, $q \in K_x$. Thus, $(\forall x \in q)(q \in K_x)$, that is, $q \in K$. To show that K is unbounded, let $t_0 \in P_U(A)$. Set $S = \{ p \in P_U(A) \mid t_0 \subseteq p \}$. We define a ternary relation R_2 on S as follows:

$$R_2(p_0, p_1, q) \text{ iff } p_0, p_1 \subseteq q \wedge q \in \bigcap_{x \in p_0} K_x.$$

By our claim, we know that $\bigcap_{x \in p} K_x$ is unbounded for each $p \in P_U(A)$. Therefore, we have

$$(\forall p_0 \in S)(\forall p_1 \in S)(\exists q \in S) R_2(p_0, p_1, q).$$

By the weak form of U-Closure Principle, there is a non-empty U-set s such that $s \subseteq S$ and

$$(\forall p_0 \in s)(\forall p_1 \in s)(\exists q \in s) R_2(p_0, p_1, q).$$

Set $t_1 = \bigcup s$. Then, $t_1 \in S$. We want to show that $t_1 \in K$. Let $x \in t_1$. Set $D_x = \{ p \in s \mid p \in K_x \text{ and } x \in p \}$. We show that D_x is a non-empty directed set and $t_1 = \bigcup D_x$. Since $x \in t_1$, there is $p \in s$ such tat $x \in p$. So there is a $q \in s$ such that $R_2(p, p, q)$, that is, $p \subseteq q$ and $q \in \bigcap_{x \in p} K_x$. Therefore, $q \in D_x$. By the same fashion, we can see that D_x is directed. To show that $t_1 = D_x$, it is enough to show that $t_1 \subseteq \bigcup D_x$. So let $p_1 \in s$. Since $x \in t_1$, there is a $p_0 \in s$ such that $x \in p_0$. Let $q \in s$ be such that $R_2(p_0, p_1, q)$, that is, $p_0, p_1 \subseteq q$ and $q \in \bigcap_{x \in p_0} K_x$. Then, $q \in D_x$ and $p_1 \subseteq q$. Thus, $t_1 \subseteq \bigcup D_x$. Since K_x is closed, we have $t_1 \in K_x$. Thus, we have shown that $(\forall x \in t_1)(t_1 \in K_x)$, that is, $t_1 \in K$.

(ii) $\Rightarrow$ (i): For the weak form of U-Closure Principle, let S be a non-empty set and let R be a ternary relation on S such that

$$(\forall x \in S)(\forall y \in S)(\exists z \in S) R(x, y, z).$$

For every $x, y \in S$, we set $K_{x,y} = \{ s \in P_U(S) \mid (\exists z \in s) R(x, y, z) \}$. It is clear that $K_{x,y}$ is a closed unbounded subset of $P_U(S)$ for each $x, y \in S$. Set $K_x = \Delta_{y \in S} K_{x,y}$ and $K = \Delta_{x \in S} K_x$. Then, K is closed unbounded by the assumption. Let $s \in K$ and $s \neq 0$. Then,

$$(\forall x \in s)(\forall y \in s)(s \in K_{x,y}),$$

that is,

$$(\forall x \in s)(\forall y \in s)(\exists z \in s) R(x, y, z).$$

For U-Collection, we must show that

$$(\forall w_0 \cdots \forall w_{n-1})(\forall s)((\forall x \in s)(\exists y)\varphi(xysw_0 \ldots w_{n-1}) \rightarrow (\exists t)(\forall x \in s)(\exists y \in t)\varphi(xysw_0 \ldots w_{n-1}))$$

Suppose that $(\forall x \in s)(\exists y)\varphi(xysw_0 \ldots w_{n-1})$. Then, by the Collection Principle, there is a set z such that $(\forall x \in s)(\exists y \in z)\varphi(xy)$. Set $A = s \cup z$. For each $x \in A$, we set $K_x = \{ t \in P_U(A) \mid (\exists y \in A)\varphi(xy) \rightarrow (\exists y \in t)\varphi(xy) \}$. Clearly, K_x is closed unbounded for each $x \in A$. Set $K = \Delta_{x \in A} K_x$. Then, K is closed unbounded by our assumption. Since K is unbounded, there is a $t \in K$ such that $s \subseteq t$. Then,

$$(\exists y \in A)\varphi(xy) \to (\exists y \in t)\varphi(xy) \quad \text{for all } x \in t.$$

Since $(\exists y \in A)\varphi(xy)$ for all $x \in s$ and $s \subseteq t$, we have $(\forall x \in s)(\exists y \in t)\varphi(xy)$. Thus, we have shown that

$$(\forall x \in s)(\exists y)\varphi(xysw_0 \ldots w_{n-1}) \to (\exists t)(\forall x \in s)(\exists y \in t)\varphi(xysw_0 \ldots w_{n-1}).$$

(ii) ⇒ (iii): First, we show that every countable set is a U-set. For every $n < \omega$, we set $K_n = \{ s \in P_U(\omega) \mid n+1 \in s \}$. Then, clearly, each K_n is closed unbounded in $P_U(\omega)$. Therefore, $K = \Delta_{n<\omega} K_n$ is closed unbounded by our assumption. Since K is unbounded, we can choose $s \in K$ such that $0 \in s$. Then, $s = \omega$. Thus, ω is a U-set. Since U-Collection holds, every countable set is a U-set. Now, let S be a non-empty set and let R be a binary relation such that for every finite sequence $\langle x_i : i < n \rangle$ from S there is an element x of S such that $\langle x_i : i < n \rangle \; R \; x$. Fix $n < \omega$. We define an F_i^n $(0 \leq i \leq n)$ recursively as follows:

$$F_0^n(x_0, \ldots, x_{n-1}) = \{ s \in P_U(S) \mid (\exists x \in s)(\langle x_i : i < n \rangle \; R \; x) \}.$$
$$F_1^n(x_0, \ldots, x_{n-2}) = \mathop{\Delta}_{y \in S} F_0^n(x_0, \ldots, x_{n-2}, y).$$

$$\vdots$$

$$F_n^n = \mathop{\Delta}_{y \in S} F_{n-1}^n(y).$$

Since $F_0^n(x_0, \ldots, x_{n-1})$ is a closed unbounded subset of $P_U(S)$, we can see that F_n^n is closed unbounded by the assumption. Since ω is a U-set and each F_n^n is closed unbounded, we have $\bigcap_{n<\omega} F_n^n$ is closed unbounded (by the Claim in the proof of (i) ⇒ (ii)). Let t be a U-set such that $t \subseteq S$. Since $\bigcap_{n<\omega} F_n^n$ is unbounded, there is an $s \in \bigcap_{n<\omega} F_n^n$ such that $t \subseteq s$. It is clear that s is a required U-set.

(iii) ⇒ (i): It is clear that the weak form of U-Closure Principle hold. For U-Collection, we must show that

$$(\forall x \in s)(\exists y)\varphi(xysw_0 \ldots w_{n-1}) \to (\exists t)(\forall x \in s)(\exists y \in t)\varphi(xysw_0 \ldots w_{n-1}).$$

Suppose that $(\forall x \in s)(\exists y)\varphi(xysw_0 \ldots w_{n-1})$ holds. By Collection, there is a set z such that

$$(\forall x \in s)(\exists y \in z)\varphi(xysw_0 \ldots w_{n-1}).$$

Let S be a set such that $s, z \subseteq S$. We define a binary relation R as follows:
Let $\langle x_i : i < n \rangle$ be a finite sequence from S and x be an element of S. Then,

$$\langle x_i : i < n \rangle \; R \; x \text{ iff } n = 0 \vee (n > 0 \wedge (\exists y \in S)\varphi(x_0 y) \to \varphi(x_0 x)).$$

Then, for every finite sequence $\langle x_i : i < n \rangle$ from S there is an element x of S such that $\langle x_i : i < n \rangle \; R \; x$. By the assumption, there is a U-set t such that $s \subseteq t$ and for every finite sequence $\langle x_i : i < n \rangle$ from t there is an element x of t such that $\langle x_i : i < n \rangle \, R \, x$. Therefore, we have

$$(\forall x \in t)(\exists y \in t)((\exists y \in S)\varphi(xy) \rightarrow \varphi(xy)).$$

Since $(\forall x \in s)(\exists y \in S)\varphi(xy)$, we have $(\forall x \in s(\exists y \in t)\varphi(xy)$. Thus, we have shown that U-Collection holds.

(iii) $\Rightarrow$ (iv): Let $\mathcal{L}'$ be a first-order language such that the set of non-logical symbols of $\mathcal{L}'$ is a U-set. Since the set of logical symbols of $\mathcal{L}'$ is countable, the set of all symbols of $\mathcal{L}'$ is a U-set. Let S be the set of all finite sequences of symbols of $\mathcal{L}'$. We define a binary relation R as follows:

Let $\langle x_i : i < n \rangle$ be a finite sequence from S and let x be an element of S.

$$\langle x_i : i < n \rangle \; R \; x \text{ iff } x \text{ is the concatenation of } \langle x_i \mid i < n \rangle.$$

Then, for every finite sequence $\langle x_i : i < n \rangle$ from S there is an element x of S such that $\langle x_i : i < n \rangle \; R \; x$. Let t be the set of all symbols of $\mathcal{L}'$. By the U-Closure Principle, there is a U-set s such that $t \subseteq s \subseteq S$ and for every finite sequence $\langle x_i : i < n \rangle$ from s there is an element x of s such that x is the concatenation of $\langle x_i : i < n \rangle$. Since a formula of $\mathcal{L}'$ is a finite sequence of symbols of $\mathcal{L}'$, s contains the set of all formulas of $\mathcal{L}'$. Therefore, the set of all formulas of $\mathcal{L}'$ is a U-set. Now, let $\mathcal{A} = \langle A; \ldots \rangle$ be a structure of $\mathcal{L}'$ and let C be a U-set such that $C \subseteq A$. Let $\varphi(x_0 \ldots x_{n-1})$ be a formula of the form $(\exists y)\psi(yx_0 \ldots x_{n-1})$ with exactly free variables $x_0, \ldots, x_{n-1}$. For each $0 \leq k \leq n$, we define a $(n-k)$-ary function F_k^φ as follows:

$$\begin{aligned} F_0^\varphi(a_0, \ldots, a_{n-1}) &= \{ s \in P_U(A) \mid \mathcal{A} \models \varphi[a_0 \ldots a_{n-1}] \\ &\qquad \text{implies } (\exists b \in s)\mathcal{A} \models \psi[b, a_0, \ldots, a_{n-1}] \} \\ F_1^\varphi(a_0, \ldots, a_{n-1}) &= \mathop{\Delta}_{a \in A} F_0^\varphi(a_0, \ldots, a_{n-2}, a) \\ &\;\vdots \\ F_n^\varphi &= \mathop{\Delta}_{a \in A} F_{n-1}^\varphi(a) \end{aligned}$$

It is clear that $F_0^\varphi(a_0, \ldots, a_{n-1})$ is closed unbounded in $P_U(A)$ for each $a_0, \ldots, a_{n-1}$. By the assumption, F_n^φ is closed unbounded. Since the set of all formulas of $\mathcal{L}'$ is a U-set, $K = \bigcap \{ F_n^\varphi \mid \varphi \text{ is a formula of the form } (\exists y)\psi \}$ is closed unbounded. Therefore, there is a U-set B such that $C \subseteq B \subseteq A$ and $B \in K$. Then, we have

$$(\forall a_0 \cdots a_{n-1} \in B)(\mathcal{A} \models \varphi[a_0 \ldots a_{n-1}] \rightarrow (\exists b \in B)\mathcal{A} \models \psi[b, a_0, \ldots, a_{n-1}])$$
$$\text{for any formula } \varphi(x_0 \ldots x_{n-1}) \text{ of the form } (\exists y)\psi(yx_0 \ldots x_{n-1}).$$

Therefore, B determines an elementary substructure $\mathcal{B}$ of $\mathcal{A}$. If A is transitive, we can see that $K_x = \{ s \in P_U(A) \mid x \text{ is a } U\text{-set} \rightarrow x \subseteq s \}$ is closed unbounded for any $x \in A$ (by using the fact that A is transitive). Therefore, $\Delta_{x \in A} K_x$ is closed unbounded, and so $K \cap \Delta_{x \in A} K_x$ is closed unbounded. Then, $B \in K \cap \Delta_{x \in A} K_x$ is a U-transitive set and determines the elementary substructure of $\mathcal{A}$.

(iv) ⇒ (v): Let $\varphi_1, \dots, \varphi_k$ be formulas of $\mathcal{L}_U$ with free variables $x_0, \dots, x_{n-1}$. Let t be a U-set. By Reflection Principle, there is a set A such that $t \subseteq A$ and each φ_k is absolute for A. By the assumption, there is a U-set s such that $t \subseteq s$ and $\langle s; \in, U \cap s\rangle \prec \langle A; \in, U \cap A\rangle$. Therefore, each φ_k is absolute for s.

(v) ⇒ (i): For U-Collection, we must show that

$$(\forall x \in t)(\exists y)\varphi(xytw_0 \dots w_{n-1}) \to (\exists s)(\forall x \in t)(\exists y \in s)\varphi(xytw_0 \dots w_{n-1}).$$

Suppose $(\forall x \in t)(\exists y)\varphi(xytw_0 \dots w_{n-1})$. Set $t' = t \cup \{w_0, \dots, w_{n-1}\} \cup \{t\}$. Then, t' is a U-set. By the assumption, there is a U-set s such that $t' \subseteq s$,

$$(\forall x)(\forall y)(\forall t)(\forall w_0 \cdots w_{n-1} \in s)(\varphi(xytw_0 \dots w_{n-1}) \leftrightarrow \varphi^{(s)}(xytw_0 \dots w_{n-1}))$$

and

$$(\forall x)(\forall y)(\forall t)(\forall w_0 \cdots w_{n-1} \in s)((\exists y)\varphi(xytw_0 \dots w_{n-1}) \\ \leftrightarrow (\exists y \in s)\varphi^{(s)}(xytw_0 \dots w_{n-1})).$$

Since $(\forall x \in t)(\exists y)\varphi(xytw_0 \dots w_{n-1})$, $t, w_0, \dots, w_{n-1} \in s$ and $t \subseteq s$, we have

$$(\forall x \in t)(\exists y \in s)\varphi^{(s)}(xytw_0 \dots w_{n-1}),$$

and so

$$(\forall x \in t)(\exists y \in s)\varphi(xytw_0 \dots w_{n-1}).$$

Thus, we have shown that

$$(\forall x \in t)(\exists y)\varphi(xytw_0 \dots w_{n-1}) \to (\exists s)(\forall x \in t)(\exists y \in s)\varphi(xytw_0 \dots w_{n-1}).$$

For the weak form of U-Closure Principle, let S be a non-empty set and let R be a ternary relation R on S such that

$$(\forall x \in S)(\forall y \in S)(\exists z \in S)R(x, y, z).$$

Let $\varphi_0(u), \varphi_1(xyzw), \varphi_2(uw)$ denote the following formulas:

$$\varphi_0(u) \equiv (\exists x)(x \in u),$$
$$\varphi_1(xyzw) \equiv \langle x, y, z\rangle \in w,$$
$$\varphi_2(uw) \equiv (\forall x \in u)(\forall y \in u)(\exists z \in u)\varphi_1(xyzw).$$

Since $\{R, S\}$ is a U-set, there is a U-set s' such that $R, S \in s'$ and $\varphi_0, \varphi_1, \varphi_2$ are absolute for s'. Set $s = s' \cap S$. Then, s is a U-set. Since $\varphi_0(S)$, we have $\varphi_0^{(s')}(S)$, that is, $s = s' \cap S \neq 0$. Since $\varphi_2(S, R)$, we have $\varphi_2^{(s's)}(S, R)$, that is,

$$(\forall x \in s)(\forall y \in s)(\exists z \in s)\varphi_1^{(s')}(xyzR).$$

Since φ_1 is absolute for s', we have

$$(\forall x \in s)(\forall y \in s)(\exists z \in s)R(x, y, z). \dashv$$

13.6 Lemma. *Assume the condition for U-sets such that every U-set is well-orderable. Then, two weak forms of U-Closure Principle for binary relations and ternary relations are equivalent under U-Union, U-Pair and U-Collection.*

PROOF. It is clear that the principle for ternary relations implies the principles for binary relations. For the converse, by the previous lemma it is enough to show that the principle for binary relations and U-Collection imply Diagonal Intersection Principle for U-sets under U-Pair and U-Union. The proof is almost the same as the proof of (i) $\Rightarrow$ (ii) in the previous lemma. Instead of the ternary relation R_1 in that proof, we define a binary relation R_1 on S as follows:

$$p\ R_1\ q \text{ iff } (\forall x \in a)(\exists r)(r \in K_x \wedge p \subseteq r \subseteq q).$$

By the same fashion as in that proof, we can get a U-set s such that $s \subseteq S$ and $(\forall p)(\exists q)p\ R_1\ q$. Since s is well-orderable, there is a function $f : \omega \to s$ so that $(\forall n < \omega)f(n)\ R_1\ f(n+1)$. Set $t_1 = \bigcup \text{range}(f)$ instead of $t_1 = \bigcup s$. For R_2, at this time, we define

$$p\ R_2\ q \text{ iff } p \subseteq q \wedge q \in \bigcap_{x \in p} K_x.$$

Then, every thing works well in the same way. ⊣

13.7 Lemma. *The following are equivalent:*

(i) *U-Pair, U-Union and U-Closure principle.*
(ii) $\mathbf{RF}_U$

PROOF. (i) $\Rightarrow$ (ii): It is clear from the "moreover" part of (iv) in Lemma 13.5 and the proof of (iv) $\Rightarrow$ (v).
(ii) $\Rightarrow$ *(i)*: By Lemma 13.5, it is enough to show that U-Pair, U-Union and Reflection Principle for U-sets. For U-pair, by $\mathbf{RF}_U$, there s a U-set s such that

$$\begin{aligned}&(\forall x \in s)(\forall y \in s)((\exists t)(x \in t \wedge y \in t) \leftrightarrow (\exists t \in s)(x \in t \wedge y \in t))\\ &\quad \wedge((\forall x)(\forall y)(\exists t)(x \in t \wedge y \in t) \leftrightarrow (\forall x \in s)(\exists y \in s)(\exists t \in s)(x \in t \wedge y \in t)).\end{aligned}$$

Since

$$x \in s \wedge y \in s \to (\exists t)(x \in t \wedge y \in t),$$

then

$$x \in s \wedge y \in s \to (\exists t \in s)(x \in t \wedge y \in t).$$

And so

$$(\forall x \in s)(\forall y \in s)(\exists t \in s)(x \in t \wedge y \in t).$$

Therefore,

$$(\forall x)(\forall y)(\exists t)(x \in t \wedge y \in t).$$

For U-Union, let φ_0 and φ_1 be as follows:

$$\varphi_0(t_0) \equiv (\exists t_1)(\forall t_2 \in t_0)(t_2 \subseteq t_1),$$
$$\varphi_1 \equiv (\forall t_0)\varphi_0(t_0).$$

By $\mathbf{RF}_U$, there is a U-transitive set s such that φ_0 andφ_1 are absolute for s. Since s is U-transitive, we have

$$t_0 \in s \rightarrow (\forall t_2 \in t_0)(t_2 \subseteq s),$$

and so

$$t_0 \in s \rightarrow (\exists t_1)(\forall t_2 \in t_0)(t_2 \subseteq t_1).$$

Since φ_0 is absolute for s, we have $\varphi_0^{(s)}(t_0)$, that is,

$$t_0 \in s \rightarrow (\exists t_1 \in s)(\forall t_2 \in t_0)(t_2 \subseteq t_1).$$

So we have

$$(\forall t_0 \in s)(\exists t_1 \in s)(\forall t_2 \in t_0)(t_0 \subseteq t_1).$$

Since φ_1 is absolute for s, we have

$$(\forall t_0)(\exists t_1))(\forall t_2 \in t_0)(t_2 \subseteq t_1).$$

For Reflection Principle for U-sets, it is easily seen that Reflection Principle for U-sets is provable in $\mathbf{ST}_U$, and so by using (d) of Th.12.3, it is provable from $\mathbf{RF}_U$. ⊣

The most interesting case for us is that

$$(\forall x)(x \text{ is a } U\text{-set} \leftrightarrow x \text{ is a countable set}).$$

In this case, U-Collection Principle is called *Countable Collection Principle.* And U-Closure Principle is called $\aleph_0$-Closure Principle. The meaning of Reflection Principle for countable sets, $\mathbf{RF}_{\aleph_0}, \ldots$ are self-evident.

13.8 Lemma. *Countable Collection Principle is equivalent to Countable Axiom of Choice.*

PROOF. Assume AC_ω (Countable Axiom Choice). We must show that

$$(\forall x \in s)(\exists y)\varphi \rightarrow (\exists t)(\forall x \in s)(\exists y \in t)\varphi.$$

Assume $(\forall x \in s)(\exists y)\varphi$. For each $x \in s$, let α_x be the least ordinal α such that $(\exists y \in R_\alpha)\varphi$. Then, $\{\{y \in R_{\alpha_x} \mid \varphi\} \mid x \in s\}$ is countable. By AC_ω, we can choose y_x from $\{y \in R_{\alpha_x} \mid \varphi\}$ for each $x \in s$. Set $t = \{y_x \mid x \in s\}$. Then, t is countable and $(\forall x \in s)(\exists y \in t)\varphi$.

Conversely, assume Countable Collection Principle. Assume $(\forall x \in s)(\exists y)(y \in x)$. Then, Countable Collection Principle, there is a countable set t such that

$$(\forall x \in s)(\exists y \in t)(y \in x).$$

Since t is countable, t has a well-ordering $\leq$. Therefore, for each x, we can choose y_x so that y_x is the least element y of t with respect to $\leq$ such that $y \in x$. ⊣

It is clear that the weak form of $\aleph_0$-Closure Principle for binary relations and DC (the Axiom of Dependent Choice) is equivalent. Since DC implies AC_ω [4] (and so Countable Collection Principle), we have the following lemma.

[4] For the proof, for example, see [4, p 23].

13.9 Lemma. *The following are equivalent.*[5]

(i) *The Axiom of Dependent Choice.*
(ii) *Diagonal Intersection Principle for countable sets.*
(iii) *Reflection Principle for countable sets.*
(iv) $\mathbf{RF}_{\aleph_0}$. ⊣

13.10 Definition. (a) $\mathbf{ZF}_U^{\mathbf{UB}}$ is a theory obtained by adding to $\mathbf{UB}_U$ all the axioms of **ZF** relative to $\mathcal{L}_U(\text{aa})$ (that is, the separation scheme and collection scheme range over all formulas of $\mathcal{L}_U(\text{aa})$).
(b) $\mathbf{ZF}_U^W(\text{aa})$ is a theory obtained by adding to $\mathbf{ST}_U$ all the axioms of **ZF** relative to $\mathcal{L}_U$.
(c) $\mathbf{ZF}_U(\text{aa})$ is a theory obtained by adding the following separation scheme to $\mathbf{ZF}_U^W(\text{aa})$.
Separation Scheme for U-sets: $(\forall s)(\exists t)(\exists x)(x \in t \leftrightarrow x \in s \wedge \varphi)$, where φ is a formula of $\mathcal{L}_U(\text{aa})$ in which t does not occur free.

Like as Lemma 4.3, we have the following lemma.

13.11 Lemma. $\mathbf{ZF}_U^{\mathbf{UB}}$ *is a conservative extension of* **ZF** + *U-Pair* + *U-Union* + *U-Collection.* ⊣

By using Lemma 13.8, we have the following corollary.

13.12 Corollary. $\mathbf{ZF}_{\aleph_0}^{\mathbf{UB}}$ *is a conservative extension of* **ZF** + *Countable Axiom of Choice.* ⊣

For $\mathbf{ZF}_U^W(\text{aa})$, by Lemma 12.3, we have the following lemma.

13.13 Lemma. $\mathbf{ZF}_U^W(\text{aa})$ *is a conservative extension of* $\mathbf{ZF} + \mathbf{RF}_U$. ⊣

As an immediate consequence of the lemma and Lemma 13.9, we have the following theorem.

13.14 Theorem. $\mathbf{ZF}_{\aleph_0}^W(\text{aa})$ *is a conservative extension of* $\mathbf{ZF} + DC$. ⊣

For $\mathbf{ZF}_{\aleph_0}(\text{aa})$, we have the same results as $\mathbf{ZF}(\text{aa})$. For example, we have the following theorem.

13.15 Theorem. $(\text{aa}\, s)(s$ *is a model of* $\mathbf{ZF})$ *is provable in* $\mathbf{ZF}_{\aleph_0}(\text{aa})$. ⊣

---

[5] The equivalence of (i) and (iii) is already known by [16, Appendix].

References

[1] K.J. Barwise, M. Kaufmann, M. Makkai, Stationary logic, *Ann Math Logic* 13 (1978) 171–224.
[2] J. Baumgartner, Ineffability properties of cardinals. I, in: Infinite and Finite Sets, Vol. 1 (North-Holland, Amsterdam, 1975) 109–130.
[3] C.C. Chang, H.J. Keisler, Model Theory (North-Holand, Amsterdam, 1973).
[4] T. Jech, The Axiom of Choice (North-Holland, Amsterdam, 1973).
[5] T. Jech, Some combinatorial problems concerning uncountable cardinals, *Ann Math Logic* 5 (1973) 165–224.
[6] T. Jech and W. Powell, Standard models of set theory with predication, *Bull Amer Math Soc* 77 (1971) 808-813.
[7] Y. Kakuda, Set theory based on the language with the additional quantifier "for almost all" I, *Math Sem Notes, Kobe University* 8 (1980) 603–609.
[8] Y. Kakuda, Set theory extracted from Cantor's theological ontology, To appear in *Ann Jap Ass Phil Sci* (1989) .
[9] S. Kamo, Unpublished Notes, (1981) .
[10] M. Kaufmann, Set theory with a filter quantifier, *J Symb Logic* 48 (1983) 263–287.
[11] M. Kaufmann and S. Shelah, A nonconservative result on global choice, *Ann Pure Appl Logic* 27 (1984) 209–214.
[12] H. J. Keisler, Logic with the quantifier "there exists uncountably many", *Ann Math Logic* 1 (1970) 1–93.
[13] E. M. Kleinberg, A combinatorial characterization of normal M-ultrafilters, *Adv Math* 30 (1978) 77–84.
[14] K. Kunen, Some applications of iterated ultrapowers in set theory, *Ann Math Logic* 1 (1970) 179–227.
[15] A Levy, Axiom schemata of strong infinity in axiomatic set theory, *Pacific J Math* 10 (1960) 223–238.
[16] A. Levy, A hierarchy of formulas in set theory, *Mem Amer Math Soc* 57 (1965) .
[17] R. Montague and R.L. Vaught, Natural models of set theory, *Fund Math* 47 (1959) 219–242.
[18] J. Schmerl, On κ-like structures which embedded stationary and closed unbounded subsets, *Ann Math Logic* 10 (1976) 289–314.

Department of Mathmtics
College of Liberal Arts
Kobe University
Nada, Kobe, Japan

Syntactical Simulation of Many-Valued Logic *

Hiroaki KATSUTANI

In order to deal with uncertain information, we often consider that the formulas representing propositions may take various values according to their degree of certainty, and try to calculate them by the methods of many-valued logic. In the traditional studies of the syntax of many-valued logic, however, the provable formulas have meant those whose values belong to a previously specified subset of the value set (that is, the set of all values which formulas can take). But we would like to know how the values of formulas develop rather than whether the values of them belong to the subset or not. Accordingly, in the present work, we extend the classical logic by incorporating into syntax the mechanism for inferring the evaluation of the values of formulas.

In the theory of many-valued logic, the value of a formula is decided uniquely by algebraic operation only from the values of its subformulas. But, if we regard the value of a formula as meaning the verisimilitude of the proposition, it would seem unnatural to handle it in the same way. For this reason, we deal with the limitation of the values of formulas, in formal systems. That is to say, we express propositions such as "the value of a proposition A is greater than or equal to that of a proposition B" and "the value of a proposition A is greater than or equal to a" as formulas, and derive the limits of the values of propositions by formal deductions.

We called the logic obtained in this way "evaluational logic". Its systems are extensions of the classical logic, and it simulates many-valued logic syntactically.

For a proposition A, the proposition "the value of A is greater than or equal to a" seems to be within a modality like the necessity. In fact, the provabilities of formulas in the normal modal logics K, D, T (that is, von Wright's logic M) and S4 are translated into provabilities in the corresponding evaluational logics. Namely, evaluational logics are conservative extensions of normal modal logics. Furthermore, evaluational logic is considered to be multi-modal logic having modalities of various grades. In other words, we can quantify modalities within it.

From Section 1 to Section 2, we present formal systems of evaluational logics by the sequential method. In Section 3, we see some theorems in each system. Section 4 is devoted to the cut-elimination theorems for them. From Section 5 to Section 6,

* This research was partially supported by Grant-in-Aid for Co-operative Research (No. 61302010), The Ministry of Education, Science and Culture.

we give an algebraic semantics for evaluational logic and prove the soundness and the completeness with respect to it. In Section 7, we state the relation between evaluational logics and normal modal logics mentioned in the preceding paragraph.

§1. Languages

First, we specify two languages $\mathsf{L_m}$ and $\mathsf{L_c}$ of evaluational logics, and discuss their meanings.

The alphabet of the language $\mathsf{L_m}$ consists of the following symbols:
1) Logical symbols: $\neg, \wedge, \vee, \supset$ and $\prec$;
2) Caliber connectives: $\otimes$ and $\oplus$;
3) Sentential variables;
4) Caliber symbols: two caliber constants 1 and 0, and caliber variables;
5) Auxiliary symbols: $($, $)$, $\rightarrow$, $\Rightarrow$, $\not\Rightarrow$ and , (comma).

The alphabet of the language $\mathsf{L_c}$ is obtained from that of $\mathsf{L_m}$ by adding another caliber connective $\sim$.

In evaluational logic we treat three sorts of expressions: calibers, formulas and sequents.

An expression is a caliber if it can be shown to be a caliber by a finite number of applications of the following two rules:

(1) A caliber symbol is a caliber;
(2) If a and b are calibers, then $(a \otimes b)$, $(a \oplus b)$ and $(\sim a)$ are calibers ($(\sim a)$ is not a caliber in $\mathsf{L_m}$ but that in $\mathsf{L_c}$).

An expression is a formula (in evaluational logic) if it can be shown to be a formula by a finite number of applications of the usual formation rules for formulas in the classical logic and of the formation rule that

if a and b are each either a caliber or a formula, then $(\mathsf{a} \prec \mathsf{b})$ is a formula.

In the formation rule above, if a or b is a formula then it is a subformula of $(\mathsf{a} \prec \mathsf{b})$. An expression is called a valuence if it is a caliber or a formula. A formula whose outermost logical symbol is $\prec$ is said to be evaluational.

Henceforth, in the descriptions of the formal expressions, the letters $\mathsf{A}, \mathsf{B}, \mathsf{C}, \ldots$ will denote formulas, the letters $\mathsf{a}, \mathsf{b}, \mathsf{c}, \ldots$ valuences, the letters $a, b, c, \ldots$ calibers, the letters $\Gamma, \Delta, \Lambda, \ldots$ finite (possibly empty) sequences of formulas separated by commas, the letters $\alpha, \beta, \gamma, \ldots$ finite (possibly empty) sequences of valuences separated by commas, unless stated otherwise.

An f-sequent (formulary sequent) is a sequent in the usual sense, i.e., an expression of the form $\Gamma \rightarrow \Sigma$. A v-sequent (valuent sequent) is an expression of the form $\Gamma \Rightarrow \alpha \not\Rightarrow \zeta$. In a v-sequent $\Gamma \Rightarrow \alpha \not\Rightarrow \zeta$, the sequence Γ of formulas is called the antecedent, and the sequences α and ζ of valuences are called the antecedental and the succedental, respectively. A sequent (in evaluational logic) is either an f-sequent or a v-sequent. We abbreviate a v-sequent $\Rightarrow \alpha \not\Rightarrow \zeta$ with the empty antecedent to $\alpha \not\Rightarrow \zeta$.

For sequences α,β of valuences, we say that α includes β if each valuence occurring in n places in β occurs at least in n places in α.

We discuss the informal meanings of the languages above in the rest of this section.

The logical symbol $\prec$ is intended to denote an order relation on the value set, and a caliber represents an element in the value set. Thus, for example, a formula $\mathsf{A}\prec\mathsf{B}$ means that the value of B is greater than or equal to that of A, and a formula $a\prec\mathsf{A}$ [$\mathsf{A}\prec a$] means that the value of A is greater [less, *respectively*] than or equal to a. The order intended by $\prec$ preserves the implication relation among propositions: If a proposition $\mathsf{A}\supset\mathsf{B}$ is valid then the proposition $\mathsf{A}\prec\mathsf{B}$ is also valid. Accordingly, for example, the evaluational logics yield the theorems

$$(\mathsf{A}\prec\mathsf{a})\supset((\mathsf{A}\wedge\mathsf{B})\prec\mathsf{a}), \qquad (\mathsf{B}\prec\mathsf{b})\supset((\mathsf{A}\wedge\mathsf{B})\prec\mathsf{b}),$$

and

$$(\mathsf{a}\prec\mathsf{A})\supset(\mathsf{a}\prec(\mathsf{A}\vee\mathsf{B})), \qquad (\mathsf{b}\prec\mathsf{B})\supset(\mathsf{b}\prec(\mathsf{A}\vee\mathsf{B})).$$

Let us regard the value of a formula as the degree of certainty of its meaning. Then, an evaluational formula $a\prec\mathsf{A}$ expresses an evaluation of degree of (positive) certainty of the proposition A, and, as a augments, it claims A with greater certainty. On the other hand, an evaluational formula $\mathsf{A}\prec a$ expresses an evaluation of degree of negative certainty of A, and, as a lessens, it claims $\neg\mathsf{A}$ with greater certainty.

When premises are not certain, an increase in premises makes a conclusion from them more uncertain. Only from the information that the degree of certainty of propositions A and B are greater than or equal to a and b respectively, what lower limit of degree of certainty of the proposition $\mathsf{A}\wedge\mathsf{B}$ we can derive in general? The caliber connective $\otimes$ is supplied to respond this question. The evaluational logics are designed to yield the theorem

$$(a\prec\mathsf{A})\wedge(b\prec\mathsf{B})\supset(a\otimes b\prec(\mathsf{A}\wedge\mathsf{B})).$$

As for the caliber connective $\oplus$, the evaluational logics yield the theorem

$$(\mathsf{A}\prec a)\wedge(\mathsf{B}\prec b)\supset((\mathsf{A}\vee\mathsf{B})\prec a\oplus b).$$

These symbols $\otimes$ and $\oplus$ are interpreted as commutative, associative operations on the value set with unit elements expressed by caliber constants 1 and 0 respectively.

The short arrow $\Rightarrow$ has the same meaning and role that the arrow $\longrightarrow$ has, and we need not distinguish it from $\longrightarrow$. It is distinguished only for the sake of appearance. A v-sequent is, so to speak, a nest of sequents in which the triangle arrow $\Rightarrow$ is the inner arrow. $\Rightarrow$ mostly has the same meaning with the logical symbol $\prec$ in the similar sense that $\longrightarrow$ has the same meaning with the logical symbol $\supset$.

§2. Deduction systems

Now we present the deduction systems of evaluational logic by the modified sequential method. Concerning the definitions of syntactical notions in sequential calculuses, we mostly follow [T]. In particular, recall that the phrase "above [below] a inference figure **I**" in a proof figure means "above [below] the lower sequent [upper sequents, *respectively*] of **I**". We begin with the system Km.

Km is a deduction system on the language L_m. It consists of the following axiom schemata and inference rules:

1) Axiom schemata:

$$D \rightarrow D, \qquad d \nrightarrow d, \qquad \nrightarrow 1, \qquad 0 \nrightarrow \ ;$$

2) Inference rules:

2-1) Structural rules:

$$\text{f-weakening}: \frac{\Gamma \rightarrow \Sigma}{D, \Gamma \rightarrow \Sigma}, \frac{\Gamma \gg \alpha \nrightarrow \zeta}{D, \Gamma \gg \alpha \nrightarrow \zeta}, \frac{\Gamma \rightarrow \Sigma}{\Gamma \rightarrow \Sigma, D};$$

$$\text{f-contraction}: \frac{D, D, \Gamma \rightarrow \Sigma}{D, \Gamma \rightarrow \Sigma}, \frac{D, D, \Gamma \gg \alpha \nrightarrow \zeta}{D, \Gamma \gg \alpha \nrightarrow \zeta}, \frac{\Gamma \rightarrow \Sigma, D, D}{\Gamma \rightarrow \Sigma, D};$$

$$\text{f-exchange}: \frac{\Gamma, D, E, \Delta \rightarrow \Sigma}{\Gamma, E, D, \Delta \rightarrow \Sigma}, \frac{\Gamma, D, E, \Delta \gg \alpha \nrightarrow \zeta}{\Gamma, E, D, \Delta \gg \alpha \nrightarrow \zeta}, \frac{\Gamma \rightarrow \Sigma, D, E, \Pi}{\Gamma \rightarrow \Sigma, E, D, \Pi};$$

$$\text{f-cut}: \frac{\Gamma \rightarrow \Sigma, D \qquad D, \Delta \rightarrow \Pi}{\Gamma, \Delta \rightarrow \Sigma, \Pi};$$

$$(1\nrightarrow): \frac{\Gamma \gg \alpha \nrightarrow \zeta}{\Gamma \gg 1, \alpha \nrightarrow \zeta}; \qquad (\nrightarrow 0): \frac{\Gamma \gg \alpha \nrightarrow \zeta}{\Gamma \gg \alpha \nrightarrow \zeta, 0};$$

$$\text{v-exchange}: \frac{\Gamma \gg \alpha, d, e, \beta \nrightarrow \zeta}{\Gamma \gg \alpha, e, d, \beta \nrightarrow \zeta}, \frac{\Gamma \gg \alpha \nrightarrow \zeta, d, e, \eta}{\Gamma \gg \alpha \nrightarrow \zeta, e, d, \eta};$$

$$\text{v-cut}: \frac{\Gamma \gg \alpha \nrightarrow \zeta, d \qquad \Delta \gg d, \beta \nrightarrow \eta}{\Gamma, \Delta \gg \alpha, \beta \nrightarrow \zeta, \eta};$$

2-2) Operational rules:

$$(\wedge\rightarrow): \frac{A, \Gamma \rightarrow \Sigma}{A \wedge B, \Gamma \rightarrow \Sigma}, \frac{B, \Gamma \rightarrow \Sigma}{A \wedge B, \Gamma \rightarrow \Sigma}; \qquad (\rightarrow\wedge): \frac{\Gamma \rightarrow \Sigma, A \qquad \Gamma \rightarrow \Sigma, B}{\Gamma \rightarrow \Sigma, A \wedge B};$$

$$(\vee\rightarrow): \frac{A, \Gamma \rightarrow \Sigma \qquad B, \Gamma \rightarrow \Sigma}{A \vee B, \Gamma \rightarrow \Sigma}; \qquad (\rightarrow\vee): \frac{\Gamma \rightarrow \Sigma, A}{\Gamma \rightarrow \Sigma, A \vee B}, \frac{\Gamma \rightarrow \Sigma, B}{\Gamma \rightarrow \Sigma, A \vee B};$$

$$(\neg\rightarrow): \frac{\Gamma \rightarrow \Sigma, A}{\neg A, \Gamma \rightarrow \Sigma}; \qquad (\rightarrow\neg): \frac{A, \Gamma \rightarrow \Sigma}{\Gamma \rightarrow \Sigma, \neg A};$$

$$(\supset\rightarrow): \frac{\Gamma \rightarrow \Sigma, A \qquad B, \Delta \rightarrow \Pi}{A \supset B, \Gamma, \Delta \rightarrow \Sigma, \Pi}; \qquad (\rightarrow\supset): \frac{A, \Gamma \rightarrow \Sigma, B}{\Gamma \rightarrow \Sigma, A \supset B};$$

$$(\otimes\nrightarrow): \frac{\Gamma \gg a, b, \alpha \nrightarrow \zeta}{\Gamma \gg a \otimes b, \alpha \nrightarrow \zeta}; \qquad (\nrightarrow\otimes): \frac{\Gamma \gg \alpha \nrightarrow \zeta, a \qquad \Gamma \gg \beta \nrightarrow \eta, b}{\Gamma \gg \alpha, \beta \nrightarrow \zeta, \eta, a \otimes b};$$

$$(\oplus\nrightarrow): \frac{\Gamma \gg a, \alpha \nrightarrow \zeta \qquad \Gamma \gg b, \beta \nrightarrow \eta}{\Gamma \gg a \oplus b, \alpha, \beta \nrightarrow \zeta, \eta}; \qquad (\nrightarrow\oplus): \frac{\Gamma \gg \alpha \nrightarrow \zeta, a, b}{\Gamma \gg \alpha \nrightarrow \zeta, a \oplus b};$$

2-3) Bridging rules:

$$\text{(total) conversion}: \frac{\Gamma \longrightarrow \Sigma}{\Gamma \nrightarrow \Sigma};$$

$$(\prec\geqslant): \frac{\Gamma \geqslant \alpha \nrightarrow \zeta, \mathsf{a} \qquad \Delta \geqslant \mathsf{b}, \beta \nrightarrow \eta}{\mathsf{a} \prec \mathsf{b}, \Gamma, \Delta \geqslant \alpha, \beta \nrightarrow \zeta, \eta}; \qquad (\longrightarrow\prec): \frac{\Gamma \geqslant \mathsf{a} \nrightarrow \mathsf{b}}{\Gamma \longrightarrow \mathsf{a} \prec \mathsf{b}}.$$

In the $(\prec\geqslant)$ figure and the $(\longrightarrow\prec)$ figure above, the formula $\mathsf{a} \prec \mathsf{b}$ is called the principal formula, and the valuences a and b are called auxiliary valuences.

The f-cut and the v-cut are generically called cuts. An inference figure whose upper sequents and lower sequent are all v-sequents is said to be valuent.

We extend the system Km by two sorts of ways. One sort of ways is to add further valuent rules, and along this line we give three deduction systems Ki, Kc and Kg.

Ki is a deduction system on the language L_m. It is obtained from Km by replacing the rules $(1\nrightarrow)$ and $(\nrightarrow 0)$ by the stronger rules

$$\text{v-weakening}: \frac{\Gamma \geqslant \alpha \nrightarrow \zeta}{\Gamma \geqslant \mathsf{d}, \alpha \nrightarrow \zeta}, \frac{\Gamma \geqslant \alpha \nrightarrow \zeta}{\Gamma \geqslant \alpha \nrightarrow \zeta, \mathsf{d}}.$$

In these inference figures, the valuence d is called the weakening valuence.

Kc is a deduction system on the language L_c. It is obtained from Km by adding the operational inference rules

$$(\sim\nrightarrow): \frac{\Gamma \geqslant \alpha \nrightarrow \zeta, a}{\Gamma \geqslant \sim a, \alpha \nrightarrow \zeta}; \qquad (\nrightarrow\sim): \frac{\Gamma \geqslant a, \alpha \nrightarrow \zeta}{\Gamma \geqslant \alpha \nrightarrow \zeta, \sim a}.$$

Kg is another deduction system on L_c. It consists of all the axiom schemata and the inference rules contained in Ki or Kc.

Since contractions are admitted neither in the antecedental nor in the succedental of a v-sequent, derivations containing only calibers are essentially those in Grišin's system GL^0 (see [G]).

The other sort of ways to extend the system Km is to add further bridging rules, and along this line we give three deduction systems Dm, Tm and S4m. All of these are systems on the language L_m.

The deduction system Dm is obtained from Km by adding the bridging rule

$$\nrightarrow\text{-vanishing}: \frac{\Gamma \geqslant \nrightarrow}{\Gamma \longrightarrow}.$$

The deduction system Tm is obtained from Dm by adding the bridging rule

$$\text{release}: \frac{\Gamma \geqslant \Delta \nrightarrow \Sigma}{\Gamma, \Delta \longrightarrow \Sigma}$$

with the proviso that in a proof figure this release figure occurs only below a conversion figure **I** such that every sequent below **I** above this release figure is a v-sequent in which the antecedental includes Δ and the succedental Σ. From the set of all such conversion figures for a release figure **J** in a proof figure, we select the rightmost figure, and call it the suspending (conversion) figure to **J**. In addition, the v-sequents occurring above **J** below the suspending figure to **J** are called suspensive v-sequents to **J**. All of them occur in one thread.

The deduction system S4m is obtained from Tm by replacing the total conversion rule by the stronger rule

$$\text{(partial) conversion}: \frac{\Gamma, \Delta \rightarrow \Sigma}{\Gamma \Rightarrow \Delta \nRightarrow \Sigma}$$

provided that every formula occurring in Γ is evaluational.

We see some properties of these deduction systems.

FACT 2.1. In any system, if a formula A occurs in the antecedental [succedental] of an upper sequent of a valuent inference figure **I**, then **I** is a $(\prec\Rightarrow)$ figure in which A is an auxiliary valuence or else A occurs in the antecedental [succedental, *respectively*] of the lower sequent of **I**.

FACT 2.2. In any system, if a formula A occurs in the antecedental [succedental] of the lower sequent of a valuent inference figure **I**, then **I** is a v-weakening figure in which A is the weakening valuence or else A occurs in the antecedental [succedental, *respectively*] of an upper sequent of **I**.

FACT 2.3. In any system, every formula occurring in the antecedent of an upper sequent of a valuent inference figure **I** occurs in the antecedent of the lower sequent of **I**.

LEMMA 2.4. For a nonevaluational formula A and a sequence Γ of formulas, let Γ^* denote the sequence of formulas obtained from Γ by deleting all the occurrences of A. In any system, a proof figure of a v-sequent $\Gamma \Rightarrow \alpha \nRightarrow \zeta$ can be transformed into a proof figure of the v-sequent $\Gamma^* \Rightarrow \alpha \nRightarrow \zeta$ without increasing the number of occurrences of inference figures of each inference rule.

PROOF. Let **P** denote a proof figure of the v-sequent $\Gamma \Rightarrow \alpha \nRightarrow \zeta$, and n be the number of occurrences of v-sequents **S** in **P** such that A occurs in the antecedent of each v-sequent from **S** down to the end-sequent in **P**. Since in **P** A is introduced to the antecedent of a v-sequent only by an f-weakening figure, the lemma is easily proved by the induction on n.

§3. Some results

Here we state what formulas are proved in each system. We omit the proof figures of theorems.

For two formulas A and B, we introduce the expression $A \equiv B$ as the abbreviation of the formula $(A \supset B) \wedge (B \supset A)$, and the expression $a \simeq b$ as that of the formula $(a \prec b) \wedge (b \prec a)$.

THEOREM 3.1. *In the system* Km, *the following formulas are provable:*
(1) $a \prec a$;

(2) $(a \prec b) \wedge (b \prec c) \supset (a \prec c)$;
(3) $1 \prec (A \vee \neg A)$;
(4) $(A \wedge \neg A) \prec 0$;
(5) $(a \prec A) \wedge (b \prec B) \supset (a \otimes b \prec (A \wedge B))$;
(6) $(A \prec a) \wedge (B \prec b) \supset ((A \vee B) \prec a \oplus b)$;
(7) $(A \prec B) \equiv (\neg B \prec \neg A)$;
(8) $(1 \prec A) \equiv (\neg A \prec 0)$.

THEOREM 3.2. *In the system* Km, *the formula* $A \prec B$ *is deduced from the formula* $A \supset B$.

THEOREM 3.3. *In the system* Ki, *the following formulas are provable:*
(1) $a \prec 1$;
(2) $0 \prec a$.

THEOREM 3.4. *In the system* Kc, *the following formulas are provable:*
(1) $1 \prec a \oplus {\sim} a$;
(2) $a \otimes {\sim} a \prec 0$;
(3) ${\sim}{\sim} a \rightleftharpoons a$;
(4) ${\sim}(a \otimes b) \rightleftharpoons {\sim} a \oplus {\sim} b$;
(5) ${\sim}(a \oplus b) \rightleftharpoons {\sim} a \otimes {\sim} b$.

THEOREM 3.5. *In the system* Dm, *the following formulas are provable:*
(1) $\neg(1 \prec 0)$;
(2) $\neg((1 \prec A) \wedge (1 \prec \neg A))$;
(3) $\neg((A \prec 0) \wedge (\neg A \prec 0))$.

THEOREM 3.6. *In the system* Tm, *the following formulas are provable:*
(1) $(1 \prec A) \supset A$;
(2) $(A \prec 0) \supset \neg A$.

THEOREM 3.7. *In the system* Tm, *the formula* B *is deduced from the formulas* A *and* $A \prec B$.

THEOREM 3.8. *In the system* S4m, *the formula*
$$(a \prec b) \supset 1 \prec (a \prec b)$$
is provable.

§ 4. Cut-elimination theorems

For the sequential calculuses presented in the preceding section, the cut-elimination theorems hold.

In order to prove the cut-elimination theorems, we introduce two new inference rules: the f-transition rule and the v-transition rule. The f-transition rule is the inference rule usually called the mix, that is the rule

$$\frac{\Gamma \rightarrow \Sigma \qquad \Delta \rightarrow \Pi}{\Gamma, \Delta^* \rightarrow \Sigma^*, \Pi} \langle \mathsf{T} \rangle$$

where both Δ and Σ contain a formula T, and Δ^* and Σ^* are the sequences of formulas obtained from Δ and Σ respectively by deleting all the occurrences of T in them. In this figure the formula T is called the transient. The v-transition rule is the inference rule

$$\frac{\Gamma \Rightarrow \alpha \not\Rightarrow \zeta, \mathsf{t}, \eta \qquad \Delta \Rightarrow \beta, \mathsf{t}, \gamma \not\Rightarrow \xi}{\Gamma, \Delta \Rightarrow \alpha, \beta, \gamma \not\Rightarrow \zeta, \eta, \xi} \langle \mathsf{t} \rangle .$$

In this figure the valuence t is called the transient. The f-transition and the v-transition are generically called transitions.

Let Km* be the deduction system obtained from the system Km by replacing the two cut rules by these two transition rules. As is easily verified, the same sequents are proved in Km and Km*. For this reason, we prove the next theorem.

THEOREM 4.1. *If a sequent is provable in the system* Km*, *then it is provable in* Km* *without a transition figure.*

PROOF. As usual, we deal only with proof figures in Km* containing just one transition figure.

Preliminarily, we define three scales, the rank, the order and the grade, for measuring the complexity of a proof figure **P** containing one transition figure **T**. If **T** is an f-transition figure, then the left [right] rank of **P** is the number of occurrences of f-sequents **S** above **T** in **P** such that the transient of **T** occurs in the succedent [antecedent] of each f-sequent from **S** down to the left [right, *respectively*] upper sequent of **T**. Likewise, if **T** is a v-transition figure, then the left [right] rank of **P** is the number of occurrences of v-sequents **S** above **T** in **P** such that the transient of **T** occurs in the succedental [antecedental] of each v-sequent from **S** down to the left [right, *respectively*] upper sequent of **T**. The rank of **P** is the sum of its left rank and right rank. The order of **P** is the number of occurrences of conversion figures above **T** in **P**. The grade of **P** is the number of occurrences of logical symbols and caliber connectives in the transient of **T**. We prove that we can eliminate a transition figure from a proof figure in Km* by triple induction on its grade, order and rank.

Let a proof figure **P** in Km* contain one transition figure **T**, and ρ_l and ρ_r be its left rank and right rank respectively. The case where at least one of the upper sequents of **T** is an axiom sequent is easily handled. Accordingly, we presume that **T** is of the form

$$\frac{\dfrac{\vdots}{\mathbf{S}_l}\,(\mathbf{I}_l) \qquad \dfrac{\vdots}{\mathbf{S}_r}\,(\mathbf{I}_r)}{\mathbf{S}_c}\,(\mathbf{T})$$

where $\mathbf{I}_l, \mathbf{I}_r$ denote inference figures, $\mathbf{S}_l, \mathbf{S}_r$ and $\mathbf{S}_c$ sequents.

First, suppose that $\rho_l = \rho_r = 1$. With this presupposition, we prove two lemmas.

Lemma. If one of $\mathbf{I}_l$ and $\mathbf{I}_r$ is a conversion figure, then so is the other.

Proof. Let $\mathbf{I}_l$ be a conversion figure. Since every valuence occurring in $\mathbf{S}_l$ is a formula, the transient of $\mathbf{T}$ is a formula. From Fact 2.2, $\mathbf{I}_r$ is not a valuent figure, hence $\mathbf{I}_r$ is a conversion figure.

Similarly, if $\mathbf{I}_r$ is a conversion figure, then so is $\mathbf{I}_l$.

Lemma. If $\mathbf{I}_l$ is a $(\rightarrow\prec)$ figure, then $\mathbf{I}_r$ is an f-weakening figure or a $(\rightarrow\prec)$ figure.

Proof. Let $\mathbf{I}_l$ be a $(\rightarrow\prec)$ figure. Then the transient of $\mathbf{T}$ is an evaluational formula. Hence $\mathbf{I}_r$ is either an f-weakening figure in which the transient is the weakening formula or a $(\rightarrow\prec)$ figure.

From these lemmas, only the following five cases can occur as to $\mathbf{I}_l$ and $\mathbf{I}_r$:

(1) Either $\mathbf{I}_l$ or $\mathbf{I}_r$ is an f-weakening figure;

(2) Both $\mathbf{I}_l$ and $\mathbf{I}_r$ are operational figures and their principal formulas are the transient of $\mathbf{T}$;

(3) Both $\mathbf{I}_l$ and $\mathbf{I}_r$ are conversion figures;

(4) $\mathbf{I}_l$ is an operational figure whose principal formula is the transient of $\mathbf{T}$, and $\mathbf{I}_r$ is a $(\rightarrow\prec)$ figure;

(5) Both $\mathbf{I}_l$ and $\mathbf{I}_r$ are $(\rightarrow\prec)$ figures.

The cases (1) and (2) are handled in the standard way.

Consider the case (3). The part of $\mathbf{P}$ on $\mathbf{T}$ is of the form

$$\dfrac{\dfrac{\Gamma \rightarrow \Sigma, \mathsf{T}, \Pi}{\Gamma \nrightarrow \Sigma, \mathsf{T}, \Pi} \qquad \dfrac{\Delta, \mathsf{T}, \Lambda \rightarrow \Xi}{\Delta, \mathsf{T}, \Lambda \nrightarrow \Xi}}{\Gamma, \Delta, \Lambda \nrightarrow \Sigma, \Pi, \Xi}\,\langle\mathsf{T}\rangle\ .$$

We transform this part into

$$\dfrac{\dfrac{\dfrac{\Gamma \rightarrow \Sigma, \mathsf{T}, \Pi \qquad \Delta, \mathsf{T}, \Lambda \rightarrow \Xi}{\Gamma, \Delta^*, \Lambda^* \rightarrow \Sigma^*, \Pi^*, \Xi}\,\langle\mathsf{T}\rangle}{\Gamma, \Delta, \Lambda \rightarrow \Sigma, \Pi, \Xi}}{\Gamma, \Delta, \Lambda \nrightarrow \Sigma, \Pi, \Xi}\ .$$

Then the order is decreased.

Consider the case (4). The part of $\mathbf{P}$ on $\mathbf{T}$ is of the form

$$\dfrac{\Gamma \rightarrow \Sigma \qquad \dfrac{\Delta > \mathsf{a} \nrightarrow \mathsf{b}}{\Delta \rightarrow \mathsf{a} \prec \mathsf{b}}}{\Gamma, \Delta^* \rightarrow \Sigma^*, \mathsf{a} \prec \mathsf{b}}\,\langle\mathsf{T}\rangle$$

where Δ and Σ contain the transient T, and T is not an evaluational formula. From Lemma 2.4, the v-sequent $\Delta^* > \mathsf{a} \nrightarrow \mathsf{b}$ is provable without a transition figure. Hence we resume the proof figure as follows:

$$\frac{\dfrac{\Delta^* \Rightarrow a \not\Rightarrow b}{\Delta^* \longrightarrow a \prec b}}{\Gamma, \Delta^* \longrightarrow \Sigma^*, a \prec b}\ .$$

Consider the case (5). The part of **P** on **T** is of the form

$$\frac{\dfrac{\Gamma \Rightarrow a \not\Rightarrow b}{\Gamma \longrightarrow a \prec b} \qquad \dfrac{\Delta \Rightarrow c \not\Rightarrow d}{\Delta \longrightarrow c \prec d}}{\Gamma, \Delta^* \longrightarrow c \prec d}\ \langle a \prec b \rangle$$

where Δ contains the formula $a \prec b$. From the induction hypothesis, the next proposition (*) holds:

> If a v-sequent is provable and among the transition figures only those whose transients are a or b are used to prove it, then it is provable without a transition figure.

For a finite sequence Λ of formulas, by $\Lambda^\times$ we denote the sequence of formulas obtained from Λ by replacing all the occurrences of the formula $a \prec b$ by the sequence Γ. Under these notations, we prove the next lemma.

LEMMA. If a v-sequent $\Lambda \Rightarrow \alpha \not\Rightarrow \zeta$ is provable without a transition figure, then the v-sequent $\Lambda^\times \Rightarrow \alpha \not\Rightarrow \zeta$ is provable without a transition figure.

PROOF. Let **Q** and **R** denote the proof figures of the v-sequent $\Gamma \Rightarrow a \not\Rightarrow b$ and a v-sequent $\Lambda \Rightarrow \alpha \not\Rightarrow \zeta$ without a transition figure respectively, n be the number of occurrences of v-sequents **S** in **R** such that $a \prec b$ occurs in the antecedent of each v-sequent from **S** down to end in **R**. We prove the lemma by the induction on n.

Assume that $n = 0$. Then Λ does not contain $a \prec b$, thus **R** itself is a proof figure of $\Lambda^\times \Rightarrow \alpha \not\Rightarrow \zeta$.

Assume that the last inference figure in **R** is of the form

$$\frac{\Phi \Rightarrow \beta \not\Rightarrow \eta, a \qquad \Psi \Rightarrow b, \gamma \not\Rightarrow \xi}{a \prec b, \Phi, \Psi \Rightarrow \beta, \gamma \not\Rightarrow \eta, \xi}\ .$$

We transform this part into

$$\cfrac{\cfrac{\dfrac{\Phi^\times \Rightarrow \beta \not\Rightarrow \eta, a \qquad \mathbf{Q}}{\Phi^\times, \Gamma \Rightarrow \beta \not\Rightarrow \eta, b}\ \langle a \rangle \qquad \Psi^\times \Rightarrow b, \gamma \not\Rightarrow \xi}{\Phi^\times, \Gamma, \Psi^\times \Rightarrow \beta, \gamma \not\Rightarrow \eta, \xi}\ \langle b \rangle}{\Gamma, \Phi^\times, \Psi^\times \Rightarrow \beta, \gamma \not\Rightarrow \eta, \xi}\ ,$$

and from the proposition (*), the v-sequent $\Gamma, \Phi^\times, \Psi^\times \Rightarrow \beta, \gamma \not\Rightarrow \eta, \xi$ is provable without a transition figure.

The cases where the last inference figure in **R** is of any other form are obvious, and the lemma is proved.

From this lemma, the v-sequent $\Delta^\times \Rightarrow c \not\Rightarrow d$ is provable without a transition figure. Hence we resume the proof figure as follows:

$$\dfrac{\dfrac{\Delta^{\times} \Rightarrow \mathsf{c} \nRightarrow \mathsf{d}}{\Delta^{\times} \rightarrow \mathsf{c} \prec \mathsf{d}}}{\Gamma, \Delta^{*} \rightarrow \mathsf{c} \prec \mathsf{d}} .$$

Secondly, suppose that $\rho_l \geq 2$ or that $\rho_r \geq 2$. In the former case, since the transient of **T** is not the principal formula of a $(\rightarrow\prec)$ figure, the left rank is decreased in the usual way. In the latter case the right rank is decreased similarly.

Now the theorem is proved.

COROLLARY 4.2. *If a sequent is provable in the system* Km, *then it is provable in* Km *without a cut figure.*

The cut-elimination theorem for the system Dm is proved in the similar way.

THEOREM 4.3. *If a sequent is provable in the system* Dm, *then it is provable in* Dm *without a cut figure.*

Let Tm* be the deduction system obtained from the system Tm by replacing the two cut rules by the two transition rules.

THEOREM 4.4. *If a sequent is provable in the system* Tm*, *then it is provable in* Tm* *without a transition figure.*

PROOF. Let a proof figure **P** in Tm* contain just one transition figure **T**. We prove that we can eliminate a transition figure from **P** by triple induction on its grade, order and rank defined in the proof of Theorem 4.1. The problem is the case where at least one of the upper sequents of **T** is the lower sequent of a release figure.

Consider the case where the left upper sequent of **T** is the lower sequent of a release figure **J**. Let **I** denote the suspending figure to **J**. Without loss of generality, we can presume that the part of **P** from **I** down to **T** is of the form

$$\begin{array}{c}
\dfrac{\Phi, \Delta \rightarrow \Sigma, \Psi}{\Phi, \Delta \nRightarrow \Sigma, \Psi}\,(\mathrm{I}) \\
\vdots \\
\overline{\Xi \rightarrow \alpha, \Delta \nRightarrow \Sigma, \zeta} \\
\vdots \\
\dfrac{\dfrac{\overline{\Gamma \Rightarrow \Delta \nRightarrow \Sigma}}{\Gamma, \Delta \rightarrow \Sigma}\,(\mathrm{J}) \qquad \Lambda \rightarrow \Pi}{\Gamma, \Delta, \Lambda^{*} \rightarrow \Sigma^{*}, \Pi}\,\langle \mathrm{T} \rangle
\end{array}$$

where Λ and Σ contain the transient T. We transform this part into

$$
\begin{array}{c}
\Phi, \Delta \rightarrow \Sigma, \Psi \quad \Lambda \rightarrow \Pi \\
\hline
\Phi, \Delta, \Lambda^* \rightarrow \Sigma^*, \Psi^*, \Pi \\
\hline\hline
\Phi, \Delta, \Lambda^* \rightarrow \Sigma^*, \Pi, \Psi \\
\hline
\Phi, \Delta, \Lambda^* \nrightarrow \Sigma^*, \Pi, \Psi \\
\hline
\vdots \\
\hline
\Xi \rightarrow \alpha, \Delta, \Lambda^* \nrightarrow \Sigma^*, \Pi, \zeta \\
\vdots \\
\hline
\Gamma > \Delta, \Lambda^* \nrightarrow \Sigma^*, \Pi \\
\hline
\Gamma, \Delta, \Lambda^* \rightarrow \Sigma^*, \Pi
\end{array}
\quad \langle \mathsf{T} \rangle
$$

Then the order is decreased.

Consider the case where the right upper sequent of **T** is the lower sequent of a release figure **J** with the suspending figure **I** to **J**. Without loss of generality, we can presume that the part of **P** from **I** down to **T** is of the form

$$
\begin{array}{cc}
 & \Phi, \Lambda \rightarrow \Pi, \Psi \\
\cline{2-2}
 & \Phi, \Lambda \nrightarrow \Pi, \Psi \quad \text{(I)} \\
\cline{2-2}
 & \vdots \\
\cline{2-2}
 & \Delta > \Lambda \nrightarrow \Pi \\
\cline{2-2}
\Gamma \rightarrow \Sigma & \Delta, \Lambda \rightarrow \Pi \quad \text{(J)} \\
\hline
\multicolumn{2}{c}{\Gamma, \Delta^*, \Lambda^* \rightarrow \Sigma^*, \Pi \quad \langle \mathsf{T} \rangle}
\end{array}
$$

where either Δ or Λ contains the transient T, and so does Σ. If Λ contains T, then, in a similar way to that in the preceding case, we transform **P** into a proof figure without a transition figure whose last part is of the form

$$
\begin{array}{c}
\Phi, \Gamma, \Lambda^* \rightarrow \Sigma^*, \Pi, \Psi \\
\hline
\Phi, \Gamma, \Lambda^* \nrightarrow \Sigma^*, \Pi, \Psi \quad (\mathbf{I}') \\
\hline
\vdots \\
\hline
\Delta > \Gamma, \Lambda^* \nrightarrow \Sigma^*, \Pi \\
\hline
\Delta, \Gamma, \Lambda^* \rightarrow \Sigma^*, \Pi \quad (\mathbf{J}')
\end{array}
$$

where $\mathbf{I}'$ is the suspending figure to the release figure $\mathbf{J}'$. If Δ contains T, we resume the proof figure as follows:

$$
\begin{array}{c}
\Gamma \rightarrow \Sigma \qquad \Delta, \Gamma, \Lambda^* \rightarrow \Sigma^*, \Pi \\
\hline
\Gamma, \Delta^*, \Gamma^*, \Lambda^* \rightarrow \Sigma^*, \Sigma^*, \Pi \quad \langle \mathsf{T} \rangle \\
\hline\hline
\Gamma, \Delta^*, \Lambda^* \rightarrow \Sigma^*, \Pi
\end{array}
\ .
$$

The transition figure occurring here is eliminated in the same way as in the proof of Theorem 4.1.

COROLLARY 4.5. *If a sequent is provable in the system* Tm, *then it is provable in* Tm *without a cut figure.*

Let S4m* be the deduction system obtained from the system S4m by replacing the two cut rules by the two transition rules.

THEOREM 4.6. *If a sequent is provable in the system* S4m*, *then it is provable in* S4m* *without a transition figure.*

PROOF. Let a proof figure **P** in S4m* contain just one transition figure **T**. We prove that we can eliminate a transition figure from **P** by triple induction on its grade, order and rank defined in the proof of Theorem 4.1. The problem is the case where the last part of **P** is of the form

$$\dfrac{\Lambda_1, \Phi_1 \to \Psi_1}{\Lambda_1 \Rightarrow \Phi_1 \nRightarrow \Psi_1}\,(\mathbf{I}_1) \quad \cdots \quad \dfrac{\Lambda_n, \Phi_n \to \Psi_n}{\Lambda_n \Rightarrow \Phi_n \nRightarrow \Psi_n}\,(\mathbf{I}_n)$$

$$\vdots \qquad\qquad \vdots$$

$$\dfrac{\dfrac{\Gamma \Rightarrow \mathsf{a} \nRightarrow \mathsf{b}}{\Gamma \to \mathsf{a} \prec \mathsf{b}} \qquad \dfrac{\Delta, \Lambda_1, \ldots, \Lambda_n \Rightarrow \alpha \nRightarrow \zeta}{\Delta, \Lambda_1, \ldots, \Lambda_n, \Xi \to \Sigma}}{\Gamma, \Delta^*, \Lambda_1^*, \ldots, \Lambda_n^*, \Xi^* \to \Sigma}\,\langle \mathsf{a} \prec \mathsf{b} \rangle$$

where at least one of Δ, $\Lambda_1, \ldots, \Lambda_n$ and Ξ contains the transient $\mathsf{a} \prec \mathsf{b}$, every formula in Λ_i ($1 \leq i \leq n$) is evaluational, and $\mathbf{I}_1, \ldots, \mathbf{I}_n$ are all the conversion figures in **P** under which no conversion figure occur. We denote by $\Gamma^{\#}$ the sequence of formulas obtained from Γ by deleting all the occurrences of nonevaluational formulas, and by $\Lambda_i^{\times}$ ($1 \leq i \leq n$) the sequence of formulas obtained from Λ_i by replacing all the occurrences of $\mathsf{a} \prec \mathsf{b}$ by $\Gamma^{\#}$. From Lemma 2.4, the proof figure of the v-sequent $\Gamma \Rightarrow \mathsf{a} \nRightarrow \mathsf{b}$ can be transformed into the proof figure of the v-sequent $\Gamma^{\#} \Rightarrow \mathsf{a} \nRightarrow \mathsf{b}$ without increasing the number of occurrences of inference figures of each inference rule. Hence, if Λ_i ($1 \leq i \leq n$) contains $\mathsf{a} \prec \mathsf{b}$, then we transform the part of **P** above into

$$\dfrac{\dfrac{\dfrac{\dfrac{\Gamma^{\#} \Rightarrow \mathsf{a} \nRightarrow \mathsf{b}}{\Gamma^{\#} \to \mathsf{a} \prec \mathsf{b}} \qquad \Lambda_i, \Phi_i \to \Psi_i}{\Gamma^{\#}, \Lambda_i^*, \Phi_i^* \to \Psi_i}\,\langle \mathsf{a} \prec \mathsf{b} \rangle}{\Lambda_i^{\times}, \Phi_i \to \Psi_i}}{\Lambda_i^{\times} \Rightarrow \Phi_i \nRightarrow \Psi_i}$$

$$\vdots$$

$$\dfrac{\dfrac{\Delta, \Lambda_1^{\times}, \ldots, \Lambda_n^{\times} \Rightarrow \alpha \nRightarrow \zeta}{\Delta, \Lambda_1^{\times}, \ldots, \Lambda_n^{\times}, \Xi \to \Sigma}}{\Gamma, \Delta, \Lambda_1^*, \ldots, \Lambda_n^*, \Xi \to \Sigma}$$

For each transition figure occurring in this figure, we can eliminate it because the order of the proof figure ending with it is decreased. If either Δ or Ξ contains $\mathsf{a} \prec \mathsf{b}$, then we resume the proof figure as follows:

$$\dfrac{\dfrac{\Gamma \rightarrow \mathsf{a} \prec \mathsf{b} \qquad \Gamma, \Delta, \Lambda_1^*, \ldots, \Lambda_n^*, \Xi \rightarrow \Sigma}{\Gamma, \Gamma^*, \Delta^*, \Lambda_1^*, \ldots, \Lambda_1^*, \Xi^* \rightarrow \Sigma}\langle \mathsf{a} \prec \mathsf{b} \rangle}{\Gamma, \Delta^*, \Lambda_1^*, \ldots, \Lambda_n^*, \Xi^* \rightarrow \Sigma} \ .$$

The transition figure occurring here is eliminated in the same ways as in the proof of Theorem 4.1 or Theorem 4.4.

COROLLARY 4.7. *If a sequent is provable in the system* S4m, *then it is provable in* S4m *without a cut figure.*

As a consequence, our logics are decidable.

§ 5. Algebraic semantics

In the present section, we give the algebraic semantics for evaluational logic.

Let $\boldsymbol{B} = \langle B, \cup, \cap, c, \bot, \top \rangle$ be a Boolean lattice. We denote by $\leq$ the canonical order of $\boldsymbol{B}$. Let V be a set, and let $\oplus, \otimes$ be two binary operations on V, $\sim$ a unary operation on V, and $0, 1$ two elements in V. Let v be a mapping from B to V, and $\lhd$ a mapping from $V \times V$ to B. For $a, b \in V$ we use the notation $a \lhd b$ instead of $\lhd(a, b)$ hereafter. Informally speaking, B is intended as a set of propositions, V as a value set, v as a valuation on B, and $\lhd$ as an order relation on V.

For an algebraic system $\boldsymbol{M} = \langle V, \oplus, \otimes, 0, 1 \rangle$, the quadruple $\langle \boldsymbol{B}, \boldsymbol{M}, v, \lhd \rangle$ is called a Km-algebra if it satisfies the following conditions:

(1) Both $\langle V, \oplus, 0 \rangle$ and $\langle V, \otimes, 1 \rangle$ are commutative monoids;

(2) $v(\bot) \lhd 0 = 1 \lhd v(\top) = \top$;

(3) For any $a, b, c \in V$,

$a \lhd a = \top$	(reflexivity of $\lhd$),
$(a \lhd b) \cap (b \lhd c) \leq (a \lhd c)$	(transitivity of $\lhd$),
$a \lhd b \leq a \oplus c \lhd b \oplus c$	(compatibility of $\lhd$ with $\oplus$),
$a \lhd b \leq a \otimes c \lhd b \otimes c$	(compatibility of $\lhd$ with $\otimes$),
$(a \oplus b) \otimes c \lhd a \oplus (b \otimes c) = \top$	(balanceability between $\oplus$ and $\otimes$);

(4) For any $p, q \in B$,

$v(p \cup q) \lhd v(p) \oplus v(q) = \top$	(subadditivity of v),
$v(p) \otimes v(q) \lhd v(p \cap q) = \top$	(supermultiplicativity of v).

A Km-algebra $\langle \boldsymbol{B}, \boldsymbol{M}, v, \lhd \rangle$ is called a Ki-algebra if it satisfies the condition

$$0 \lhd a = a \lhd 1 = \top \quad \text{for any } a \in V .$$

For an algebraic system $\boldsymbol{C} = \langle V, \oplus, \otimes, \sim, 0, 1 \rangle$, the quadruple $\langle \boldsymbol{B}, \boldsymbol{C}, v, \lhd \rangle$ is called a Kc-algebra if the quadruple $\langle \boldsymbol{B}, \langle V, \oplus, \otimes, 0, 1 \rangle, v, \lhd \rangle$ is a Km-algebra and the condition

$$1 \lhd a \oplus \sim a = a \otimes \sim a \lhd 0 = \top \quad \text{for any } a \in V$$

holds.

A Kc-algebra $\langle \boldsymbol{B}, \langle V, \oplus, \otimes, \sim, 0, 1 \rangle, v, \lhd \rangle$ is called a Kg-algebra if the quadruple $\langle \boldsymbol{B}, \langle V, \oplus, \otimes, 0, 1 \rangle, v, \lhd \rangle$ is a Ki-algebra.

A Km-algebra $\langle \boldsymbol{B}, \boldsymbol{M}, v, \triangleleft \rangle$ is called a Dm-algebra if it satisfies the condition

$$1 \triangleleft 0 = \bot .$$

A Dm-algebra $\langle \boldsymbol{B}, \boldsymbol{M}, v, \triangleleft \rangle$ is called a Tm-algebra if it satisfies the condition

$$1 \triangleleft v(p) \leq p \quad \text{for any } p \in B .$$

A Tm-algebra $\langle \boldsymbol{B}, \boldsymbol{M}, v, \triangleleft \rangle$ is called an S4m-algebra if it satisfies the condition

$$a \triangleleft b \leq 1 \triangleleft v(a \triangleleft b) \quad \text{for any } a, b \in V .$$

Next, we define the interpretations. An interpretation on the language L_m into a Km-algebra $\langle \boldsymbol{B}, \boldsymbol{M}, v, \triangleleft \rangle$ is a mapping I which maps a formula in L_m to an element in B and a caliber in L_m to an element in V and satisfies the conditions

(1)	$I(0) = 0$,	(2)	$I(1) = 1$,
(3)	$I(\mathsf{A} \wedge \mathsf{B}) = I(\mathsf{A}) \cap I(\mathsf{B})$,	(4)	$I(\mathsf{A} \vee \mathsf{B}) = I(\mathsf{A}) \cup I(\mathsf{B})$,
(5)	$I(\neg \mathsf{A}) = I(\mathsf{A})^c$,	(6)	$I(\mathsf{A} \supset \mathsf{B}) = I(\mathsf{A})^c \cup I(\mathsf{B})$,
(7)	$I(a \otimes b) = I(a) \otimes I(b)$,	(8)	$I(a \oplus b) = I(a) \oplus I(b)$,
(9)	$I(\mathsf{A} \prec \mathsf{B}) = v(I(\mathsf{A})) \triangleleft v(I(\mathsf{B}))$,	(10)	$I(a \prec \mathsf{A}) = I(a) \triangleleft v(I(\mathsf{A}))$,
(11)	$I(\mathsf{A} \prec a) = v(I(\mathsf{A})) \triangleleft I(a)$,	(12)	$I(a \prec b) = I(a) \triangleleft I(b)$,

for any formulas A, B and any calibers a, b in L_m. An interpretation on the language L_c into a Kc-algebra $\langle \boldsymbol{B}, \boldsymbol{C}, v, \triangleleft \rangle$ is a mapping I which maps a formula in L_c to an element in B and a caliber in L_c to an element in V and satisfies the conditions from (1) to (12) above and the condition

$$I(\sim a) = \sim I(a)$$

for any formulas A, B and any calibers a, b in L_c.

A formula A in L_m is said to be valid under an interpretation I into a Km-algebra $\langle \boldsymbol{B}, \boldsymbol{M}, v, \triangleleft \rangle$ if $I(\mathsf{A}) = \top$. A formula A in L_c is said to be valid under an interpretation I into a Kc-algebra $\langle \boldsymbol{B}, \boldsymbol{C}, v, \triangleleft \rangle$ if $I(\mathsf{A}) = \top$.

A formula A in L_m is said to be Km-valid if A is valid under any interpretation into any Km-algebra. A formula A in L_m [L_c, L_c] is said to be Ki-valid [Kc-valid, Kg-valid] if A is valid under any interpretation into any Ki-algebra [Kc-algebra, Kg-algebra, *respectively*]. A formula A in L_m is said to be Dm-valid [Tm-valid, S4m-valid] if A is valid under any interpretation into any Dm-algebra [Tm-algebra, S4m-algebra, *respectively*].

LEMMA 5.1. Let $\langle \boldsymbol{B}, \boldsymbol{M}, v, \triangleleft \rangle$ be a Km-algebra. For any $p, q \in B$, if $p \leq q$ then $v(p) \triangleleft v(q) = \top$.

PROOF. If $p \leq q$, i.e., $q \cup p^c = \top$, then

$$\begin{aligned} \top &= 1 \triangleleft v(q \cup p^c) \\ &= 1 \triangleleft v(q) \oplus v(p^c) \\ &= v(p) \triangleleft (v(q) \oplus v(p^c)) \otimes v(p) \\ &= v(p) \triangleleft v(q) \oplus (v(p^c) \otimes v(p)) \\ &= v(p) \triangleleft v(q) \oplus v(\bot) \\ &= v(p) \triangleleft v(q) . \end{aligned}$$

§6. Soundness and completeness

We first prove the soundness theorem and the completeness theorem for the system Km with respect to the algebraic semantics.

THEOREM 6.1. *If a formula in* $\mathrm{L_m}$ *is provable in the system* Km, *then it is* Km-*valid.*

PROOF. Let us take a Km-algebra $\mathcal{F} = (\langle B, \cup, \cap, c, \bot, \top\rangle, \langle V, \oplus, \otimes, 0, 1\rangle, v, \triangleleft)$ and an interpretation I into $\mathcal{F}$, and denote by $\leq$ the canonical order on B. Preliminarily, we introduce some notations. For a finite (possibly empty) sequence Γ of formulas

$$\mathsf{A}_1, \ldots, \mathsf{A}_n$$

where $n \geq 0$, we define two elements $[\![\Gamma]\!]^{\cap}$ and $[\![\Gamma]\!]^{\cup}$ in B by

$$[\![\Gamma]\!]^{\cap} = \begin{cases} I(\mathsf{A}_1)\cap\cdots\cap I(\mathsf{A}_n) & \text{if } n \geq 1, \\ \top & \text{if } n = 0; \end{cases}$$

$$[\![\Gamma]\!]^{\cup} = \begin{cases} I(\mathsf{A}_1)\cup\cdots\cup I(\mathsf{A}_n) & \text{if } n \geq 1, \\ \bot & \text{if } n = 0. \end{cases}$$

For a valuence a, we define an element [a] in V by

$$[\mathsf{a}] = \begin{cases} v(I(\mathsf{a})) & \text{if a is a formula,} \\ I(\mathsf{a}) & \text{if a is a caliber.} \end{cases}$$

Note that $I(\mathsf{a} \prec \mathsf{b}) = [\mathsf{a}] \triangleleft [\mathsf{b}]$ for any valuences a, b. In addition, for any calibers a, b, $[a] \oplus [b] = [a \oplus b]$ and $[a] \otimes [b] = [a \otimes b]$. For a finite (possibly empty) sequence α of valuences

$$\mathsf{a}_1, \ldots, \mathsf{a}_n$$

where $n \geq 0$, we define two elements $[\alpha]^{\oplus}$ and $[\alpha]^{\otimes}$ in V by

$$[\alpha]^{\oplus} = \begin{cases} [\mathsf{a}_1]\oplus\cdots\oplus[\mathsf{a}_n] & \text{if } n \geq 1, \\ 0 & \text{if } n = 0; \end{cases}$$

$$[\alpha]^{\otimes} = \begin{cases} [\mathsf{a}_1]\otimes\cdots\otimes[\mathsf{a}_n] & \text{if } n \geq 1, \\ 1 & \text{if } n = 0. \end{cases}$$

Under these notations, we define the validity of sequents as follows:

(1) An f-sequent $\Gamma \rightarrow \Sigma$ is said to be valid under I if $[\![\Gamma]\!]^{\cap} \leq [\![\Sigma]\!]^{\cup}$;

(2) A v-sequent $\Gamma \Rightarrow \alpha \not\Rightarrow \zeta$ is said to be valid under I if $[\![\Gamma]\!]^{\cap} \leq ([\alpha]^{\otimes} \triangleleft [\zeta]^{\oplus})$.

It suffices to prove that every axiom sequent is valid and that if the upper sequents of each inference figure in Km are valid then its lower sequent is also valid. These are verified by straightforward calculation, so we see only a few cases.

First, consider a $(\prec\Rightarrow)$ figure

$$\frac{\Gamma \Rightarrow \alpha \not\Rightarrow \zeta, \mathsf{a} \qquad \Delta \Rightarrow \mathsf{b}, \beta \not\Rightarrow \eta}{\mathsf{a} \prec \mathsf{b}, \Gamma, \Delta \Rightarrow \alpha, \beta \not\Rightarrow \zeta, \eta}.$$

If $[\![\Gamma]\!]^{\cap} \leq [\alpha]^{\otimes} \triangleleft [\zeta, \mathsf{a}]^{\oplus}$ and $[\![\Delta]\!]^{\cap} \leq [\mathsf{b}, \beta]^{\otimes} \triangleleft [\eta]^{\oplus}$, then

$$\begin{aligned}
[\![\mathsf{a}\prec\mathsf{b},\Gamma,\Delta]\!]^{\cap} &\le ([\mathsf{a}]\lhd[\mathsf{b}])\cap([\alpha]^{\otimes}\lhd[\zeta,\mathsf{a}]^{\oplus})\cap([\mathsf{b},\beta]^{\otimes}\lhd[\eta]^{\oplus})\\
&\le ([\zeta]^{\oplus}\oplus[\mathsf{a}]\lhd[\zeta]^{\oplus}\oplus[\mathsf{b}])\cap([\alpha]^{\otimes}\lhd[\zeta]^{\oplus}\oplus[\mathsf{a}])\cap([\mathsf{b}]\otimes[\beta]^{\otimes}\lhd[\eta]^{\oplus})\\
&\le ([\alpha]^{\otimes}\lhd[\zeta]^{\oplus}\oplus[\mathsf{b}])\cap([\mathsf{b}]\otimes[\beta]^{\otimes}\lhd[\eta]^{\oplus})\\
&\le ([\alpha]^{\otimes}\otimes[\beta]^{\otimes}\lhd([\zeta]^{\oplus}\oplus[\mathsf{b}])\otimes[\beta]^{\otimes})\cap([\zeta]^{\oplus}\oplus([\mathsf{b}]\otimes[\beta]^{\otimes})\lhd[\zeta]^{\oplus}\oplus[\eta]^{\oplus})\\
&\le ([\alpha,\beta]^{\otimes}\lhd[\zeta]^{\oplus}\oplus([\mathsf{b}]\otimes[\beta]^{\otimes}))\cap([\zeta]^{\oplus}\oplus([\mathsf{b}]\otimes[\beta]^{\otimes})\lhd[\zeta,\eta]^{\oplus})\\
&\le [\alpha,\beta]^{\otimes}\lhd[\zeta,\eta]^{\oplus}\,.
\end{aligned}$$

Hence a $(\prec\gg)$ figure preserves the validity of sequents.

Next, consider a $(\not\gg\otimes)$ figure

$$\frac{\Gamma \gg \alpha \not\gg \zeta, a \qquad \Gamma \gg \beta \not\gg \eta, b}{\Gamma \gg \alpha,\beta \not\gg \zeta,\eta, a\otimes b}\,.$$

If $[\![\Gamma]\!]^{\cap} \le [\alpha]^{\otimes}\lhd[\zeta,a]^{\oplus}$ and $[\![\Gamma]\!]^{\cap} \le [\beta]^{\otimes}\lhd[\eta,b]^{\oplus}$, then

$$\begin{aligned}
[\![\Gamma]\!]^{\cap} &\le ([\alpha]^{\otimes}\lhd[\zeta,a]^{\oplus})\cap([\beta]^{\otimes}\lhd[\eta,b]^{\oplus})\\
&\le [\alpha]^{\otimes}\otimes[\beta]^{\otimes}\lhd[\zeta,a]^{\oplus}\otimes[\eta,b]^{\oplus}\\
&\le [\alpha,\beta]^{\otimes}\lhd([\zeta]^{\oplus}\oplus[a])\otimes([\eta]^{\oplus}\oplus[b])\\
&\le [\alpha,\beta]^{\otimes}\lhd[\zeta]^{\oplus}\oplus(([\eta]^{\oplus}\oplus[b])\otimes[a])\\
&\le [\alpha,\beta]^{\otimes}\lhd[\zeta]^{\oplus}\oplus[\eta]^{\oplus}\oplus([a]\otimes[b])\\
&\le [\alpha,\beta]^{\otimes}\lhd[\zeta,\eta,a\otimes b]^{\oplus}\,.
\end{aligned}$$

Hence a $(\not\gg\otimes)$ figure preserves the validity of sequents.

A total conversion figure preserves the validity of sequents, which is shown on the basis of Lemma 5.1.

It is shown similarly that other inference figures in Km also preserve the validity of sequents, and the proof of the theorem is completed.

In order to prove the completeness theorem for the system Km, we extend the system Km on the language $\mathrm{L_m}$ to a system $\mathrm{Km^v}$ on a language $\mathrm{L_m^v}$. The alphabet of the language $\mathrm{L_m^v}$ consists of the symbols in $\mathrm{L_m}$ and another symbol v of a special sort. The calibers and the formulas in $\mathrm{L_m^v}$ are defined by mutual induction as follows:

(1) An expression is a caliber if it can be shown to be a caliber by a finite number of applications of the formation rules for calibers in $\mathrm{L_m}$ and of the formation rule that

if A is a formula then $\mathsf{v}(\mathsf{A})$ is a caliber;

(2) An expression is a formula if it can be shown to be a formula by a finite number of applications of the formation rules for formulas in $\mathrm{L_m}$.

The deduction system $\mathrm{Km^v}$ is obtained from Km by adding the inference rules

$$\text{valuation}: \frac{\Gamma \gg \mathsf{A}, \alpha \not\gg \zeta}{\Gamma \gg \mathsf{v}(\mathsf{A}), \alpha \not\gg \zeta}, \frac{\Gamma \gg \alpha \not\gg \zeta, \mathsf{A}}{\Gamma \gg \alpha \not\gg \zeta, \mathsf{v}(\mathsf{A})}\,.$$

LEMMA 6.2. The system $\mathrm{Km^v}$ is a conservative extension of the system Km.

PROOF. Similarly to Km, the cut-elimination theorem for $\mathrm{Km^v}$ is proved.

THEOREM 6.3. *If a formula in* L_m *is Km-valid, then it is provable in the system* Km.

PROOF. We construct a Km-algebra $\mathcal{F} = \langle\langle B, \cup, \cap, c, \bot, \top\rangle, \langle V, \oplus, \otimes, 0, 1\rangle, v, \triangleleft\rangle$ and an interpretation I into $\mathcal{F}$ such that if a formula in L_m is valid under I then it is provable in Km.

Let F be the set of all formulas in L_m^v, $\top$ the set of all provable formulas in Km^v, and $\bot$ the set of all disprovable formulas in Km^v. We define an equivalence relation $\cong$ on F by

$$A \cong B \quad \text{if and only if} \quad (A \supset B) \in \top \text{ and } (B \supset A) \in \top \qquad \text{for any } A, B \in F.$$

Let B be the quotient set $F/\cong$ of F with respect to $\cong$. We denote the equivalence class of $A \in F$ by $[\![A]\!]$. As is well known, we can define a join $\cup$, a meet $\cap$ and a complementation c on B by

$$[\![A]\!] \cup [\![B]\!] = [\![A \vee B]\!],\ [\![A]\!] \cap [\![B]\!] = [\![A \wedge B]\!],\ [\![A]\!]^c = [\![\neg A]\!] \quad \text{for any } A, B \in F,$$

and $\langle B, \cup, \cap, c, \bot, \top\rangle$ is a Boolean lattice.

Let C be the set of all calibers in L_m^v. We define a relation $\simeq$ on C by

$$a \simeq b \quad \text{if and only if} \quad (a \prec b) \in \top \text{ and } (b \prec a) \in \top \qquad \text{for any } a, b \in C.$$

This relation $\simeq$ on C is an equivalence relation. Let V be the quotient set $C/\simeq$ of C with respect to $\simeq$. We denote the equivalence class of $a \in C$ by $[a]$, and $[0]$ by 0, $[1]$ by 1.

From the fact that

$$\text{for any } A, B \in F, \text{ if } A \cong B \text{ then } v(A) \simeq v(B),$$

we can define a mapping v from B to V by

$$v([\![A]\!]) = [v(A)] \quad \text{for any } A \in F.$$

From the fact that

for any $a, b, c, d \in C$,

if $a \simeq b$ and $c \simeq d$, then $a \oplus c \simeq b \oplus d$, $a \otimes c \simeq b \otimes d$, and $a \prec c \cong b \prec d$,

we can define binary operations $\oplus$ and $\otimes$ on V and a mapping $\triangleleft$ from $V \times V$ to B by

$$[a] \oplus [b] = [a \oplus b],\ [a] \otimes [b] = [a \otimes b],\ [a] \triangleleft [b] = [\![a \prec b]\!] \quad \text{for any } a, b \in V.$$

Then, $\mathcal{F} = \langle\langle B, \cup, \cap, c, \bot, \top\rangle, \langle V, \oplus, \otimes, 0, 1\rangle, v, \triangleleft\rangle$ defined above is a Km-algebra, which is verified in a straightforward way. For instance, the balanceability between $\oplus$ and $\otimes$ is due to

$$([a] \oplus [b]) \otimes [c] \triangleleft [a] \oplus ([b] \otimes [c]) = [\![(a \oplus b) \otimes c \prec a \oplus (b \otimes c)]\!]$$

and

$$\dfrac{\dfrac{\dfrac{\dfrac{a \mathrel{+\!\!\!\triangleright} a \qquad b \mathrel{+\!\!\!\triangleright} b}{a \oplus b \mathrel{+\!\!\!\triangleright} a, b} \qquad c \mathrel{+\!\!\!\triangleright} c}{a \oplus b, c \mathrel{+\!\!\!\triangleright} a, b \otimes c}}{(a \oplus b) \otimes c \mathrel{+\!\!\!\triangleright} a \oplus (b \otimes c)}}{\rightarrow (a \oplus b) \otimes c \prec a \oplus (b \otimes c)}\ .$$

Finally, we define a mapping I defined for every valuence in L_m as follows:

(1) $I(A) = [\![A]\!]$ for any formula A in L_m;

(2) $I(a) = [a]$ for any caliber a in L_m.

I is an interpretation on L_m into $\mathcal{F}$, and if a formula in L_m is valid under I then it is provable in Km^v. From Lemma 6.2, if a formula in L_m is valid under I then it is provable in Km.

Now the proof of the theorem is completed.

The soundness and the completeness for other systems are proved in similar ways.

THEOREM 6.4.

1) A formula in L_m *is* Ki*-valid if and only if it is provable in the system* Ki.

2) A formula in L_c *is* Kc*-valid if and only if it is provable in the system* Kc.

3) A formula in L_c *is* Kg*-valid if and only if it is provable in the system* Kg.

THEOREM 6.5. *A formula in* L_m *is* Dm*-valid if and only if it is provable in the system* Dm.

THEOREM 6.6. *A formula in* L_m *is* Tm*-valid if and only if it is provable in the system* Tm.

PROOF. We handle the "if" part only. Let us take a Tm-algebra $\mathcal{F} = \langle\langle B, \cup, \cap, c, \bot, \top\rangle, \langle V, \oplus, \otimes, 0, 1\rangle, v, \triangleleft\rangle$ and an interpretation I into $\mathcal{F}$, and denote the canonical order on B by $\leq$. We use the notations given in the proof of Theorem 6.1. Moreover, for sequences α, β of valuences, if α includes β, by $\alpha - \beta$ we denote any sequence of valuences obtained from α by deleting as many occurrences of a as those in β for each valuence a in β.

We keep the definition of the validity of f-sequents given in the proof of Theorem 6.1, but modify that of the validity of v-sequents as follows: In a proof figure,

(1) a v-sequent $\Gamma \Rrightarrow \alpha \nRrightarrow \zeta$ which is not suspensive is said to be valid under I if $[\![\Gamma]\!]^{\cap} \leq [\alpha]^{\otimes} \triangleleft [\zeta]^{\oplus}$;

(2) a suspensive v-sequent $\Gamma \Rrightarrow \alpha \nRrightarrow \zeta$ to a release figure

$$\frac{\Delta \Rrightarrow \Lambda \nRrightarrow \Sigma}{\Delta, \Lambda \rightarrow \Sigma}$$

is said to be valid under I if $[\![\Gamma]\!]^{\cap} \leq [\alpha - \Lambda]^{\otimes} \triangleleft ([\zeta - \Sigma]^{\oplus} \oplus v(([\![\Lambda]\!]^{\cap})^{c} \cup [\![\Sigma]\!]^{\cup}))$.

It is easily verified that in a proof figure in Tm every inference figure preserves the validity of sequents defined above.

THEOREM 6.7. *A formula in* L_m *is* S4m*-valid if and only if it is provable in the system* S4m.

PROOF. We handle the "if" part only. Let us take an S4m-algebra $\mathcal{F} = \langle\langle B, \cup, \cap, c, \bot, \top\rangle, \langle V, \oplus, \otimes, 0, 1\rangle, v, \triangleleft\rangle$ and an interpretation I into $\mathcal{F}$, and denote the canonical order on B by $\leq$. In a proof figure in S4m every inference figure preserves the validity of sequents defined in the proof of Theorem 6.6, which is verified straightforwardly.

For instance, under the notations given in the proof of Theorem 6.1, consider a partial conversion figure which is not a suspending figure

$$\frac{A_1, \ldots, A_m, \Delta \rightarrow \Sigma}{A_1, \ldots, A_m \Rightarrow \Delta \not\Rightarrow \Sigma}$$

where $m \geq 0$ and $A_1, \ldots, A_m$ are evaluational formulas. If $[\![A_1, \ldots, A_m, \Delta]\!]^{\cap} \leq [\![\Sigma]\!]^{\cup}$, then

$$[A_1] \otimes \cdots \otimes [A_m] \otimes [\Delta]^{\otimes} \triangleleft [\Sigma]^{\oplus} = \top,$$

and since $I(A_i) \leq 1 \triangleleft [A_i]$ for any $i \leq m$,

$$\begin{aligned} [\![A_1, \ldots, A_m]\!]^{\cap} &= I(A_1) \cap \cdots \cap I(A_m) \\ &\leq (1 \triangleleft [A_1]) \cap \cdots \cap (1 \triangleleft [A_m]) \\ &\leq 1 \triangleleft ([A_1] \otimes \cdots \otimes [A_m]) \\ &\leq [\Delta]^{\otimes} \triangleleft ([A_1] \otimes \cdots \otimes [A_m] \otimes [\Delta]^{\otimes}) \\ &\leq [\Delta]^{\otimes} \triangleleft [\Sigma]^{\oplus} . \end{aligned}$$

Hence a partial conversion figure which is not a suspending figure preserves the validity of sequents.

§7. Relation to the normal modal logics

We show that the evaluational logics are conservative extensions of the normal modal logics.

Let $L_{\Box}$ be the language of standard sentential modal logics, i.e., the language whose logical symbols are $\wedge, \vee, \neg, \supset$ and the necessity operator $\Box$, and have the same sentential variables that the language L_m has. We define a transformation φ which transforms a formula in $L_{\Box}$ into a formula in L_m inductively as follows: For any formulas P, Q in $L_{\Box}$,

(1) $\varphi(P) = P$ if P is a sentential variables;

(2) $\varphi(P * Q) = \varphi(P) * \varphi(Q)$, here $*$ stands for $\wedge, \vee$ or $\supset$;

(3) $\varphi(\neg P) = \neg\varphi(P)$;

(4) $\varphi(\Box P) = 1 \prec \varphi(P)$.

Clearly we can construct a transformation ψ such that ψ is defined for any formula in L_m whose evaluational subformulas are of the form $1 \prec A$ and $\psi(\varphi(P)) = P$ for any formula P in $L_{\Box}$.

In the following, for a sequence Γ of formulas in L_m, $1 \prec \Gamma$ [$\psi(\Gamma)$] will denote the sequence of formulas in L_m [$L_{\Box}$] obtained from Γ by replacing all the occurrences of each formula A in Γ by $1 \prec A$ [$\psi(A)$, *respectively*]. Likewise, for a sequence Γ of formulas in $L_{\Box}$, $\Box\Gamma$ [$\varphi(\Gamma)$] will denote the sequence of formulas in $L_{\Box}$ [L_m] obtained from Γ by replacing all the occurrences of each formula P in Γ by $\Box P$ [$\varphi(P)$, *respectively*].

We postulate that the modal logics K, D, T and S4 are formalized by the sequential method. For example, K is the deduction system obtained from the deduction system LK of the classical logic by adding the inference rule

$$\frac{\Gamma \rightarrow \mathsf{P}}{\Box\Gamma \rightarrow \Box\mathsf{P}}\,.$$

THEOREM 7.1. *For a formula* P *in* $L_\Box$, *if* P *is provable in the system* K *then* $\varphi(\mathsf{P})$ *is provable in the system* Km.

PROOF. Since Km yields the derivation figure

$$\begin{array}{c} \Delta \rightarrow \mathsf{A} \\ \hline\hline \\ \hline 1 \prec \Delta \rightarrow 1 \prec \mathsf{A} \end{array}\,,$$

if a sequent $\Gamma \rightarrow \Sigma$ in $L_\Box$ is provable in K, then the f-sequent $\varphi(\Gamma) \rightarrow \varphi(\Sigma)$ is provable in Km.

THEOREM 7.2. *For a formula* P *in* $L_\Box$, *if* $\varphi(\mathsf{P})$ *is provable in the system* Kg *then* P *is provable in the system* K.

PROOF. Let Γ, Σ denote sequences of formulas in L_m whose evaluational subformulas are of the form $1 \prec \mathsf{A}$. We prove that a proof figure **P** of the f-sequent $\Gamma \rightarrow \Sigma$ in Kg can be transformed into a proof figure of the sequent $\psi(\Gamma) \rightarrow \psi(\Sigma)$ in K by the induction on the number of occurrences of f-sequents in **P**. On account of the cut-elimination theorem (Theorem 4.1), we presuppose that no cut figure occurs in **P**. Hence, 1 is the only caliber that can occur in **P**, and every evaluational formula occurring in **P** is of the form $1 \prec \mathsf{A}$.

We consider the case where the last inference figure in **P** is a $(\rightarrow\prec)$ figure **J**. Let **R** denote the rightmost thread in **P**. Since every valuent inference figure occurring in **P** is a structural figure or a $(\prec\geqslant)$ figure, if a sequent **S** in **R** and the sequent **S**$'$ just below **S** are v-sequents, then the succedental of **S**$'$ includes that of **S**. If every sequent occurring above **J** in **R** is a v-sequent, then the initial sequent in **R** is the sequent $1 \nrightarrow 1$ or the sequent $\nrightarrow 1$, thus the succedental of each v-sequent above **J** in **R** contains the caliber 1. But **J** is of the form

$$\frac{\Gamma \geqslant 1 \nrightarrow \mathsf{A}}{\Gamma \rightarrow 1 \prec \mathsf{A}}\,,$$

hence at least one f-sequent occurs above **J** in **R**. Let **I** denote the lowermost conversion figure in **R**. The part of **R** from **I** down to **J** is of the form

$$\left.\begin{array}{c} \Delta \rightarrow \\ \hline \Delta \nrightarrow \\ \hline \vdots \\ \hline \Gamma \geqslant 1 \nrightarrow \mathsf{A} \end{array}\right\} \text{only v-sequents occur} \qquad \begin{array}{l}(\mathbf{I}) \\ \\ \\ (\mathbf{J})\end{array}$$

$$\frac{\Gamma \geqslant 1 \nrightarrow \mathsf{A}}{\Gamma \rightarrow 1 \prec \mathsf{A}}\;(\mathbf{J})$$

or

$$\cfrac{\cfrac{\Delta \rightarrow \mathsf{A}}{\Delta \not\Rightarrow \mathsf{A}}\,(\mathrm{I})}{\vdots}\quad\Bigg\}\ \text{only v-sequents occur}$$

$$\cfrac{\Gamma \Rightarrow 1 \not\Rightarrow \mathsf{A}}{\Gamma \rightarrow 1 \prec \mathsf{A}}\,(\mathrm{J})$$

For each formula B contained in Δ, since B is not contained in the antecedental of the upper sequent of **J**, from Fact 2.1 B becomes an auxiliary valuence of a $(\prec\Rightarrow)$ figure below **I** above **J**, which is of the form

$$\frac{\Phi \Rightarrow \alpha \not\Rightarrow \zeta, 1 \qquad \Psi \Rightarrow \mathsf{B}, \beta \not\Rightarrow \eta}{1 \prec \mathsf{B}, \Phi, \Psi \Rightarrow \alpha, \beta \not\Rightarrow \zeta, \eta},$$

and from Fact 2.3 the formula $1 \prec \mathsf{B}$ is contained in Γ. Thus, every formula contained in the sequence $\Box\psi(\Delta) = \psi(1 \prec \Delta)$ is contained in the sequence $\psi(\Gamma)$. Therefore, since the sequent $\psi(\Delta) \rightarrow$ or $\psi(\Delta) \rightarrow \psi(\mathsf{A})$ is provable in K from the induction hypothesis, we resume the proof figure in K as follows:

$$\cfrac{\cfrac{\cfrac{\psi(\Delta) \rightarrow}{\psi(\Delta) \rightarrow \psi(\mathsf{A})}}{\Box\psi(\Delta) \rightarrow \Box\psi(\mathsf{A})}}{\psi(\Gamma) \rightarrow \psi(1 \prec \mathsf{A})}\ \text{some structural figures}$$

or

$$\cfrac{\cfrac{\psi(\Delta) \rightarrow \psi(\mathsf{A})}{\Box\psi(\Delta) \rightarrow \Box\psi(\mathsf{A})}}{\psi(\Gamma) \rightarrow \psi(1 \prec \mathsf{A})}\ \text{some structural figures}\ .$$

The cases where the last inference figure in **P** is not a $(\rightarrow\prec)$ figure are straightforward.

Now the theorem is proved.

COROLLARY 7.3. *For a formula* P *in* $L_\Box$, *the following five conditions are equivalent:*
(1) P *is provable in the system* K;
(2) $\varphi(\mathsf{P})$ *is provable in the system* Km;
(3) $\varphi(\mathsf{P})$ *is provable in the system* Ki;
(4) $\varphi(\mathsf{P})$ *is provable in the system* Kc;
(5) $\varphi(\mathsf{P})$ *is provable in the system* Kg.

THEOREM 7.4. *For a formula* P *in* $L_\Box$, P *is provable in the system* D *if and only if* $\varphi(\mathsf{P})$ *is provable in the system* Dm.

PROOF. Similar to the proofs of previous theorems.

Theorem 7.5. *For a formula* P *in* $L_\Box$, P *is provable in the system* T *if and only if* $\varphi(P)$ *is provable in the system* Tm.

Proof. We prove the "if" part only by the induction similar to that in the proof of Theorem 7.2. The problem is the case where a proof figure **P** in Tm without a cut figure ends with a release figure **J**:

$$\frac{\Gamma \Rightarrow \Delta \not\Rightarrow \Sigma}{\Gamma, \Delta \rightarrow \Sigma}.$$

Let **I** denote the suspending conversion figure to **J**. Without loss of generality, we can presume that **I** is of the form

$$\frac{\Lambda, \Delta \rightarrow \Sigma, \Pi}{\Lambda, \Delta \not\Rightarrow \Sigma, \Pi}.$$

From Fact 2.1, each formula contained in Λ or Π becomes an auxiliary valuence of a $(\prec\Rightarrow)$ figure below **I** above **J**. However, a $(\prec\Rightarrow)$ figure occurs in **P** only in the form

$$\frac{\Phi \Rightarrow \alpha \not\Rightarrow \zeta, 1 \qquad \Psi \Rightarrow A, \beta \not\Rightarrow \eta}{1 \prec A, \Phi, \Psi \Rightarrow \alpha, \beta \not\Rightarrow \zeta, \eta}.$$

Hence the sequence Π of formulas is empty. Furthermore, from Fact 2.3, for each formula A contained in Λ the formula $1 \prec A$ is contained in Γ. Thus, every formula contained in the sequence $\Box\psi(\Lambda) = \psi(1 \prec \Lambda)$ is contained in the sequence $\psi(\Gamma)$. Therefore, since the sequent $\psi(\Lambda), \psi(\Delta) \rightarrow \psi(\Sigma)$ is provable in T from the induction hypothesis, we resume the proof figure in T as follows:

$$\cfrac{\cfrac{\psi(\Lambda), \psi(\Delta) \rightarrow \psi(\Sigma)}{\Box\psi(\Lambda), \psi(\Delta) \rightarrow \psi(\Sigma)}}{\psi(\Gamma), \psi(\Delta) \rightarrow \psi(\Sigma)} \text{ some structural figures}.$$

Theorem 7.6. *For a formula* P *in* $L_\Box$, P *is provable in the system* S4 *if and only if* $\varphi(P)$ *is provable in the system* S4m.

Proof. Similar to the proof of previous theorems.

References

[G] V. N. Grišin, *Predicate and set-theoretic calculi based on logic without contractions*, ***Math. USSR Izvestija***, Vol. 18 (1982), No. 1.

[T] G. Takeuti, ***Proof Theory***, 2nd edition, Studies in Logic and the Foundations of Mathematics, North-Holland, 1987.

Division of System Science,
The Graduate School of Science and Technology,
Kobe University,
Nada-ku, Kobe 657, Japan

Consistency of Beeson's Formal System RPS and Some Related Results

SATOSHI KOBAYASHI
Department of Mathematics
University of Tokyo

ABSTRACT. We prove the consistency of Beeson's formal system **RPS** and its extensions. The relationships between **RPS** and various principles of constructive mathematics are investigated.

In 1980, P. Aczel[**2**] introduced the concept of Frege structures to analyze the Frege's logical notion of set. As is well-known, Frege's set theory is inconsistent. It contains the Russel paradox. But Aczel reconstructed the theory in a (weak but) consistent form.

The principal idea of Aczel[**2**] is to define a set as a propositional function. An object x belongs to a propositional function f iff $f(x)$ is a true proposition. If $\phi(x)$ is a proposition for every object x, we can define the "comprehension term" $\{ x \mid \phi(x) \}$ as $\lambda x.\phi(x)$. Since $\lambda x.\neg(x(x))$ is not a propositional function, the Russell paradox is avoided.

Aczel considered the theory only semantically, but some people have axiomatized it. For example, J. Smith[**11**] proposed the formal system "**LT**" and used it to give an interpretation of Martin-Löf's type theory.

Beeson[**3**] also gave a formalization of the theory of Frege structures called "**F**". He thought that Aczel's theory was "a significant contribution to the theory of rules and proofs in constructive mathematics"[**3**, P. 410]. But Aczel himself did not develop the theory of proofs in the framework of [**2**]. And **F** has no axioms for proofs. So Beeson extended **F** by adding the axioms for constructive proofs and proposed a new theory "**RPS**"—a theory of Rules, Proofs, and Sets. But he did not give a consistency proof of **RPS**. He stated "it is not known whether **RPS** is consistent" [**3**, P. 416].

In this paper we introduce a new theory which contains **RPS** as a subsystem, and prove its consistency. We also investigate the relationships between our theory and various principles of constructive mathematics, such as existence properties, choice principles, church's thesis, Markov's principle, and so on.

1. INTRODUCTION OF RPS

The author thinks that most of the readers are unfamiliar to **F** and **RPS**. They are defined as follows:

F *is* **EON** *(theory of rules)* + *axioms about propositions*

RPS *is* **F** + *axioms about proofs*

We first introduce **EON**.

1.1 EON.

EON (Elementary theory of Operations and Numbers) is a variant of Heyting arithmetic.

The underlying logic of **RPS** is **LPT**—the Logic of Partial Terms. It is a variant of first order intuitionistic predicate calculus. A special feature of **LPT** is that it admit the formation of terms which do not necessarily denote anything, such as $\{e\}(x)$. The formation rules of terms are as usual. The formation rules of formulae are as usual, except there is one more rule: if t is a term, then $t\downarrow$ (read, "t is defined", or "t denotes") is an atomic formula. The propositional axioms and rules of inference are as usual. The quantifier axioms and rules are as follows:

(Q1) $$\frac{B \to A}{B \to \forall x A} \quad (x \textit{ not free in } B)$$

(Q2) $$\frac{A \to B}{\exists x A \to B} \quad (x \textit{ not free in } B)$$

(Q3) $$\forall x A \,\&\, t\downarrow \to A[t/x]$$

(Q4) $$A[t/x] \,\&\, t\downarrow \to \exists x A$$

The equality axioms are as follows:

(E1) $$x = x \,\&\, (x = y \to y = x)$$

(E2) $$t \simeq s \,\&\, \phi(t) \to \phi(s)$$

(E3) $$t = s \to t\downarrow \,\&\, s\downarrow$$

Here t and s are terms, x and y are variables, and $t \simeq s$ is an abbreviation of $t\downarrow \vee s\downarrow \to t = s$. The axiom **E3** is actually a special case of a more general axiom,

(S1) $$R(t_1, \ldots, t_n) \to t_1\downarrow \,\&\, \cdots \,\&\, t_n\downarrow \quad (R \textit{ an atomic formula})$$

The next axiom is a special case of **S1**:

(*) $$f(t_1, \ldots, t_n)\downarrow \to t_1\downarrow \,\&\, \cdots \,\&\, t_n\downarrow$$

We also require

(S2) $c\downarrow$ *(every constant symbol c)*

(S3) $x\downarrow$ *(every variable x)*

The intended semantics of this logic is as follows. Variables, constants and predicate symbols are interpreted as usual, but function symbols are interpreted as partial functions. That is to say, the language is interpreted in *partial structure*. The interpretation

$\tilde{t}$ of t and the truth of $t\downarrow$ are defined by induction on the complexity of the term t as follows. If t is a constant or variable, then $t\downarrow$ is true and $\tilde{t}$ is the element of the model which interprets t. If t is of the form $f(t_1, \ldots, t_n)$ and $\tilde{f}$ is the partial function which interprets f, then $t\downarrow$ is true iff $\tilde{f}(\tilde{t_1}, \ldots, \tilde{t_n})$ is defined, and in that case $\tilde{t} = \tilde{f}(\tilde{t_1}, \ldots, \tilde{t_n})$. For details, see Beeson[**3**, Chapter VI].

Language of EON:

Constant symbols of **EON** are $\mathbf{k}$, $\mathbf{s}$, $\mathbf{p}$, $\mathbf{p}_0$, $\mathbf{p}_1$, $\mathbf{d}$, $\mathbf{s_N}$, $\mathbf{p_N}$ and 0. Predicate symbols are $=$ and N. **EON** contains **Ap** as a function symbol. $\mathbf{Ap}(x, y)$ is usually written as $(x \cdot y)$ or xy. We use the conventions of association to the left, e.g. xyz means $(xy)z$. We often write $f(t_1, \ldots, t_n)$ for $ft_1 \ldots t_n$, and $\langle x, y\rangle$ for $\mathbf{p}xy$. The numeral $\overline{m}$ which corresponds to natural number m is defined as usual, i.e. $\overline{0}$ is 0, and $\overline{n+1}$ is $\mathbf{s_N}\overline{n}$.

Axioms of EON:

(EON1) $\quad \mathbf{k}xy = x \quad$ (*$\mathbf{k}$ combinator*)

(EON2) $\quad \mathbf{s}xyz \simeq xz(yz) \mathbin{\&} \mathbf{s}xy\downarrow \quad$ (*$\mathbf{s}$ combinator*)

(EON3) $\quad \mathbf{k} \neq \mathbf{s}$

(EON4) $\quad \mathbf{p}xy\downarrow \mathbin{\&} \mathbf{p}_0(\mathbf{p}xy) = x \mathbin{\&} \mathbf{p}_1(\mathbf{p}xy) = y \quad$ (*pairing*)

(EON5), (EON6) *Axioms for natural numbers:*

$$N(0) \mathbin{\&} \forall x(N(x) \to [N(\mathbf{s_N}(x)) \mathbin{\&} \mathbf{p_N}(\mathbf{s_N}x) = x \mathbin{\&} \mathbf{s_N}x \neq 0])$$
$$\forall x(N(x) \mathbin{\&} x \neq 0 \to N(\mathbf{p_N}x) \mathbin{\&} \mathbf{s_N}(\mathbf{p_N}x) = x)$$

(EON7), (EON8) *Definition by integer cases:*

$$N(a) \mathbin{\&} N(b) \mathbin{\&} a = b \to \mathbf{d}(a, b, x, y) = x$$
$$N(a) \mathbin{\&} N(b) \mathbin{\&} a \neq b \to \mathbf{d}(a, b, x, y) = y$$

(EON9) *Induction for all formulae:*

$$\phi(0) \mathbin{\&} \forall x(N(x) \mathbin{\&} \phi(x) \to \phi(\mathbf{s_N}x)) \to \forall x(N(x) \to \phi(x))$$

Introduction of λ-abstraction.

In **EON**, one can define λ-abstraction as a syntactical operation. For each term t and variable x, we define $\lambda x.t$ by induction on the complexity of the term t. $\lambda x.x$ is **skk**. If t is a constant or a variable other than x, $\lambda x.t$ is $\mathbf{k}t$. If t is of the form uv, $\lambda x.t$ is $\mathbf{s}(\lambda x.u)(\lambda x.v)$. It is easy to see that the free variables of $\lambda x.t$ are those of t, excluding x, and that $\mathbf{EON} \vdash \lambda x.t\downarrow \mathbin{\&} (\lambda x.t)(x) \simeq t$. We often write $\lambda x_1 \ldots x_n.t$ for $\lambda x_1. \cdots \lambda x_n.t$.

N.B. λ-abstraction defined above does not commute with substitution. For example, $(\lambda y.x)[\mathbf{kk}/x] \equiv (\mathbf{k}x)[\mathbf{kk}/x] \equiv \mathbf{k}(\mathbf{kk})$, but $(\lambda y.\mathbf{kk}) \equiv \mathbf{s}(\mathbf{kk})(\mathbf{kk})$.

The Recursion Theorem.

THEOREM. *Let* $t \equiv \lambda yx.f(yy)x$ *and* $\mathbf{Y} \equiv \lambda f.tt$. *Then* **EON** *proves*

$$\mathbf{Y}f\downarrow \,\&\, [g = \mathbf{Y}f \rightarrow \forall x(gx \simeq fgx)].$$

PROOF: Since $\mathbf{Y}f \simeq tt \simeq (\lambda x.f(yy)x)[t/y]$, we have $\mathbf{Y}f\downarrow$ and $gx \simeq f(tt)x \simeq fgx$.

N.B. Exactly speaking, **Y** is not a "fixpoint operator". That is, $\mathbf{Y}f \simeq f(\mathbf{Y}f)$ is not necessarily true. For example, let $\omega \equiv (\lambda x.xx)(\lambda x.xx)$ and $f \equiv \lambda y.\omega$. Then $\mathbf{EON} \vdash \mathbf{Y}f\downarrow$ but $\mathbf{EON} \not\vdash f(\mathbf{Y}f)$.

Models of EON.

EON has various models, e.g. Kleene's first model K_1 (indices of the partial recursive functions) and second model K_2, Scott's D_∞ models, graph models of Plotkin-Scott, term models, and so on. Here we treat the models K_1' and TT, which we shall use later.

(1) K_1'. The universe of K_1' is $\mathbf{N} \cup \{-1\}$. The application operation is defined by

$$x \cdot y = \begin{cases} \{x+1\}(y+1) - 1, & \text{if } \{x+1\}(y+1) \text{ is defined} \\ -1, & \text{otherwise.} \end{cases}$$

where $\{x\}(y)$ is the partial recursive function application which is the application operation of K_1. Predicate N is interpreted by **N**.

It is easy to see that this is actually a model. For example, if $\tilde{\mathbf{k}}$ and $\tilde{\mathbf{s}}$ interpret **k** and **s** in K_1, then $(\tilde{\mathbf{k}} - 1)xy = x$ and $(\tilde{\mathbf{s}} - 1)xyz \simeq xz(yz)$ hold in K_1'. Moreover, K_1' is a model of

$$\textbf{(TA)} \qquad \forall x, y(xy\downarrow)$$

where "**TA**" stands for "Total Application".

We can assume that $\{0\}(n)$ is undefined for all natural number n. Then $(-1) \cdot x = -1$ for all x in K_1'. Hence $\forall f \in \mathbf{N}^{\mathbf{N}} f \in \mathbf{N}$ is true in K_1', where $f \in \mathbf{N}^{\mathbf{N}}$ means $\forall n \in \mathbf{N}\, f \cdot n \in \mathbf{N}$.

(2) TT (Total Term model). TT is also a model of **EON** + **TA**. The universe of TT is the set of all terms of **EON**. Reduction relation is introduced to TT in the obvious way (the definition is found in [**3**, P.111]), and it satisfies the Church-Rosser property. Of course one may add new constant symbols to **EON**. We define $t \sim u$ iff t and u have a common reduct. By the Church-Rosser property, "$\sim$" is a equivalence relation. Application $t \cdot u$ is defined to be the term tu. We interpret the constant symbols as themselves, the predicate N as the set of numerals, and equality as the equivalence relation "$\sim$". TT becomes a model of **EON** under this interpretation (see [**3**, Theorem 6.2.1]).

Both K_1' and TT have standard integers, i.e. the structure of the set of natural numbers of the model is isomorphic to the usual structure of the positive integers.

1.2 The Theory F.

We introduce a formal system **F** (**F** is for "Frege structure").

LANGUAGE OF **F**. *That of* **EON**, *plus two new unary predicates* Ω *and* Δ, *plus new constant symbols* $\dot{\&}$, $\dot{\vee}$, $\dot{\rightarrow}$, $\dot{\bot}$, $\dot{\forall}$, $\dot{\exists}$, $\dot{=}$, $\dot{\mathbf{N}}$.

Here $\Omega(a)$ means that "a is a proposition", and $\Delta(a)$ means that "a is a true proposition". For the sake of readability, we write e.g. $a \mathbin{\dot{\&}} b$ instead of $\dot{\&}ab$.

AXIOMS OF **F**. *Those of* **EON**, *plus induction for all formulae of the new language, plus the following axioms about propositions and truths:*

(F1) $$\forall x(\Delta(x) \rightarrow \Omega(x))$$

(F2) $$\forall x \forall y(\Omega(x \mathbin{\dot{=}} y) \mathbin{\&} (\Delta(x \mathbin{\dot{=}} y) \leftrightarrow x = y))$$

(F3) $$\forall x(\Omega(\dot{\mathbf{N}}x) \mathbin{\&} (\Delta(\dot{\mathbf{N}}x) \leftrightarrow N(x)))$$

(F4) $$\forall x[\Omega(a \mathbin{\dot{\&}} b) \leftrightarrow \Omega(a) \mathbin{\&} \Omega(b)] \\ \mathbin{\&} \forall a, b[\Delta(a \mathbin{\dot{\&}} b) \leftrightarrow \Delta(a) \mathbin{\&} \Delta(b)]$$

(F5) $$\forall a, b[\Omega(a \mathbin{\dot{\vee}} b) \leftrightarrow \Omega(a) \mathbin{\&} \Omega(b)] \\ \mathbin{\&} \forall a, b[\Omega(a \mathbin{\dot{\vee}} b) \rightarrow (\Delta(a \mathbin{\dot{\vee}} b) \leftrightarrow \Delta(a) \vee \Delta(b))]$$

(F6) $$\Omega(\dot{\bot}) \mathbin{\&} (\Delta(\dot{\bot}) \leftrightarrow \bot)$$

(F7) $$\forall a, b[\Omega(a) \mathbin{\&} (\Delta(a) \rightarrow \Omega(b)) \\ \rightarrow \Omega(a \mathbin{\dot{\rightarrow}} b) \mathbin{\&} (\Delta(a \mathbin{\dot{\rightarrow}} b) \leftrightarrow (\Delta(a) \rightarrow \Delta(b)))]$$

(F8) $$\forall x(\Omega(fx)) \rightarrow \Omega(\dot{\forall} f) \mathbin{\&} (\Delta(\dot{\forall} f) \leftrightarrow \forall x(\Delta(fx)))$$

(F9) $$\forall x(\Omega(fx)) \rightarrow \Omega(\dot{\exists} f) \mathbin{\&} (\Delta(\dot{\exists} f) \leftrightarrow \exists x(\Delta(fx)))$$

NOTATION. *We write* $x \dot{\in} X$ *for* Xx, $S(X)$ *for* $\forall x \Omega(x \dot{\in} X)$, *and* $x \in X$ *for* $\Delta(x \dot{\in} X)$.

$S(X)$ means that "X is a set".

Aczel's extended conjunction. We define "Aczel's extended conjunction" $a \mathbin{\dot{\&}\!\supset} b$ by $a \mathbin{\dot{\&}\!\supset} b \equiv a \mathbin{\dot{\&}} (a \mathbin{\dot{\rightarrow}} b)$. It is easy to see that

$$\forall a, b(\Omega(a \mathbin{\dot{\&}\!\supset} b) \leftrightarrow \Omega(a) \mathbin{\&} (\Delta(a) \rightarrow \Omega(b))) \\ \mathbin{\&} \forall a, b(\Delta(a \mathbin{\dot{\&}\!\supset} b) \leftrightarrow \Delta(a) \mathbin{\&} \Delta(b)).$$

NOTATION. *We adopt the following abbreviations:*

$$\dot{\forall} x.a \equiv \dot{\forall} \lambda x.a$$
$$\dot{\exists} x.a \equiv \dot{\exists} \lambda x.a$$
$$\dot{\forall} x \dot{\in} X.a \equiv \dot{\forall} x.(x \dot{\in} X \mathbin{\dot{\rightarrow}} a)$$
$$\dot{\exists} x \dot{\in} X.a \equiv \dot{\exists} x.(x \dot{\in} X \mathbin{\dot{\&}\!\supset} a).$$

Defining sets in F. We can define null set $\emptyset$, finite set $\{a_1, \ldots, a_n\}$, product $X \times Y$, disjoint sum $X + Y$, function space Y^X, dependent product $\prod_{x \dot{\in} X} Y_x$, and dependent sum $\sum_{x \dot{\in} X} Y_x$ as follows:

$$\begin{aligned}
\{x \mid \phi\} &\equiv \lambda x.\phi \\
\emptyset &\equiv \{x \mid \dot{\bot}\} \\
\{a_1, \ldots, a_n\} &\equiv \{x \mid x \doteq a_1 \dot{\vee} \cdots \dot{\vee} x \doteq a_n\} \\
X \times Y &\equiv \{x \mid \mathbf{p}_0 x \dot{\in} X \mathbin{\dot{\&}} \mathbf{p}_1 x \dot{\in} Y\} \\
X + Y &\equiv \{x \mid \mathbf{p}_0 x \doteq 0 \mathbin{\dot{\&}} \mathbf{p}_1 x \dot{\in} X \dot{\vee} \mathbf{p}_0 x \doteq 1 \mathbin{\dot{\&}} \mathbf{p}_1 x \dot{\in} Y\} \\
Y^X &\equiv \{f \mid \dot{\forall} x \dot{\in} X. f x \dot{\in} Y\} \\
\prod_{x \dot{\in} X} Y_x &\equiv \{f \mid \dot{\forall} x \dot{\in} X. f x \dot{\in} Y x\} \\
\sum_{x \dot{\in} X} Y_x &\equiv \{x \mid \mathbf{p}_0 x \dot{\in} X \mathbin{\dot{\&}\!\supset} \mathbf{p}_1 x \dot{\in} Y(\mathbf{p}_0 x)\}.
\end{aligned}$$

1.3 RPS.

Beeson[3] introduced **RPS** as an extension of formal system **PS** (**PS** is for Propositions and Sets). But **PS** is only a definitional extension of **F**, so we introduce **RPS** as a direct extension of **F**.

LANGUAGE OF **RPS**. *That of* **F**, *plus a new binary predicate symbol P and a new constant $\dot{P}$.*

The intended interpretation of $P(u, x)$ is that "*x is a proposition and u is a proof of x*".

AXIOMS OF **RPS**. *Those of* **F**, *plus induction for all formulae of the new language, plus the following axioms about proofs:*

(RPS1) $\Omega(\alpha \mathbin{\dot{\&}} \beta) \rightarrow [P(u, \alpha \mathbin{\dot{\&}} \beta) \leftrightarrow P(\mathbf{p}_0 u, \alpha) \mathbin{\&} P(\mathbf{p}_1 u, \beta)]$

(RPS2) $\Omega(\alpha \dot{\vee} \beta) \rightarrow [P(e, \alpha \dot{\vee} \beta) \leftrightarrow (\mathbf{p}_0 e = 0 \mathbin{\&} P(\mathbf{p}_1 e, \alpha)) \vee (\mathbf{p}_0 e = 1 \mathbin{\&} P(\mathbf{p}_1 e, \beta))]$

(RPS3) $\forall u \neg P(u, \dot{\bot})$

(RPS4) $\Omega(\dot{\exists}\alpha) \rightarrow [P(e, \dot{\exists}\alpha) \leftrightarrow P(\mathbf{p}_1 e, \alpha(\mathbf{p}_0 e))]$

(RPS5) $\Omega(\dot{\forall}\alpha) \rightarrow [P(e, \dot{\forall}\alpha) \leftrightarrow \forall x (P(\mathbf{p}_0 e x, \alpha x) \mathbin{\&} P(\mathbf{p}_1 e, \dot{\forall} x. \dot{P}(\mathbf{p}_0 e x, \alpha x))]$

(RPS6) $\Omega(\alpha \dot{\rightarrow} \beta) \rightarrow [P(e, \alpha \dot{\rightarrow} \beta) \leftrightarrow \forall q (P(q, \alpha) \rightarrow P(\mathbf{p}_0 e q, \beta)) \mathbin{\&} P(\mathbf{p}_1 e, \dot{\forall} q. (\dot{P}(q, \alpha) \dot{\rightarrow} \dot{P}(\mathbf{p}_0 e q, \beta)))]$

(RPS7) $\Delta(\alpha) \leftrightarrow \exists u P(u, \alpha)$ *(to assert is to prove)*

(RPS8) $P(u, \alpha) \leftrightarrow \Delta(\dot{P} u \alpha)$

Axioms RPS1–RPS6 look like the definition of realizability, but RPS5 and RPS6 are much more complicated than realizability interpretations of $\forall x A$ and $A \rightarrow B$. These axioms express Beeson's ideas on constructive proofs. RPS5 says that *"e proves $\forall x\phi(x)$ iff e is a pair $\langle u, v\rangle$ where u is a rule such that for each object x, ux is a proof of $\phi(x)$, and v is a supplementary data which proves this property"*. RPS6 says that *"e proves $A \rightarrow B$ iff e is a pair $\langle u, v\rangle$ where u is a rule which transforms any proof q of A into a proof uq of B, and v is a supplementary data which proves this property"*.

These complicated axioms make the consistency proof difficult. If we delete the "second clauses" of these axioms, we can easily prove the consistency. This is a point of our consistency proof. We show that we can delete the second clauses. That is, we do not need "supplementary datas".

2. Consistency Proof

In this section we give a consistency proof of **RPS**. First we make a convention as follows. That is, we delete the predicate symbol Δ from our list of primitive symbols and define $\Delta(\alpha)$ as an abbreviation of $\exists x P(x, \alpha)$. Then axiom RPS7 becomes unnecessary, so we remove it from the list of axioms.

Outline of the proof. First we introduce two formal systems $\mathbf{RPS}'$ and $\mathbf{RPS}^+$. $\mathbf{RPS}'$ is a modification of **RPS**, and $\mathbf{RPS}^+$ is defined as an extension of $\mathbf{RPS}'$. Secondly we prove that **RPS** is a subsystem of $\mathbf{RPS}^+ + \mathbf{TA}$. Next we interpret $\mathbf{RPS}^+ + \mathbf{TA}$ in $\mathbf{RPS}' + \mathbf{TA}$ using realizability interpretation. Thus the consistency of **RPS** is reduced to that of $\mathbf{RPS}' + \mathbf{TA}$. Finally, we construct a model of $\mathbf{RPS}' + \mathbf{TA}$. This ends the proof.

2.1 $\mathbf{RPS}'$ and $\mathbf{RPS}^+$.

We define $\mathbf{RPS}'$ by modifying **RPS**. The modification is as follows:

(1) Replace the axioms F2–F9 by the following F2′–F6′, F7′a, F8′a and F9′ respectively:

(F2′) $$\forall x \forall y(\Omega(x \doteq y))$$

(F3′) $$\forall x(\Omega(\dot{\mathrm{N}} x))$$

(F4′) $$\forall a, b[\Omega(a \mathbin{\dot{\&}} b) \leftrightarrow \Omega(a) \mathbin{\&} \Omega(b)]$$

(F5′) $$\forall a, b[\Omega(a \mathbin{\dot{\vee}} b) \leftrightarrow \Omega(a) \mathbin{\&} \Omega(b)]$$

(F6′) $$\Omega(\dot{\bot})$$

(F7′a) $$\forall a, b[\Omega(a) \mathbin{\&} (\Delta(a) \rightarrow \Omega(a)) \leftrightarrow \Omega(a \mathbin{\dot{\rightarrow}} b)]$$

(F8′a) $$\forall x(\Omega(fx)) \leftrightarrow \Omega(\dot{\forall} f)$$

(F9′) $$\forall x(\Omega(fx)) \leftrightarrow \Omega(\dot{\exists} f).$$

(2) Replace the axioms RPS1, RPS5 and RPS6 by the following RPS1′, RPS5′ and RPS6′ respectively:

(RPS1$'$) $$P(u, \alpha \mathbin{\dot{\&}} \beta) \leftrightarrow P(\mathbf{p}_0 u, \alpha) \mathbin{\&} P(\mathbf{p}_1 u, \beta)$$

(RPS5$'$) $$P(e, \dot{\forall}\alpha) \leftrightarrow \forall x P(\mathbf{p}_0 e x, \alpha x)$$

(RPS6$'$) $$\Omega(\alpha \mathbin{\dot{\rightarrow}} \beta) \rightarrow [P(e, \alpha \mathbin{\dot{\rightarrow}} \beta) \leftrightarrow \forall q(P(q, \alpha) \rightarrow P(\mathbf{p}_0 e q, \beta))].$$

(3) Remove RPS8.

(4) **RPS** does not describe when $\dot{P}xy$ becomes a proposition and what the proofs of $x \doteq y$, $\dot{\mathbf{N}}x$ and $\dot{P}xy$ are. So we add the following four axioms:

($\Omega\dot{P}$) $$\forall x, y(\Omega(\dot{P}xy) \leftrightarrow \Omega(y))$$

($P \doteq$) $$\forall e(P(e, x \doteq y) \leftrightarrow x = y)$$

($P\dot{\mathbf{N}}$) $$\forall e(P(e, \dot{\mathbf{N}}x) \leftrightarrow N(x))$$

($P\dot{P}$) $$\forall x, y(P(x, \dot{P}(y, \alpha)) \leftrightarrow P(y, \alpha)).$$

Note that we have deleted the "second clauses" of RPS5 and RPS6.

$\mathbf{RPS}^+$ is defined as $\mathbf{RPS}' +$ the following two axioms:

(F7$'$b) $$\Omega(a \mathbin{\dot{\rightarrow}} b) \rightarrow (\Delta(a \mathbin{\dot{\rightarrow}} b) \leftrightarrow (\Delta(a) \rightarrow \Delta(b)))$$

(F8$'$b) $$\Delta(\dot{\forall} f) \leftrightarrow \forall x(\Delta(fx)).$$

2.2 THEOREM. **RPS** *is a subsystem of* $\mathbf{RPS}^+ + \mathbf{TA}$.

PROOF: It is enough to show that F2–F9, RPS1, RPS5, RPS6 and RPS8 are derivable in $\mathbf{RPS}^+ + \mathbf{TA}$.

We can easily obtain F2 from F2$'$ and $P \doteq$; F3 from F3$'$ and $P\dot{\mathbf{N}}$; F4 from F4$'$ and RPS1$'$; F5 from F5$'$ and RPS2; F6 from F6$'$ and RPS3; F7 from F7$'$a and F7$'$b; F8 from F8$'$a and F8$'$b respectively. We show that F9, RPS5, RPS6 are derivable.

Ad F9: Suppose $\forall x \Omega(fx)$. First we check $\Delta(\dot{\exists} f) \rightarrow \exists x(\Delta(fx))$. Suppose $P(e, \dot{\exists} f)$. Then by RPS4 we have $P(\mathbf{p}_1 e, \alpha(\mathbf{p}_0 e))$. So we have $\exists x \exists u P(u, fx)$. Hence $\Delta(\dot{\exists} f) \rightarrow \exists x(\Delta(fx))$.

Conversely, suppose $P(u, fx)$. Then by RPS4 we have $P(\langle x, u\rangle, \dot{\exists} f)$. Hence we obtain $\exists x(\Delta(fx)) \rightarrow \Delta(\dot{\exists} f)$. Thus we have derived F9.

Ad RPS5: It is sufficient to show

$$\forall x P(\mathbf{p}_0 e x, \alpha x) \leftrightarrow P(\mathbf{p}_1 e, \dot{\forall} x.\dot{P}(\mathbf{p}_0 e x, \alpha x)).$$

But we have

$$\begin{aligned} & P(\mathbf{p}_1 e, \dot{\forall} x.\dot{P}(\mathbf{p}_0 e x, \alpha x)) \\ & \quad \leftrightarrow \forall x P(\mathbf{p}_0(\mathbf{p}_1 e) x, \dot{P}(\mathbf{p}_0 e x, \alpha x)) \qquad \text{by RPS5}' \end{aligned}$$

$$\leftrightarrow \forall x P(\mathbf{p}_0 ex, \alpha x) \qquad \text{by } P\dot{P} \text{ and } \mathbf{TA}.$$

Ad RPS6: Suppose $\Omega(a \dot{\rightarrow} b)$. It is enough to show

$$P(\mathbf{p}_1 e, \dot{\forall} q.(\dot{P}(q,\alpha) \dot{\rightarrow} \dot{P}(\mathbf{p}_0 eq, \beta))) \leftrightarrow \forall q(P(q,\alpha) \rightarrow P(\mathbf{p}_0 eq, \beta)).$$

But we have

$$\begin{aligned}
&P(\mathbf{p}_1 e, \dot{\forall} q.(\dot{P}(q,\alpha) \dot{\rightarrow} \dot{P}(\mathbf{p}_0 eq, \beta))) \\
&\quad \leftrightarrow \forall q P(\mathbf{p}_0(\mathbf{p}_1 e)q, \dot{P}(q,\alpha) \dot{\rightarrow} \dot{P}(\mathbf{p}_0 eq, \beta)) && \text{by RPS5}' \\
&\quad \leftrightarrow \forall q \forall x (P(x, \dot{P}(q,\alpha)) \\
&\qquad\qquad \rightarrow P(\mathbf{p}_0(\mathbf{p}_0(\mathbf{p}_1 e)q)x, \dot{P}(\mathbf{p}_0 eq, \beta))) && \text{by RPS6}' \\
&\quad \leftrightarrow \forall q(P(q,\alpha) \rightarrow P(\mathbf{p}_0 eq, \beta)) && \text{by } P\dot{P} \text{ and } \mathbf{TA}.
\end{aligned}$$

∎

2.3 Realizability Interpretation.

We shall define for each formula A a new formula $e \mathbf{r} A$ (read "e *(*$\mathbf{r}$*-)realizes* A"), such that the free variables of $e \mathbf{r} A$ are among those of A and e.

DEFINITION.

$e \mathbf{r} A$	*is* A *for atomic* A
$e \mathbf{r} (A \rightarrow B)$	*is* $\forall a(a \mathbf{r} A \rightarrow ea\downarrow \,\&\, ea \mathbf{r} B)$
$e \mathbf{r} \exists x A$	*is* $\mathbf{p}_0 e \mathbf{r} A(\mathbf{p}_1 e)$
$e \mathbf{r} \forall x A$	*is* $\forall x(ex\downarrow \,\&\, ex \mathbf{r} A)$
$e \mathbf{r} A \vee B$	*is* $N(\mathbf{p}_0 e) \,\&\, (\mathbf{p}_0 e = 0 \rightarrow \mathbf{p}_1 e \mathbf{r} A)$ $\&\, (\mathbf{p}_0 e \neq 0 \rightarrow \mathbf{p}_1 e \mathbf{r} B)$
$e \mathbf{r} A \,\&\, B$	*is* $\mathbf{p}_0 e \mathbf{r} A \,\&\, \mathbf{p}_1 e \mathbf{r} B$

Remark. According to our conventions, $\Delta(\alpha)$ is no longer an atomic formula but an abbreviation of $\exists x P(x,\alpha)$. So, by the above definition,

$$\begin{aligned}
e \mathbf{r} \Delta(\alpha) &\equiv e \mathbf{r} \exists x P(x,\alpha) \\
&\equiv \mathbf{p}_0 e \mathbf{r} P(\mathbf{p}_1 e, \alpha) \\
&\equiv P(\mathbf{p}_1 e, \alpha).
\end{aligned}$$

We say that a formula A is *(provably) realized* in a theory T, if $T \vdash e \mathbf{r} A$ for some term e with free variables among those of A.

DEFINITION. *The formula A is called self-realizing, if there is a term $\mathbf{j}_A$ of* **EON** *such that $FV(\mathbf{j}_A) \subseteq FV(A)$ and* **EON** *proves*

(i) $A \to \mathbf{j}_A\downarrow \,\&\, \mathbf{j}_A \mathbf{\ r\ } A$

(ii) $(e \mathbf{\ r\ } A) \to A$.

DEFINITION. *The formula A is called negative if it contains no $\vee$'s and no $\exists$'s.*

2.3.1 LEMMA. *Every negative formula is self-realizing.*

PROOF: First we define $\mathbf{j}_A$ inductively as follows. We take $\mathbf{j}_A \equiv 0$ for atomic A. If A is $B \,\&\, C$, then $\mathbf{j}_A$ is $\langle \mathbf{j}_B, \mathbf{j}_C \rangle$. If A is $B \to C$, then $\mathbf{j}_A$ is $\lambda x.\mathbf{j}_C$. If A is $\forall x B$, then $\mathbf{j}_A$ is $\lambda x.\mathbf{j}_B$.

Now we prove (i) and (ii) by induction on the complexity of the formula A. It is clear for atomic A. Consider the case that A is $B \to C$. We check (i). Suppose $B \to C$ and $a \mathbf{\ r\ } B$. By (ii) for B, we have B. Then we get C and hence, by the induction hypothesis, $\mathbf{j}_C \mathbf{\ r\ } C$. But since $\mathbf{j}_A a$ is $\mathbf{j}_C$, we have $\mathbf{j}_A a \mathbf{\ r\ } C$. Hence $\mathbf{j}_A \mathbf{\ r\ } B \to C$. Now we check (ii). Suppose $e \mathbf{\ r\ } B \to C$. Then $\forall a(a \mathbf{\ r\ } B \to ea\downarrow \,\&\, ea \mathbf{\ r\ } C)$. We have to show $B \to C$. Suppose B. Then by (i) for B, $\mathbf{j}_B \mathbf{\ r\ } B$. So $e\mathbf{j}_B \mathbf{\ r\ } C$. Hence by (ii) for C, we have C.

The other cases are easy and we omit the proof. ∎

2.3.2 LEMMA.

(1) *If the hypotheses of the inference rules are provably realized, so are the conclusions.*

(2) *Let Γ be a set of formulae which are provably realized in* **EON**. *Suppose* **EON** + $\Gamma \vdash A$. *Then A is provably realized in* **EON**.

(3) *Let Γ be a set of formulae which are provably realized in* **RPS$'$**. *Suppose* **RPS$'$** + $\Gamma \vdash A$. *Then A is provably realized in* **RPS$'$**.

PROOF: (1), (2) See [**3**, Chapter VII, Theorem 1.6].

(3) By (1) and (2), it is sufficient to verify that the axioms of **RPS$'$** which are not among those of **EON** are provably realized in **RPS$'$**. But all these axioms except for F1 and RPS2 are negative formulae and hence self-realizing. So we have only to check that F1 and RPS2 are provably realized.

But F1 is logically equivalent to a negative formula $\forall x \forall y(P(y,x) \to \Omega(x))$, and **EON** proves that RPS2 is equivalent to a negative formula

$$\begin{aligned}\Omega(\alpha \dot{\vee} \beta) \to [P(e, \alpha \dot{\vee} \beta) \leftrightarrow\ & \\ & N(\mathbf{p}_0 e) \,\&\, \mathbf{d}(\mathbf{p}_0 e, 0, 0, \mathbf{d}(\mathbf{p}_0 e, 1, 0, 1)) = 0 \\ & \&\, (\mathbf{p}_0 e = 0 \to P(\mathbf{p}_1 e, \alpha)) \,\&\, (\mathbf{p}_0 e = 1 \to P(\mathbf{p}_1 e, \beta))].\end{aligned}$$

Hence these axioms are also provably realized. ∎

2.3.3 THEOREM. *Let Γ be a set of formulae which are provably realized in* $\mathbf{RPS}' + \mathbf{TA}$. *Suppose* $\mathbf{RPS}^+ + \mathbf{TA} + \Gamma \vdash A$. *Then A is provably realized in* $\mathbf{RPS}' + \mathbf{TA}$.

PROOF: By the above lemma, all axioms of $\mathbf{RPS}'$ are provably realized in $\mathbf{RPS}' + \mathbf{TA}$. We show that the remaining axioms of $\mathbf{RPS}^+ + \mathbf{TA}$ are also provably realized.

Consider the axiom **TA**. This is a negative formula and hence self-realizing. So **TA** is provably realized in $\mathbf{RPS}' + \mathbf{TA}$.

Next consider the axiom F7′b. We must show the following two formulae are provably realized:

$$\text{(1)} \qquad \Omega(a \dot{\rightarrow} b) \rightarrow (\Delta(a \dot{\rightarrow} b) \rightarrow (\Delta(a) \rightarrow \Delta(b)))$$
$$\text{(2)} \qquad \Omega(a \dot{\rightarrow} b) \rightarrow ((\Delta(a) \rightarrow \Delta(b)) \rightarrow \Delta(a \dot{\rightarrow} b))$$

We show that (1) is a theorem of $\mathbf{RPS}'$. Remember that $\Delta(a)$ is an abbreviation of $\exists x P(x, a)$. Suppose $\Omega(a \dot{\rightarrow} b)$ and $P(u, a \dot{\rightarrow} b)$. Then by RPS6′, we have $\forall q(P(q, a) \rightarrow P(\mathbf{p}_0 uq, b))$, and hence $\Delta(a) \rightarrow \Delta(b)$. This proves (1). So (1) is provably realized.

Now we verify that (2) is realized. Let $Z \equiv \langle 0, \langle \lambda q.\mathbf{p}_1(x\langle 0, q\rangle), 0\rangle\rangle$. We claim $\lambda e.\lambda x.Z$ realizes (2). Suppose $e \ \mathbf{r} \ \Omega(a \dot{\rightarrow} b)$ and $x \ \mathbf{r} \ (\Delta(a) \rightarrow \Delta(b))$. Then we have

$$\text{(3)} \qquad \forall y(P(\mathbf{p}_1 y, a) \rightarrow P(\mathbf{p}_1(xy), b)).$$

Suppose $P(q, a)$, then $P(\mathbf{p}_1\langle 0, q\rangle, a)$. So, by using (3), we obtain $P(\mathbf{p}_1(x\langle 0, q\rangle), b)$ and this is equivalent to $P(\mathbf{p}_0(\mathbf{p}_1 Z)q, b))$. Hence we get $\forall q(P(q, a) \rightarrow P(\mathbf{p}_0(\mathbf{p}_1 Z)q, b))$. By RPS6′, we have $P(\mathbf{p}_1 Z, a \dot{\rightarrow} b)$, which is equivalent to $Z \ \mathbf{r} \ \Delta(a \dot{\rightarrow} b)$. This proves our claim.

Finally consider the axiom F8′b. We must show that the following two formulae are provably realized:

$$\text{(4)} \qquad \Delta(\dot{\forall} f) \rightarrow \forall x(\Delta(fx))$$
$$\text{(5)} \qquad \forall x(\Delta(fx)) \rightarrow \Delta(\dot{\forall} f)$$

We show that (4) is a theorem of $\mathbf{RPS}'$. Suppose $P(e, \dot{\forall} f)$. Then, by RPS5′, we have $\forall x P(\mathbf{p}_0 ex, fx)$, which implies $\forall x(\Delta(fx))$. So (4) is a theorem of $\mathbf{RPS}'$, and hence provably realized in $\mathbf{RPS}'$.

Consider (5). Let $Z \equiv \langle 0, \langle \lambda x.\mathbf{p}_1(yx), 0\rangle\rangle$. We claim that $\lambda y.Z$ realizes (5). Suppose $y \ \mathbf{r} \ \forall x(\Delta(fx))$. Then we have $\forall x P(\mathbf{p}_1(yx), fx)$, and this is equivalent to $\forall x P(\mathbf{p}_0(\mathbf{p}_1 Z)x, fx)$. By RPS5′, we get $P(\mathbf{p}_1 Z, \dot{\forall} f)$. So $Z \ \mathbf{r} \ \Delta(\dot{\forall} f)$. Hence $\lambda y.Z$ realizes (5).

This completes the proof. ∎

2.3.4 COROLLARY. *Let* Γ *be a set of formulae which are provably realized in* **RPS'**. *If* **RPS'** + **TA** *is consistent, so is* $\mathbf{RPS}^+ + \mathbf{TA} + \Gamma$.

PROOF: Suppose $\mathbf{RPS}^+ + \mathbf{TA} + \Gamma \vdash \bot$. Then $\mathbf{RPS}' + \mathbf{TA} \vdash e \ \mathbf{r} \ \bot$ for some term e, i.e. $\mathbf{RPS}' + \mathbf{TA} \vdash \bot$. Contradiction. ∎

2.4 Models of RPS' + TA.

Let M be a model of **EON** + **TA** which has standard integers. We expand the model M to a model of **RPS'** + **TA**.

Assume that the elements $\bar{\mathbf{k}}$, $\bar{\mathbf{s}}$, $\bar{\mathbf{p}}$, ... of M interprets the constant symbols **k**, **s**, **p**, ... of **EON**. Define $\overline{ts} = \bar{t} \cdot \bar{s}$, then $\bar{t}$ is determined for each term t of **EON**. We expand "‾" to all the terms of **RPS'**. We define the interpretations of the other constant symbols of **RPS'** as follows: if M is the total term model TT, constant symbols are interpreted as themselves. For the other models, we define

$$\begin{array}{lll} \bar{\dot{\&}} = \overline{\lambda xy.\langle 1, x, y\rangle} & \bar{\dot{\vee}} = \overline{\lambda xy.\langle 2, x, y\rangle} & \bar{\dot{\rightarrow}} = \overline{\lambda xy.\langle 3, x, y\rangle} \\ \bar{\dot{\bot}} = \overline{\langle 4, 0\rangle} & \bar{\dot{\forall}} = \overline{\lambda x.\langle 5, x\rangle} & \bar{\dot{\exists}} = \overline{\lambda x.\langle 6, x\rangle} \\ \bar{\dot{\doteq}} = \overline{\lambda xy.\langle 7, x, y\rangle} & \bar{\dot{\mathbf{N}}} = \overline{\lambda x.\langle 8, x\rangle} & \bar{\dot{P}} = \overline{\lambda xy.\langle 9, x, y\rangle} \end{array}$$

where $\langle x, y, z\rangle$ is an abbreviation of $\langle x, \langle y, z\rangle\rangle$. Note that the right hand sides of these clauses are already defined. In the following, we write t for $\bar{t}$, if no confusion is feared.

Next we give the interpretation of the predicate symbols Ω and P. First we define two binary predicates $\mathcal{P}$ and $\bar{\mathcal{P}}$ on M by the following inductive clauses:

$$\begin{cases} \mathcal{P}(u, \alpha) \leftrightarrow A(\mathcal{P}, \bar{\mathcal{P}}, u, \alpha) \\ \bar{\mathcal{P}}(u, \alpha) \leftrightarrow B(\mathcal{P}, \bar{\mathcal{P}}, u, \alpha) \end{cases}$$

Here $A(\mathcal{P}, \bar{\mathcal{P}}, u, \alpha)$ is

$$\begin{array}{l} \exists a, b(\alpha = (a \doteq b) \ \& \ a = b) \\ \vee \exists a(\alpha = \dot{\mathbf{N}}a \ \& \ N(a)) \\ \vee \exists a, b(\alpha = (a \mathbin{\dot{\&}} b) \ \& \ \mathcal{P}(\mathbf{p}_0 u, a) \ \& \ \mathcal{P}(\mathbf{p}_1 u, b)) \\ \vee \exists a, b(\alpha = (a \mathbin{\dot{\vee}} b) \ \& \ (\mathbf{p}_0 u = 0 \ \& \ \mathcal{P}(\mathbf{p}_1 u, a) \vee \mathbf{p}_0 u = 1 \ \& \ \mathcal{P}(\mathbf{p}_1 u, b))) \\ \vee \exists a(\alpha = \dot{\exists} a \ \& \ \mathcal{P}(\mathbf{p}_1 u, a(\mathbf{p}_0 u))) \\ \vee \exists a(\alpha = \dot{\forall} a \ \& \ \forall x \mathcal{P}(\mathbf{p}_0 u x, a x)) \\ \vee \exists a, b(\alpha = (a \mathbin{\dot{\rightarrow}} b) \ \& \ \forall x(\neg \bar{\mathcal{P}}(x, a) \rightarrow \mathcal{P}(\mathbf{p}_0 u x, b))) \\ \vee \exists a, b(\alpha = (\dot{P} ab) \ \& \ \mathcal{P}(a, b)) \end{array}$$

and $B(\mathcal{P},\bar{\mathcal{P}},u,\alpha)$ is

$$
\begin{aligned}
&\exists a,b(\alpha = (a \dot{=} b) \,\&\, \neg a = b)\\
&\vee \exists a(\alpha = \dot{\mathbf{N}}a \,\&\, \neg N(a))\\
&\vee \alpha = \dot{\bot}\\
&\vee \exists a,b(\alpha = (a \,\dot{\&}\, b) \,\&.\neg(\neg\bar{\mathcal{P}}(\mathbf{p}_0 u, a) \,\&\, \neg\bar{\mathcal{P}}(\mathbf{p}_1 u, b)))\\
&\vee \exists a,b(\alpha = (a \,\dot{\vee}\, b)\\
&\qquad\qquad \& \neg(\mathbf{p}_0 u = 0 \,\&\, \neg\bar{\mathcal{P}}(\mathbf{p}_1 u, a) \vee \mathbf{p}_0 u = 1 \,\&\, \neg\bar{\mathcal{P}}(\mathbf{p}_1 u, b)))\\
&\vee \exists a(\alpha = \dot{\exists} a \,\&\, \bar{\mathcal{P}}(\mathbf{p}_1 u, a(\mathbf{p}_0 u)))\\
&\vee \exists a(\alpha = \dot{\forall} a \,\&\, \neg\forall x \neg\bar{\mathcal{P}}(\mathbf{p}_0 u x, ax))\\
&\vee \exists a,b(\alpha = (a \dot{\rightarrow} b) \,\&\, \neg\forall x(\mathcal{P}(x,a) \rightarrow \neg\bar{\mathcal{P}}(\mathbf{p}_0 u x, b)))\\
&\vee \exists a,b(\alpha = (\dot{P}ab) \,\&\, \bar{\mathcal{P}}(a,b)).
\end{aligned}
$$

Note that $\mathcal{P}$ and $\bar{\mathcal{P}}$ occur in these clauses only positively. So there is a solution.

Using this solution we define a unary predicate $\mathcal{Q}$ on M by the following inductive clause:

$$
\begin{aligned}
&\mathcal{Q}(\alpha)\\
&\leftrightarrow \forall u(\mathcal{P}(u,\alpha) \leftrightarrow \neg\bar{\mathcal{P}}(u,\alpha))\\
&\quad \& \ [\exists a,b(\alpha = (a \dot{=} b))\\
&\qquad \vee \exists a(\alpha = \dot{\mathbf{N}}a)\\
&\qquad \vee \alpha = \dot{\bot}\\
&\qquad \vee \exists a,b(\alpha = (a \,\dot{\&}\, b) \,\&\, \mathcal{Q}(a) \,\&\, \mathcal{Q}(b))\\
&\qquad \vee \exists a,b(\alpha = (a \,\dot{\vee}\, b) \,\&\, \mathcal{Q}(a) \,\&\, \mathcal{Q}(b))\\
&\qquad \vee \exists a(\alpha = (\dot{\exists} a) \,\&\, \forall x \mathcal{Q}(ax))\\
&\qquad \vee \exists a(\alpha = (\dot{\forall} a) \,\&\, \forall x \mathcal{Q}(ax))\\
&\qquad \vee \exists a,b(\alpha = (a \dot{\rightarrow} b) \,\&\, \mathcal{Q}(a) \,\&\, \forall x(\mathcal{P}(x,a) \rightarrow \mathcal{Q}(b)))\\
&\qquad \vee \exists a,b(\alpha = (\dot{P}ab) \,\&\, \mathcal{Q}(b))]
\end{aligned}
$$

This clause contains only positive occurrences of $\mathcal{Q}$. Hence it has a solution $\mathcal{Q}$.

Now we define a binary predicate $\mathcal{P}'$ by

$$\mathcal{P}'(x,\alpha) \equiv \mathcal{P}(x,\alpha) \,\&\, \mathcal{Q}(\alpha).$$

We interpret $\Omega(x)$ and $P(x,y)$ as $\mathcal{Q}(x)$ and $\mathcal{P}'(x,y)$ respectively. Then we have the following theorem:

2.4.1 THEOREM. *$\langle M, \mathcal{Q}, \mathcal{P}' \rangle$ is a model of* **RPS′ + TA** *+ classical logic.*

PROOF: We show that all the axioms of **RPS′** are true under the interpretation. Since M is a model of **EON + TA** and has standard integers, the axioms of **EON + TA** are all true.

Consider the other axioms. First we prove the following formula:

$$
\begin{aligned}
(1) \qquad \mathcal{Q}(\alpha) \leftrightarrow & \exists a, b(\alpha = (a \doteq b)) \\
& \vee \exists a(\alpha = \dot{\mathrm{N}}a) \\
& \vee \alpha = \bot \\
& \vee \exists a, b(\alpha = (a \mathbin{\dot{\&}} b) \mathbin{\&} \mathcal{Q}(a) \mathbin{\&} \mathcal{Q}(b)) \\
& \vee \exists a, b(\alpha = (a \mathbin{\dot{\vee}} b) \mathbin{\&} \mathcal{Q}(a) \mathbin{\&} \mathcal{Q}(b)) \\
& \vee \exists a(\alpha = (\dot{\exists} a) \mathbin{\&} \forall x \mathcal{Q}(ax)) \\
& \vee \exists a(\alpha = (\dot{\forall} a) \mathbin{\&} \forall x \mathcal{Q}(ax)) \\
& \vee \exists a, b(\alpha = (a \mathbin{\dot{\rightarrow}} b) \mathbin{\&} \mathcal{Q}(a) \mathbin{\&} \forall x(\mathcal{P}'(x, a) \rightarrow \mathcal{Q}(b))) \\
& \vee \exists a, b(\alpha = (\dot{P}ab) \mathbin{\&} \mathcal{Q}(b)).
\end{aligned}
$$

The proof is straightforward. The "only if" part is trivial. We show the "if" part. Consider the case of $\alpha = a \mathbin{\dot{\rightarrow}} b \mathbin{\&} \mathcal{Q}(a) \mathbin{\&} \forall x(\mathcal{P}'(x, a) \rightarrow \mathcal{Q}(b))$. We must show $\mathcal{Q}(\alpha)$. It is enough to show $\forall u(\mathcal{P}(u, a \mathbin{\dot{\rightarrow}} b) \leftrightarrow \neg\bar{\mathcal{P}}(u, a \mathbin{\dot{\rightarrow}} b))$ and $\mathcal{Q}(a) \mathbin{\&} \forall x(\mathcal{P}(x, a) \rightarrow \mathcal{Q}(b))$. The latter is clear. The former is proved as follows:

$$
\begin{aligned}
\mathcal{P}(u, a \mathbin{\dot{\rightarrow}} b) & \leftrightarrow \forall x(\neg\bar{\mathcal{P}}(x, a) \rightarrow \mathcal{P}(\mathbf{p}_0 ux, b)) \\
& \leftrightarrow \forall x(\mathcal{P}(x, a) \rightarrow \mathcal{P}(\mathbf{p}_0 ux, b)) && \text{by } \mathcal{Q}(a) \\
& \leftrightarrow \forall x(\mathcal{P}(x, a) \rightarrow \neg\bar{\mathcal{P}}(\mathbf{p}_0 ux, b)) && \text{by } \mathcal{P}'(x, a) \rightarrow \mathcal{Q}(b) \\
& \leftrightarrow \neg\bar{\mathcal{P}}(u, a \mathbin{\dot{\rightarrow}} b).
\end{aligned}
$$

The other cases are similar.

By (1), axioms F1′–F6′, F7′a, F8′a, F9′, $\Omega\dot{P}$ are clearly true. The cases of RPS1′, RPS2, RPS3, $P \doteq$, $P\dot{\mathrm{N}}$, $P\dot{P}$ are also clear. The remaining axioms are RPS4, RPS5′, RPS6′.

Ad RPS4: Suppose $\mathcal{Q}(\dot{\exists} a)$, then by (1) we have $\forall x \mathcal{Q}(ax)$. Hence,

$$
\begin{aligned}
\mathcal{P}'(u, \dot{\exists} a) & \leftrightarrow \mathcal{Q}(\dot{\exists} a) \mathbin{\&} \mathcal{P}(u, \dot{\exists} a) \\
& \leftrightarrow \mathcal{P}(\mathbf{p}_1 u, a(\mathbf{p}_0 u)) \\
& \leftrightarrow \mathcal{P}'(\mathbf{p}_1 u, a(\mathbf{p}_0 u)) && (\text{by } \forall x \mathcal{Q}(ax)).
\end{aligned}
$$

So RPS4 is true.

Ad RPS5′:

$$\begin{aligned}\mathcal{P}'(u, \dot{\forall} a) &\leftrightarrow \forall x \mathcal{P}(\mathbf{p}_0 ux, ax) \,\&\, \mathcal{Q}(\dot{\forall} a)\\ &\leftrightarrow \forall x(\mathcal{P}(\mathbf{p}_0 ux, ax) \,\&\, \mathcal{Q}(ax))\\ &\equiv \forall x \mathcal{P}'(\mathbf{p}_0 ux, ax).\end{aligned}$$

Ad RPS6′: Suppose $\mathcal{Q}(\alpha \dot{\rightarrow} \beta)$. We have to show that the following two formulae are true:

$$\mathcal{P}'(u, \alpha \dot{\rightarrow} \beta) \rightarrow \forall x(\mathcal{P}'(x, \alpha) \rightarrow \mathcal{P}'(\mathbf{p}_0 ux, \beta)) \tag{2}$$

$$\forall x(\mathcal{P}'(x, \alpha) \rightarrow \mathcal{P}'(\mathbf{p}_0 ux, \beta)) \rightarrow \mathcal{P}'(u, \alpha \dot{\rightarrow} \beta). \tag{3}$$

We verify (2). Suppose $\mathcal{P}'(u, \alpha \dot{\rightarrow} \beta)$ and $\mathcal{P}'(x, \alpha)$. Then we have the following (4)–(7):

$$\mathcal{Q}(\alpha) \tag{4}$$

$$\forall u(\mathcal{P}(u, \alpha) \rightarrow \mathcal{Q}(\beta)) \tag{5}$$

$$\forall x(\neg \bar{\mathcal{P}}(x, \alpha) \rightarrow \mathcal{P}(\mathbf{p}_0 ux, \beta)) \tag{6}$$

$$\mathcal{P}(x, \alpha) \tag{7}$$

By (4) and (7) we have $\neg \bar{\mathcal{P}}(x, \alpha)$. So by (6) we have $\mathcal{P}(\mathbf{p}_0 ux, \beta)$. And by (5) and (7), $\mathcal{Q}(\beta)$ holds. Hence $\mathcal{P}'(\mathbf{p}_0 ux, \beta)$.

Now we verify (3). Suppose

$$\forall x(\mathcal{P}'(x, \alpha) \rightarrow \mathcal{P}'(\mathbf{p}_0 ux, \beta)) \tag{8}$$

is true. Note that by $\mathcal{Q}(\alpha \dot{\rightarrow} \beta)$ we have $\mathcal{Q}(\alpha)$. Suppose $\neg \bar{\mathcal{P}}(x, \alpha)$, then $\mathcal{P}(x, \alpha)$. Hence $\mathcal{P}'(x, \alpha)$. Hence by (8) we have $\mathcal{P}(\mathbf{p}_0 ux, \beta)$. Thus we have derived $\forall x(\neg \bar{\mathcal{P}}(x, \alpha) \rightarrow \mathcal{P}(\mathbf{p}_0 ux, \beta)) \,\&\, \mathcal{Q}(\alpha \dot{\rightarrow} \beta)$. Hence $\mathcal{P}'(u, \alpha \dot{\rightarrow} \beta)$. ∎

2.4.2 Corollary. **RPS⁺ + TA** *is consistent and hence so is* **RPS**.

Proof: By the above theorem, **RPS′ + TA** is consistent. Hence, by Corollary 2.3.4, **RPS⁺ + TA** is consistent. Since **RPS** is a subsystem of **RPS⁺ + TA**, **RPS** is also consistent, of course. ∎

The following theorem is a refinement of the above result:

2.4.3 Theorem. $\mathbf{ID}_1^- \leq \mathbf{F} \leq \mathbf{RPS} \leq \mathbf{RPS}^+ + \mathbf{TA} \equiv \mathbf{RPS}' + \mathbf{TA} \leq \mathbf{ID}_2^-$.

Here $T_1 \leq T_2$ means that the proof-theoretical strength of T_1 is not stronger than that of T_2, and $T_1 \equiv T_2$ means that T_1 and T_2 have the same proof-theoretical strength. $\mathbf{ID}_n^-$ is a weak version of the theory of n times iterated inductive definitions. It is weak in the sense that you can not use the schema of transfinite induction in it. For details, see Aczel[**1**] or Beeson[**3**].

PROOF: As is seen in the proof of Theorem 2.4.1, predicates $\mathcal{P}$, $\bar{\mathcal{P}}$ and $\mathcal{Q}$ need not to be the least solutions of their defining equations. So our construction of the model $\langle M, \mathcal{Q}, \mathcal{P}' \rangle$ is formalizable in $\mathbf{ID}_2^-$. Hence $\mathbf{RPS}' + \mathbf{TA} \le \mathbf{ID}_2^-$. Next, **RPS** is an extension of **F** and Beeson[**3**, P.414] showed that $\mathbf{ID}_1^- \le \mathbf{F}$, so we have $\mathbf{ID}_1^- \le \mathbf{F} \le \mathbf{RPS}$. Finally, Theorem 2.2 shows that $\mathbf{RPS} \le \mathbf{RPS}^+ + \mathbf{TA}$ and Theorem 2.3.3 shows that $\mathbf{RPS}^+ + \mathbf{TA} \equiv \mathbf{RPS}' + \mathbf{TA}$. ∎

Remark. The exact strength of **F**, **RPS**, **RPS′** and **RPS⁺** are unknown. Theorem 9.4 of Beeson[**3**] states that $\mathbf{F} \le \mathbf{ID}_1^-$, but his proof contains a gap. One difficulty is to validate $\Omega(a \mathbin{\dot{\&}} b) \to \Omega(a) \mathbin{\&} \Omega(b)$ and $\Omega(a \mathbin{\dot{\vee}} b) \to \Omega(a) \mathbin{\&} \Omega(b)$.

3. VARIOUS PRINCIPLES OF CONSTRUCTIVE MATHEMATICS

In this section we shall investigate the relationships between our systems and various principles of constructive mathematics. In the following, "$\vdash A$" will mean "$\mathbf{RPS}^+ + \mathbf{TA} \vdash A$", unless othewise stated.

First we introduce a useful tool called "q-realizability interpretation".

3.1 q-realizability interpretation. We shall associate to each formula A a new formula $e \mathbf{\ q\ } A$ (read "*e* (q-)*realizes A*"), such that the free variables of $e \mathbf{\ q\ } A$ are among those of A and e.

DEFINITION.

$e \mathbf{\ q\ } A$	*is A for atomic A*
$e \mathbf{\ q\ } (A \to B)$	*is* $\forall a(A \mathbin{\&} a \mathbf{\ q\ } A \to ea\downarrow \mathbin{\&} ea \mathbf{\ q\ } B)$
$e \mathbf{\ q\ } \exists x A$	*is* $A(\mathbf{p}_1 e) \mathbin{\&} \mathbf{p}_0 e \mathbf{\ q\ } A(\mathbf{p}_1 e)$
$e \mathbf{\ q\ } \forall x A$	*is* $\forall x(ex\downarrow \mathbin{\&} ex \mathbf{\ q\ } A)$
$e \mathbf{\ q\ } A \vee B$	*is* $N(\mathbf{p}_0 e) \mathbin{\&} (\mathbf{p}_0 e = 0 \to A \mathbin{\&} \mathbf{p}_1 e \mathbf{\ q\ } A)$ $\mathbin{\&} (\mathbf{p}_0 e \neq 0 \to B \mathbin{\&} \mathbf{p}_1 e \mathbf{\ q\ } B)$
$e \mathbf{\ q\ } A \mathbin{\&} B$	*is* $\mathbf{p}_0 e \mathbf{\ q\ } A \mathbin{\&} \mathbf{p}_1 e \mathbf{\ q\ } B$

We say that a formula A is *(provably)* **q**-*realized* in a theory T, if $T \vdash e \mathbf{\ q\ } A$ for some term e with free variables among those of A. We define the concept of "self-q-realizing formula" similarly as "self-realizing formula" simply replacing **r** by **q**.

3.1.1 LEMMA. *Every negetive formula is self-*q*-realizing.*

PROOF: Similar to Lemma 2.3.1. The definition of $\mathbf{j}_A$ is exactly the same. ∎

3.1.2 THEOREM. *Let* Γ *be a set of formulae which are provably* **q**-*realized in* $\mathbf{RPS}^+ + \mathbf{TA}$. *Suppose* $\mathbf{RPS}^+ + \mathbf{TA} + \Gamma \vdash A$. *Then A is provably realized in* $\mathbf{RPS}^+ + \mathbf{TA} + \Gamma$.

PROOF: Similar to Theorem 2.3.3. ∎

3.2 Existence Properties.

3.2.1 THEOREM. *(Term exitence property)*

RPS$^+$ + **TA** *has the term existence property. That is, if* $\vdash \exists x A(x)$*, then for some term* t *with free variables among those of* $\exists x A(x)$*, we have* $\vdash A(t)$.

PROOF: Suppose $\vdash \exists x A(x)$. Then by Theorem 3.1.2 we can find a term t such that $FV(e) \subseteq FV(\exists x A(x))$ and $\vdash e$ **q** $\exists x A(x)$. Hence by definition of e **q** $\exists x A(x)$ we have $\vdash A(\mathbf{p}_1 e)$ and $FV(\mathbf{p}_1 e) \subseteq FV(\exists x A(x))$. ∎

Remark. **RPS** does not have the term existence property. For example, **RPS** proves $\exists x P(x, 0 \doteq 0)$ but we can not find a term t such that **RPS** proves $P(t, 0 \doteq 0)$.

3.2.2 THEOREM. *(Evaluation of numerical terms)*

If t *is a closed term and* $\vdash N(t)$ *then there exists a numeral* $\overline{m}$ *such that* **EON** $\vdash t = \overline{m}$.

PROOF: Suppose $\vdash N(t)$. Then by Theorem 3.1.2 there exists a term t such that **RPS**$'$ + **TA** $\vdash e$ **q** $N(t)$, that is **RPS**$'$ + **TA** $\vdash N(t)$. Let us take $M \equiv TT$ and construct the model $\langle M, \mathcal{Q}, \mathcal{P}' \rangle$ of **RPS**$'$ + **TA**. Then $N(t)$ is true in this model. Since t is a closed term, it is interpreted as itself. Hence t is a natural number of TT, i.e. t is reduced to some numeral $\overline{m}$. So by induction on the length of the reduction we have **EON** $\vdash t = \overline{m}$. ∎

3.2.3 THEOREM. *(Numerical existence property)*

If $\exists n(N(n) \& A(n))$ *is a closed formula and provable in* **RPS**$^+$ + **TA**, *then for some numeral* $\overline{m}$, *we have* $\vdash A(\overline{m})$.

PROOF: Suppose $\vdash \exists n(N(n) \& A(n))$, then by Theorem 3.2.1 we can find a closed term t such that $\vdash N(t) \& A(t)$. Hence by the above theorem we have $\vdash t = \overline{m}$ for some numeral $\overline{m}$. Hence $\vdash A(\overline{m})$. ∎

3.2.4 COROLLARY. *(Disjunction property)*

RPS$^+$ + **TA** *has the disjunction property, that is, if* $\vdash A \vee B$ *and* $A \vee B$ *is closed, then* $\vdash A$ *or* $\vdash B$.

PROOF: Note that $A \vee B$ is equivalent to $\exists x(N(x) \& (x = 0 \to A) \& (x \neq 0 \to B))$. By the above theorem, there exists a numeral $\overline{m}$ such that $\vdash (\overline{m} = 0 \to A) \& (\overline{m} \neq 0 \to B)$. If $m = 0$, we have $\vdash A$, and if $m \neq 0$, we have $\vdash B$. ∎

3.3 Choice Principles.

3.3.1 THEOREM. *(Rule of choice)*

Suppose $\vdash A(x) \to \exists y B(x, y)$*, where* $A(x)$ *is negative. Then for some closed term* f*, we have* $\vdash A(x) \to B(x, fx)$.

PROOF: Suppose $\vdash A(x) \to \exists y B(x, y)$. By Theorem 3.1.2 we can find a term t with no free variables other than x such that e **q** $(A(x) \to \exists y B(x, y))$, i.e. $\forall a(A(x) \& a$ **q** $A(x) \to B(x, \mathbf{p}_1(ea)))$. Since $A(x)$ is negative, we have $A(x) \to \mathbf{j}_{A(x)}$ **q** $A(x)$. Hence we have $A(x) \to B(x, \mathbf{p}_1(e\mathbf{j}_{A(x)}))$. Let $f = \lambda x.\mathbf{p}_1(e\mathbf{j}_{A(x)})$. Then f is a closed term and we have $A(x) \to B(x, fx)$. ∎

3.3.2 Dependent Choice.

We state the axiom of dependent choice as

$$\textbf{(DC)} \quad \begin{aligned} &\Delta(a\dot{\in}X \mathbin{\&} \dot{\forall}x\dot{\in}X.\dot{\exists}y\dot{\in}X.(\langle x,y\rangle\dot{\in}W)) \\ &\to \Delta(\dot{\exists}f\dot{\in}X^{\dot{\mathbf{N}}}(f(0)\doteq a \mathbin{\&} \dot{\forall}n\dot{\in}\dot{\mathbf{N}}(\langle fn, f(\mathbf{s_N}n)\rangle\dot{\in}W))). \end{aligned}$$

THEOREM. **RPS** $\vdash$ **DC**.

PROOF: Suppose $\Delta(a\dot{\in}X \mathbin{\&} \dot{\forall}x\dot{\in}X.\dot{\exists}y\dot{\in}X.(\langle x,y\rangle\dot{\in}W))$. Then we have $P(u, a\dot{\in}X)$ and $P(v, \dot{\forall}x\dot{\in}X.\dot{\exists}y\dot{\in}X.(\langle x,y\rangle\dot{\in}W))$ for some u and v. After a short calculation we obtain

$$\begin{aligned} \forall x, q(P(q, x\dot{\in}X) \to P(\mathbf{p}_0(G(x,q)), F(x,q)\dot{\in}X) \\ \mathbin{\&} P(\mathbf{p}_1(G(x,q)), \langle x, F(x,q)\rangle\dot{\in}W)) \end{aligned}$$

where $F(x,q)$ is $\mathbf{p}_0(\mathbf{p}_0 vxq)$ and $G(x,q)$ is $\mathbf{p}_1(\mathbf{p}_0 vxq)$. Now we define f and g by

$$\begin{cases} f(0) = a \\ f(\mathbf{s_N}n) = F(fn, hn) \\ h(0) = u \\ h(\mathbf{s_N}n) = \mathbf{p}_0(G(fn, hn)). \end{cases}$$

It is easy to verify by induction that

$$\forall n(N(n) \to P(hn, fn\dot{\in}X) \mathbin{\&} P(\mathbf{p}_1(G(fn,hn)), \langle fn, f(\mathbf{s_N}n)\rangle\dot{\in}W)).$$

Hence we have

$$\Delta(\dot{\exists}f\dot{\in}X^{\dot{\mathbf{N}}}(f(0)\doteq a \mathbin{\&} \dot{\forall}n\dot{\in}\dot{\mathbf{N}}(\langle fn, f(\mathbf{s_N}n)\rangle\dot{\in}W))).$$

Thus we have derived **DC**. ∎

3.3.3 Presentation Axiom of Choice.

For each term X, we define a term X^+ by $X^+ \equiv \lambda x.\dot{P}(\mathbf{p}_1 x, \mathbf{p}_0 x\dot{\in}X)$. We state the presentation axiom of choice as follows:

$$\textbf{(PAC)} \quad \begin{aligned} &(S(X) \to S(X^+)) \mathbin{\&} \forall u(u\in X \leftrightarrow \exists v(\langle u,v\rangle \in X^+)) \\ &\mathbin{\&} [\Delta(\dot{\forall}a\dot{\in}X^+.\dot{\exists}b\dot{\in}Y.\alpha(a,b)) \to \Delta(\dot{\exists}f.\dot{\forall}a\dot{\in}X^+.(fa\dot{\in}Y \mathbin{\&} \alpha(a,fa)))]. \end{aligned}$$

THEOREM. $\mathbf{RPS^+} \vdash$ **PAC**.

PROOF: (i) $S(X)\to S(X^+)$: Suppose $S(X)$, i.e. $\forall x\Omega(x\dot{\in}X)$. By the axiom $\Omega\dot{P}$ we have $\forall x\Omega(\dot{P}(\mathbf{p}_1 x, \mathbf{p}_0 x\dot{\in}X))$. Hence $\forall x\Omega(X^+x)$, that is $S(X^+)$.

(ii) $\forall u(u \in X \leftrightarrow \exists v(\langle u, v\rangle \in X^+))$: Since

$$X^+\langle u, v\rangle = \dot{P}(\mathbf{p}_1\langle u, v\rangle, \mathbf{p}_0\langle u, v\rangle \dot{\in} X) = \dot{P}(v, u\dot{\in} X),$$

we have

$$\begin{aligned} u \in X &\leftrightarrow \Delta(u\dot{\in}X) \\ &\leftrightarrow \exists v P(v, u\dot{\in}X) \\ &\leftrightarrow \exists v \Delta(\dot{P}(v, u\dot{\in}X)) \\ &\leftrightarrow \exists v \Delta(X^+\langle u, v\rangle) \\ &\leftrightarrow \exists v \, \langle u, v\rangle \in X^+. \end{aligned}$$

(iii) $\Delta(\dot{\forall} a \dot{\in} X^+.\dot{\exists} b \dot{\in} Y.\alpha(a,b)) \to \Delta(\dot{\exists} f.\dot{\forall} a \dot{\in} X^+.(fa\dot{\in} Y \mathbin{\dot{\&}} \alpha(a, fa)))$: Suppose $P(u, \dot{\forall} a \dot{\in} X^+.\dot{\exists} b \dot{\in} Y.\alpha(a,b))$. Then we have

(1) $$\forall a, q(P(q, a\dot{\in}X^+) \to \Delta(F(a,q)\dot{\in}Y \mathbin{\dot{\&}} \alpha(a, F(a,q))))$$

where $F(a,q)$ is $\mathbf{p}_0(\mathbf{p}_0 uaq)$. Let $f \equiv \lambda a.F(a,0)$. We claim

$$\Delta(\dot{\forall} a \dot{\in} X^+.(fa\dot{\in}Y \mathbin{\dot{\&}} \alpha(a, fa))).$$

Suppose $\Delta(a\dot{\in}X^+)$. Then we have $P(q, \dot{P}(\mathbf{p}_1 a, \mathbf{p}_0 a \dot{\in} X))$ for some q. So by the axiom $P\dot{P}$ we have $P(0, \dot{P}(\mathbf{p}_1 a, \mathbf{p}_0 a\dot{\in}X))$, i.e. $P(0, a\dot{\in}X^+)$. Then by (1) we get $\Delta(F(a,0)\dot{\in}Y \mathbin{\dot{\&}} \alpha(a, F(a,0)))$, i.e. $\Delta(fa\dot{\in}Y \mathbin{\dot{\&}} \alpha(a, fa))$. Hence our claim is true. Moreover, it is easy to check that

$$\forall f \Omega(\dot{\forall} a \dot{\in} X^+(fa\dot{\in}Y \mathbin{\dot{\&}} \alpha(a, fa))).$$

Hence $\Delta(\dot{\exists} f.\dot{\forall} a\dot{\in}X^+.(fa\dot{\in}Y \mathbin{\dot{\&}} \alpha(a, fa)))$. ∎

For the meaning of the axiom **PAC**, see Beeson[**3**, Chapter X, §13].

3.4 Church's Thesis and Markov's Principle.

We shall consider Church's thesis (**CT**) and Markov's principle (**MP**). These two are the foundational principles of Russian constructive mathematics. We make the convention that m and n range over the natural numbers.

CT and **MP** are stated as follows:

(CT) $$\forall f \in \mathbf{N}^{\mathbf{N}} \exists e \in \mathbf{N}\, \forall n \in \mathbf{N}\, (\{e\}(n) = f(n))$$

(MP) $$\forall x[\forall n(A(n,x) \vee \neg A(n,x)) \mathbin{\&} \neg\neg\exists n A(n,x) \to \exists n A(n,x)].$$

THEOREM. $\mathbf{RPS}^+ + \mathbf{TA} + \mathbf{CT} + \mathbf{MP}$ *is consistent.*

PROOF: Let $T \equiv \mathbf{RPS}' + \mathbf{TA}$+classical logic+$\forall f \in \mathbf{N}^{\mathbf{N}}[f \in \mathbf{N} \mathbin{\&} \forall n(f(n) = \{f+1\}(n+1) - 1)]$. It is enough to show that (i) T is consistent, and that (ii) if $\mathbf{RPS}^+ + \mathbf{TA} + \mathbf{CT} + \mathbf{MP} \vdash A$, then A is provably (**r**-)realized in T.

(i) Take $M = K_1'$ and construct the model $\langle M, \mathcal{Q}, \mathcal{P}' \rangle$. Clearly this is a model of T.

(ii) It is sufficient to prove that **CT** and **MP** are provably realized in T.

Ad **CT**: Let $u \equiv \lambda fx.\langle\langle 0, \lambda ny.0\rangle, \Lambda n.(\{f+1\}(n+1)-1)\rangle$. Here we used Kleene's Λ-notation: $\Lambda n.g(n)$ is an index e such that $\forall n\, \{e\}(n) = g(n)$. Then $u \mathbf{\ r\ } \mathbf{CT}$ easily follows from $\forall f \in \mathbf{N}^{\mathbf{N}}[f \in \mathbf{N} \,\&\, \forall n(f(n) = \{f+1\}(n+1)-1)]$.

Ad **MP**: Let $F(x,y) \equiv \mathbf{p}_0(\mathbf{p}_0 xy0)$ and $u \equiv \lambda x.\mu n(F(x,n) = 0)$, where $\mu n(F(x,y) = 0))$ is the least n such that $F(x,n) = 0$. Suppose

$$e \mathbf{\ r\ } \forall n(A(n,x) \vee \neg A(n,x)) \,\&\, \neg\neg\exists n A(n,x).$$

Then we have

$$\begin{aligned}\forall n[N(F(e,n)) \,\&\, (F(e,n) = 0 \to G(e,n) \mathbf{\ r\ } A(n,x))\\ \&\, (F(e,n) \neq 0 \to \&G(e,n) \mathbf{\ r\ } \neg A(n,x))\end{aligned}$$

and $\mathbf{p}_1 e \mathbf{\ r\ } \neg\neg\exists n A(n,x)$, where $G(e,n)$ is $\mathbf{p}_1(\mathbf{p}_0 en0)$. If $\forall n F(e,n) \neq 0$, then $\lambda nv.G(e,n) \mathbf{\ r\ } \forall n \neg A(n,x)$. But this contradicts $\mathbf{p}_1 e \mathbf{\ r\ } \neg\neg\exists n A(n,x)$. So we have $\neg\forall n F(e,n) \neq 0$. Then, by classical logic, we have $\exists n F(e,n) = 0$. Hence $u(e)$ has a natural number value. By the definition of u, we have $F(e, u(e)) = 0$. Hence $\lambda xe.\langle G(e,u(e)), u(e)\rangle$ realizes **MP**. ∎

Acknowledgements.

I would like to thank Prof. Kanji Namba for good advice and much encouragement. I am grateful to Mr. Tatsuya Shimura and Mr. Makoto Tatsuta for pointing out the errors in the earlier versions of the consistency proof.

References

1. P. Aczel, *The type theoretic interpretation of constructive set theory*, in "Logic Colloquim '77 (eds. A. MacIntyre, L. Pacholski, and J. Paris)," North-Holland, Amsterdam, 1979.
2. ———, *Frege structures and the notions of proposition, truth, and set*, in "The Kleene Symposium (eds. J. Barwise, H. J. Keisler, and K. Kunen)," North-Holland, Amsterdam, 1980, pp. 31–60.
3. M. Beeson, "Foundations of Constructive Mathematics," Springer, 1985.
4. E. Bishop, "Foundations of Constructive Analysis," McGraw-Hill, New York, 1967.
5. E. Bishop, D. Bridges, "Constructive Analysis," Springer.
6. S. Feferman, *A language and axioms for explicit mathematics*, in "Algebra and Logic, Lecture Notes in Mathematics No. 450," Springer, Berlin, 1975, pp. 87–139.
7. ———, *Constructive theories of functions and classes*, in "Logic Colloquium '78: Proceedings of the Logic Colloquium at Mons, 1978 (eds. M. Boffa, D. van Dalen, and K. McAloon)," North-Holland, Amsterdam, 1979, pp. 159–224.
8. P. Martin-Löf, *An Intuitionistic Theory of Types, Predicative Part*, in "Logic Colloquium '73," North-Holland, Amsterdam, 1975, pp. 73–118.
9. ———, *Constructive mathematics and computer programming*, in "Logic, Methodology, and Philosophy of Science VI (eds. L. J. Cohen, J. Los, H. Pfeiffer, and K. P. Podewski)," North-Holland, Amsterdam, 1982, pp. 153–179.
10. F. Richman (ed.), "Constructive Mathematics. Lecture Notes in Mathematics 873," Springer, 1981.
11. J. Smith, *An interpretation of Martin-Löf's type theory in a type-free theory of propositions*, J. Symbolic Logic **49** (1984), 730–753.

12. A. S. Troelstra, D. van Dalen (eds.), "The L. E. J. Brouwer Centenary Symposium," North-Holland, Amsterdam, 1982.

Keywords. constructive mathematics

7-3-1 Hongo Bunkyo-ku Tokyo 113 Japan

ELEMENTARY PROPERTIES OF A SYSTEM OF FUNDAMENTAL SEQUENCES FOR Γ_0

Mamoru SHIMODA[1]
Shimonoseki City College, Shimonoseki, Japan

Introduction

In this paper we take up a system of fundamental sequences for Γ_0, and we give constructive proofs of some basic properties of the system. This is a slight modification of the system in Kadota and Aoyama[1], which is also a slight modification of the system in Schmidt[5]. So these three systems are quite similar to each other, and any of these may be called a standard system of fundamental sequences for Γ_0. Most of the properties are essentially proved in [1] by using transfinite induction, but we prove them within PA, Peano Arithmeric. In Ketonen and Solovay[2] and Kurata[3], some properties of the canonical system of fundamental sequences for ε_0 are investigated, and we extend them to the system for Γ_0. For some aspects of applications to proof theory, it is necessary to prove them constructively. An example of the application will be found in [4].

In §1, we introduce some concepts about the ordinals below Γ_0 by using the notational system of Schütte[6]. Throughout this paper we consider only the ordinals below Γ_0 (the ordinal terms in [6]).

In §2, we define the system of fundamental sequences for Γ_0 and prove that $T(\gamma, n)$ is finite for every γ by using only mathematical induction. For each number we define a kind of rank of ordinals, by using rapidly growing functions, in order to avoid using transfinite induction.

In §3, we prove in PA that the system is (1)-built-up, which is

---

1) This research was partially supported by Grand-in-Aid for Co-operative Research (No. 61302010), The Ministry of Education, Science and Culture of Japan.

proved in [1] by transfinite induction. The notion "α-normal" plays a key role in the proof.

Finally in § 4, we prove that for all α and β if $\alpha > \beta$ then there is a number n such that $\alpha \xrightarrow[n]{} \beta$. This is called property A in [1]. We extend the results in 2.1 and 2.2 of [3], and show that the number n is effectively determined from α and β. This fact enables us to prove in PA the mutual equivalence of large set principles, well-founded principles, and transfinite inductions up to Γ_0. See [4], for details.

The author would like to thank R. Kurata and N. Kadota for helpful comments and suggestions.

§ 1. Preliminaries

For the representation of the ordinals below Γ_0, we adopt the notational system of [6], Chapter V. There is a primitive recursive well-ordering of order type Γ_0 on the set of all natural numbers, and we occasionally identify them with the corresponding natural numbers.

We assume the familiarity with Chapter V of [6]. We frequently use the function symbols $\phi\alpha\beta$ and $L\alpha$. The function ϕ_α $(=\lambda\beta\phi\alpha\beta)$ is the ordering function of the set of common fixed points of all functions ϕ_ξ with $\xi < \alpha$, and $L\alpha$ means the length of α.

Proposition 1.1. ([6], Theorems 14.7 and 14.8)

For every $\gamma > 0$, there exist unique $\alpha_1, \cdots, \alpha_k$ and $\beta_1, \cdots, \beta_k$ such that $\gamma = \phi\alpha_1\beta_1 + \cdots + \phi\alpha_k\beta_k$, where $\phi\alpha_1\beta_1 \geqq \cdots \geqq \phi\alpha_k\beta_k$, $\beta_i < \phi\alpha_i\beta_i$, $\alpha_i, \beta_i < \gamma$, and $L\alpha_i, L\beta_i < L\gamma$ $(i = 1, \cdots, k)$.

In this case, $\phi\alpha_1\beta_1 + \cdots + \phi\alpha_k\beta_k$ is called the normal form of γ. γ is called principal if $\gamma = \phi\alpha\beta$ for some α and β. If γ is principal and the normal form of γ is $\phi\alpha\beta$, then $\alpha, \beta < \gamma$ and $L\alpha, L\beta < L\gamma$.

Let $\alpha_1 + \cdots + \alpha_k$ and $\beta_1 + \cdots + \beta_m$ be normal forms of α and β. If $\alpha_k \geqq \beta_1$, then we say α meshes with β.

Definition 1.2. Define $\omega_k(\alpha,\beta)$ and $\zeta_k(\alpha,\beta)$ by induction on k;
(1) $\omega_0(\alpha,\beta)=\beta$, and $\omega_{k+1}(\alpha,\beta)=\phi(\alpha,\omega_k(\alpha,\beta))$.
(2) $\zeta_0(\alpha,\beta)=\alpha$, and $\zeta_{k+1}(\alpha,\beta)=\phi(\zeta_k(\alpha,\beta),\beta)$.

Define $\omega_k(\beta)=\omega_k(0,\beta)$ and $\zeta_k(\alpha)=\zeta_k(\alpha,0)$. Note that $\omega_k=\omega_k(1)$ and $\zeta_k=\zeta_k(1)$.

Lemma 1.3. (1) If $\beta<\phi\alpha\beta$, then $\omega_k(\alpha,\beta)<\omega_{k+1}(\alpha,\beta)$.
(2) If $\beta<\phi(\alpha+1,\gamma)$, then $\omega_k(\alpha,\beta)<\phi(\alpha+1,\gamma)$ for all k.

§ 2. A system of fundamental sequences for Γ_0

The following definition is essentially owing to N. Kadota (Cf. [1], § 3), and R. Kurata suggested the use of the function $\phi\alpha\beta$.

Definition 2.1. For γ and n, define $\gamma[n]$ by induction on $L\gamma$.
1. γ is principal, i.e. $\gamma=\phi\alpha\beta$ for some $\alpha,\beta<\gamma$, $L\alpha, L\beta<L\gamma$;
(1) If $\gamma=\phi 00=1$, then $\gamma[n]=0$.
(2) If $\gamma=\phi(\delta+1,0)$, then $\gamma[n]=\omega_n(\delta,\phi\delta 0)$.
(3) If $\gamma=\phi\alpha 0$, α is limit, then $\gamma[n]=\phi(\alpha[n],0)$.
(4) If $\gamma=\phi(0,\eta+1)$, then $\gamma[n]=\phi 0\eta\cdot(n+1)$.
(5) If $\gamma=\phi(\delta+1,\eta+1)$, then $\gamma[n]=\omega_n(\delta,\phi(\delta+1,\eta)+1)$.
(6) If $\gamma=\phi(\alpha,\eta+1)$, α is limit, then $\gamma[n]=\phi(\alpha[n],\phi\alpha\eta+1)$.
(7) If $\gamma=\phi\alpha\beta$, β is limit, then $\gamma[n]=\phi(\alpha,\beta[n])$.
2. γ is not principal;
(1) If $\gamma=0$, then $\gamma[n]=0$.
(2) If $\gamma=\gamma_1+\cdots+\gamma_k$ $(k\geqq 2)$, γ_j is principal $(j=1,\cdots,k)$, then $\gamma[n]=\gamma_1+\cdots+\gamma_{k-1}+\gamma_k[n]$.

Lemma 2.2. Let γ be a limit ordinal. Then
(1) $\gamma[n]<\gamma[n+1]<\gamma$ for all $n<\omega$.
(2) For all $\lambda<\gamma$, there exists $n<\omega$ such that $\lambda<\gamma[n]$, hence $\lim_{n<\omega}\gamma[n]=\gamma$.

Proof. (1) By induction on $L\gamma$ or by Lemma 1.3.
(2) By induction on $L\gamma$ and on $L\lambda$.

Hence for limit γ, $\{\gamma[n]\}_{n<\omega}$ is a fundamental sequence for γ.

Definition 2.3. ([3], Definition 2.2.1)
(1) $\alpha \xrightarrow[n]{} \beta$ if $\alpha[n]=\gamma_1$, $\gamma_1[n]=\gamma_2, \cdots, \gamma_k[n]=\beta$ for some $\gamma_1, \cdots, \gamma_k$.
(2) $\alpha \xRightarrow[n]{} \beta$ if $\alpha \xrightarrow[n]{} \beta$ or $\alpha=\beta$.
(3) $T(\alpha, n)=\{\beta ; \alpha \xrightarrow[n]{} \beta\}$.

Lemma 2.4. (1) If $\alpha \xrightarrow[n]{} \beta$, $\alpha \xrightarrow[n]{} \gamma$, and $\beta>\gamma$, then $\beta \xrightarrow[n]{} \gamma$.
(2) If $\alpha \xrightarrow[n]{} \beta$ and $\beta \xrightarrow[n]{} \gamma$, then $\alpha \xrightarrow[n]{} \gamma$.

Lemma 2.5. Let λ mesh with α. Then
(1) If $\alpha \xrightarrow[n]{} \beta$, then $\lambda+\alpha \xrightarrow[n]{} \lambda+\beta$.
(2) $\lambda+\alpha \xrightarrow[n]{} 0$ if and only if $\alpha \xrightarrow[n]{} 0$.

As usual, the m-times iteration of a function f is denoted by f^m; $f^0(x)=x$, $f^{m+1}(x)=f(f^m(x))$.

Definition 2.6. For a number m, define $f_k=f_k\langle m\rangle : \omega \longrightarrow \omega$ by induction on k; $f_0(x)=m^x$, $f_{k+1}(x)=f_k{}^{mx+m}(1)$.

Note that $f_{k+1}(0)=f_k^m(1)$ and $f_{k+1}(x+1)=f_k^m(f_{k+1}(x))$.

Lemma 2.7. Let $m \geqq 2$ and $f_k=f_k\langle m\rangle$. Then for all x and k,
(1) $x<f_k(x)<f_k(x+1)$.
(2) $k<f_k(x)<f_{k+1}(x)$.
(3) If $n<m$, then $f_k^n(f_k(0))<f_{k+1}(0)$.
(4) If $n<m$, then $f_k^n(f_{k+1}(x)+1)<f_{k+1}(x+1)$.
Proof. By induction on k and/or on x.

Definition 2.8. Let $n<\omega$ and $f_k=f_k\langle n+2\rangle$ $(k<\omega)$. For an ordinal γ, the number $r(\gamma)=r_n(\gamma)$ is defined by induction on $L\gamma$;

1. If $\gamma=0$, then $r(\gamma)=0$.
2. If γ is principal and its normal form is $\phi\alpha\beta$, then $r(\gamma)=f_{r(\alpha)}(r(\beta))$.
3. If $\gamma=\gamma_1+\cdots+\gamma_k$ $(k\geqq 2)$, γ_j is principal $(j=1,\cdots,k)$, then $r(\gamma)=r(\gamma_1)+\cdots+r(\gamma_k)$.

Lemma 2.9. $r(\omega_n(\alpha,\beta))=f^n_{r(\alpha)}(r(\beta))$.

Proposition 2.10. Let $n<\omega$ and $r=r_n$. If $\gamma>0$, then $r(\gamma[n])<r(\gamma)$

Proof. By induction on $L\gamma$.

1. γ is principal, i.e. $\gamma=\phi\alpha\beta$ for some $\alpha,\beta<\gamma$, $L\alpha, L\beta<L\gamma$;

(1) $\gamma=1$; trivial.

(2) $\gamma=\phi(\delta+1,0)$; use Lemmas 2.9 and 2.7(3).

(3) $\gamma=\phi\alpha 0$, α is limit. Since $L\alpha<L\gamma$, by induction hypothesis, $r(\alpha[n])<r(\alpha)$. Then use Lemma 2.7(2).

(4) $\gamma=\phi(0,\eta+1)$. Since $\gamma[n]=\phi 0\eta\cdot(n+1)$ and $r(0)=0$,
$$r(\gamma[n])=(n+1)\cdot r(\phi 0\eta)=(n+1)\cdot f_0(r(\eta))=(n+1)\cdot(n+2)^{r(\eta)}$$
$$<(n+2)^{r(\eta)+1}=f_0(r(\eta+1))=r(\phi(0,\eta+1))=r(\gamma).$$

(5) $\gamma=\phi(\delta+1,\eta+1)$; use Lemmas 2.9 and 2.7(4).

(6) $\gamma=\phi(\alpha,\eta+1)$, α is limit. As in (3), by induction hypothesis, $r(\alpha[n])<r(\alpha)$. Let $r(\alpha)=k+1$. Then by (2) and (4) of Lemma 2.7,
$$r(\gamma[n])=r(\phi(\alpha[n],\phi\alpha\eta+1))=f_{r(\alpha[n])}(f_{r(\alpha)}(r(\eta))+1)$$
$$\leqq f_k(f_{k+1}(r(\eta))+1)<f_{k+1}(r(\eta)+1)=f_{r(\alpha)}(r(\eta+1))=r(\gamma).$$

(7) $\gamma=\phi\alpha\beta$, β is limit. Since $L\beta<L\gamma$, by induction hypothesis, $r(\beta[n])<r(\beta)$. Then by Lemma 2.7(1),
$$r(\gamma[n])=r(\phi(\alpha,\beta[n]))=f_{r(\alpha)}(r(\beta[n]))<f_{r(\alpha)}(r(\beta))=r(\gamma).$$

2. γ is not principal. For some $\lambda,\kappa>0$, $\gamma=\lambda+\kappa$ and λ meshes with κ. Since $L\kappa<L\gamma$, by induction hypothrsis, $r(\kappa[n])<r(\kappa)$. Hence $r(\gamma[n])=r(\lambda+\kappa[n])=r(\lambda)+r(\kappa[n])<r(\lambda)+r(\kappa)=r(\gamma)$.

Theorem 2.11. (Cf. [2], Proposition 2.9, and [3], Proposition 2.2.5)

Let $n<\omega$ and $r=r_n$. If $\gamma>0$, then $T(\gamma, n)$ has cardinality at most $r(\gamma)$, i.e. $|T(\gamma, n)| \leqq r(\gamma)$.

Hence for all γ, $T(\gamma, n)$ is finite and $\gamma \underset{n}{\longrightarrow} 0$.

Proof. By induction on $r(\gamma)$.

If $\gamma=1$, then $T(\gamma, n)=\{0\}$ and $r(\gamma)=r(\phi 00)=f_0(0)=1$.

If $\gamma>1$, then $\gamma[n]>0$, so by Proposition 2.10, $r(\gamma[n])<r(\gamma)$. Then by induction hypothesis, $|T(\gamma[n], n)| \leqq r(\gamma[n])$.
Since $T(\gamma, n)=\{\gamma[n]\} \cup T(\gamma[n], n)$,

$|T(\gamma, n)| \leqq |T(\gamma[n], n)|+1 \leqq r(\gamma[n])+1 \leqq r(\gamma)$.

§ 3. (1)-built-upness of the system.

Here we prove in PA that the sytstem of fundamental sequences is (n)-built-up for all $n \geqq 1$. The system is called (n)-built-up if for every limit γ and every $i<\omega$, $\gamma[i+1] \underset{n}{\longrightarrow} \gamma[i]$ holds (Cf. [1], § 3).

Definition 3.1. (1) β is α-normal if $\beta<\phi\alpha\beta$.
(2) $\beta \overset{\alpha}{\underset{n}{\longrightarrow}} \gamma$ if for some $\beta_1, \cdots, \beta_k$, β_i is α-normal $(i=1, \cdots, k)$ and $\beta=\beta_1$, $\beta_1[n]=\beta_2, \cdots, \beta_k[n]=\gamma$.
(3) $\beta \overset{\alpha}{\underset{n}{\Longrightarrow}} \gamma$ if either $\beta \overset{\alpha}{\underset{n}{\longrightarrow}} \gamma$ or $\beta=\gamma$ and β is α-normal.

If $\beta \leqq \phi\alpha 0$, then β is α-normal. Every non-principal ordinal is α-normal for all α. If β is α-normal and $\alpha \leqq \tau$, then β is τ-normal. If β is α-normal, then $\phi\alpha\beta$ is also α-normal. Moreover, if $\beta \underset{n}{\longrightarrow} \gamma$ and λ meshes with β, then $\lambda+\beta \overset{\alpha}{\underset{n}{\longrightarrow}} \lambda+\gamma$ for all α.

The following lemmas are immediate consequences of the above facts.

Lemma 3.2. $\phi\alpha 0 \overset{\alpha}{\underset{n}{\longrightarrow}} 0$ for all α.

Lemma 3.3. (1) If $\beta \overset{\alpha}{\underset{n}{\longrightarrow}} \gamma$ and $\alpha \leqq \tau$, then $\beta \overset{\tau}{\underset{n}{\longrightarrow}} \gamma$.
(2) $\beta \overset{\alpha}{\underset{n}{\longrightarrow}} \gamma$ and $\gamma \overset{\alpha}{\underset{n}{\longrightarrow}} \delta$ implies $\beta \overset{\alpha}{\underset{n}{\longrightarrow}} \delta$.

Proposition 3.4. Let $n \geqq 1$ and $\beta > 0$.

If $\beta \xrightarrow[n]{\alpha} \gamma$, then $\omega_k(\alpha, \beta) \xrightarrow[n]{\alpha} \omega_k(\alpha, \gamma)$ for all k.

In particular, $\beta \xrightarrow[n]{\alpha} \gamma$ implies $\phi\alpha\beta \xrightarrow[n]{\alpha} \phi\alpha\gamma$.

Proof. By induction on $r(\alpha)$, $r = r_n$.

By Lemma 3.3(2), we may assume that $\beta[n] = \gamma$ and β is α-normal.

1. $k = 0$; trivial.

2. $k = 1$.

(1) β is limit; $(\phi\alpha\beta)[n] = \phi(\alpha, \beta[n]) = \phi\alpha\gamma$.

Since $\phi\alpha\beta$ is α-normal, $\phi\alpha\beta \xrightarrow[n]{\alpha} \phi\alpha\gamma$.

(2) $\beta = \eta + 1$; $\gamma = \beta[n] = \eta$.

(a) $\alpha = 0$. By Theorem 2.11 and Lemma 2.5(2),

$$(\phi\alpha\beta)[n] = \phi(0, \eta + 1)[n] = \phi 0\eta \cdot (n+1) \xrightarrow[n]{} \phi 0\eta = \phi\alpha\gamma.$$

If $\phi 0\eta \cdot (n+1) \overset{}{\underset{n}{\Longrightarrow}} \lambda \xrightarrow[n]{} \phi 0\eta$, then λ is not principal and so λ is α-normal. Hence $\phi\alpha\beta \xrightarrow[n]{\alpha} \phi\alpha\gamma$.

(b) $\alpha = \delta + 1$; $(\phi\alpha\beta)[n] = \omega_n(\delta, \phi\alpha\eta + 1)$.

Since $r(\delta) < r(\alpha)$ and $\phi\alpha\eta + 1 \xrightarrow[n]{\delta} \phi\alpha\eta$, by induction hypothesis,

$$\omega_n(\delta, \phi\alpha\eta + 1) \xrightarrow[n]{\delta} \omega_n(\delta, \phi\alpha\eta) = \phi\alpha\eta = \phi\alpha\gamma.$$

Hence by Lemma 3.3, $\phi\alpha\beta \xrightarrow[n]{\alpha} \phi\alpha\gamma$.

(c) α is limit. Let $\delta = \alpha[n]$. By Proposition 2.10, $r(\delta) < r(\alpha)$.

As in (b), by induction hypothesis,

$$(\phi\alpha\beta)[n] = \phi(\delta, \phi\alpha\eta + 1) \xrightarrow[n]{\delta} \phi(\delta, \phi\alpha\eta) = \phi\alpha\gamma.$$

Since $\delta < \alpha$, by Lemma 3.3, we have $\phi\alpha\beta \xrightarrow[n]{\alpha} \phi\alpha\gamma$.

3. $k > 1$. By induction on k.

Assume $\omega_k(\alpha, \beta) \xrightarrow[n]{\alpha} \omega_k(\alpha, \gamma)$, then by 2 above,

$$\omega_{k+1}(\alpha, \beta) = \phi(\alpha, \omega_k(\alpha, \beta)) \xrightarrow[n]{\alpha} \phi(\alpha, \omega_k(\alpha, \gamma)) = \omega_{k+1}(\alpha, \gamma).$$

Lemma 3.5. Let $n \geqq 1$. If $\phi\alpha\beta \xrightarrow[n]{\alpha} \beta$, then for all $k \geqq 1$,

$\omega_{k+1}(\alpha, \beta) \xrightarrow[n]{\alpha} \omega_k(\alpha, \beta) \xrightarrow[n]{\alpha} \beta$.

Proof. By induction on k. Use Proposition 3.4 and Lemma 3.3(2).

Lemma 3.6. Let $n \geqq 1$. Then $\phi(\alpha, \beta + 1) \xrightarrow[n]{\alpha} \phi\alpha\beta \cdot (n+1) \overset{\alpha}{\underset{n}{\Longrightarrow}} \phi\alpha\beta + 1$.

Proof. By induction on $r(\alpha)$, $r = r_n$. Use Lemmas 3.5 and 3.3.

<u>Lemma 3.7.</u> Let $n \geqq 1$ and $\alpha > 0$. If $\alpha \xrightarrow[n]{} \beta$, then $\phi\alpha 0 \xrightarrow[n]{\alpha} \phi\beta 0$.

<u>Proof.</u> Use Lemmas 3.2, 3.5, and 3.3.

By Lemmas 3.5 and 3.7, $\omega_{k+1} \xrightarrow[n]{} \omega_k$ and $\zeta_{k+1} \xrightarrow[n]{} \zeta_k$ holds for all n and k.

<u>Lemma 3.8.</u> Let $n \geqq 1$ and $\alpha > 0$. If $\alpha \xrightarrow[n]{} \beta$ and $\phi\alpha\gamma = \gamma$, then $\phi(\alpha, \gamma+1) \xrightarrow[n]{\alpha} \phi(\beta, \gamma+1)$.

<u>Proof.</u> Use Lemmas 3.6, 3.5, and 3.3.

<u>Theorem 3.9.</u> Let $n \geqq 1$.

If γ is limit and τ-normal, then $\gamma[i+1] \xrightarrow[n]{\tau} \gamma[i]$ for all $i < \omega$. Hence the system is (n)-built-up.

<u>Proof.</u> By induction on $L\gamma$.

1. γ is principal, i.e. for some $\alpha, \beta < \gamma$, $\phi\alpha\beta$ is the normal form of γ and $L\alpha, L\beta < L\gamma$. Then we have $\alpha \leqq \tau$, otherwise γ is not τ-normal. So by Lemma 3.3(1), it suffices to show $\gamma[i+1] \xrightarrow[n]{\alpha} \gamma[i]$.

(1) $\gamma = \phi(\delta+1, 0)$; use Lemmas 3.2, 3.5, and 3.3(1).

(2) $\gamma = \phi\alpha 0$, α is limit. Since $L\alpha < L\gamma$, by induction hypothesis, $\alpha[i+1] \xrightarrow[n]{\alpha} \alpha[i]$. Then use Lemmas 3.7 and 3.3(1).

(3) $\gamma = \phi(0, \eta+1)$. By Theorem 2.11 and Lemma 2.5(2), for all τ

$$\gamma[i+1] = \phi 0\eta \cdot (i+2) \xrightarrow[n]{\tau} \phi 0\eta \cdot (i+1) = \gamma[i].$$

(4) $\gamma = \phi(\delta+1, \eta+1)$; use Lemmas 3.6, 3.5, and 3.3(1).

(5) $\gamma = \phi(\alpha, \eta+1)$; α is limit. As in (2), by induction hypothesis, $\alpha[i+1] \xrightarrow[n]{\alpha} \alpha[i]$. Then use Lemmas 3.8 and 3.3(1).

(6) $\gamma = \phi\alpha\beta$, β is limit. Since β is α-normal and $L\beta < L\gamma$, by induction hypothesis, $\beta[i+1] \xrightarrow[n]{\alpha} \beta[i]$. Then use Proposition 3.4.

2. γ is not principal. There are λ and κ such that $\gamma = \lambda + \kappa$, λ meshes with κ, and κ is principal and limit. Let $\phi\alpha\beta$ be the normal form of κ. Since κ is α-normal and $L\kappa < L\gamma$, by induction hypothesis, $\kappa[i+1] \xrightarrow[n]{\alpha} \kappa[i]$. Then by Lemma 2.5(1), for all τ

$$\gamma[i+1] = \lambda + \kappa[i+1] \xrightarrow[n]{\tau} \lambda + \kappa[i] = \gamma[i].$$

Proposition 3.10. If $m<n<\omega$ and $\alpha \underset{m}{\longrightarrow} \beta$, then $\alpha \underset{n}{\longrightarrow} \beta$.

Proof. We may assume $\alpha[m]=\beta$ and $\alpha>0$.

If α is a successor, then $\alpha=\beta+1$ and $\alpha[n]=\beta$.

If α is limit, then since $n>m$, by Theorem 3.9, $\alpha[n] \underset{n}{\longrightarrow} \alpha[m]$. Hence by Lemma 2.4(2), $\alpha \underset{n}{\longrightarrow} \beta$.

§4. A constructive proof of property A

The following proposition is conjectured by R. Kurata. In the following we frequently use Lemma 2.4(2) without mention.

Proposition 4.1. Let $n \geq 1$, $\gamma=\phi\alpha\beta$, $\alpha>0$, and $\lambda=\phi\mu\nu$. Then $\alpha \underset{n}{\longrightarrow} \mu$ and $\gamma \underset{n}{\longrightarrow} \nu$ implies $\gamma \underset{n+1}{\longrightarrow} \lambda$.

Proof. By induction on $r(\gamma)$, $r=r_{n+1}$.

1. β is limit; $\gamma[n]=\phi(\alpha,\beta[n]) \underset{n}{\Longrightarrow} \nu$.

(a) $\gamma[n]=\nu$; $\lambda=\phi\mu\nu=\phi(\mu,\phi(\alpha,\beta[n]))=\phi(\alpha,\beta[n])=\gamma[n]$. So $\gamma \underset{n}{\longrightarrow} \lambda$, and by Proposition 3.10, $\gamma \underset{n+1}{\longrightarrow} \lambda$.

(b) $\gamma[n] \underset{n}{\longrightarrow} \nu$. By Theorem 3.9, $\gamma[n+1] \underset{n+1}{\longrightarrow} \gamma[n]$, hence $\gamma \underset{n+1}{\longrightarrow} \gamma[n]$. Then by Proposition 2.10, $r(\gamma[n])<r(\gamma)$. Since $\alpha \underset{n}{\longrightarrow} \mu$ and $\gamma[n] \underset{n}{\longrightarrow} \nu$, by induction hypothesis, $\gamma[n] \underset{n+1}{\longrightarrow} \lambda$. Therefore, $\gamma \underset{n+1}{\longrightarrow} \lambda$.

2. $\alpha=\delta+1$ and $\beta=0$; $\gamma=\phi(\delta+1,0)$, $\gamma[n+1]=\phi(\delta,\gamma[n])$, and $\gamma[n]=\omega_n(\delta,\phi\delta 0)=\omega_{n+1}(\delta,0)$. Since $\alpha[n]=\delta$, $\delta \underset{n}{\Longrightarrow} \mu$.

(1) $\delta=\mu$.

(a) $\gamma[n]=\nu$; $\lambda=\phi\mu\nu=\phi(\delta,\gamma[n]))=\gamma[n+1]$.

(b) $\gamma[n] \underset{n}{\longrightarrow} \nu$. By Lemmas 3.2 and 3.5, $\omega_{n+1}(\delta,0) \overset{\delta}{\underset{n}{\longrightarrow}} 0$. Hence $\gamma[n]=\omega_{n+1}(\delta,0) \overset{\delta}{\underset{n}{\longrightarrow}} \nu$. Then by Proposition 3.4, $\gamma[n+1]=\phi(\delta,\gamma[n]) \overset{\delta}{\underset{n}{\longrightarrow}} \phi\delta\nu=\phi\mu\nu=\lambda$.

Therefore by Proposition 3.10, $\gamma \underset{n+1}{\longrightarrow} \lambda$.

(2) $\delta \underset{n}{\longrightarrow} \mu$. By Theorem 3.9, $\gamma[n+1] \underset{n}{\longrightarrow} \gamma[n]$, so $\gamma[n+1] \underset{n}{\longrightarrow} \nu$. By Proposition 2.10, $r(\gamma[n+1])<r(\gamma)$. Then by induction hypothesis, $\gamma[n+1] \underset{n+1}{\longrightarrow} \lambda$. Hence $\gamma \underset{n+1}{\longrightarrow} \lambda$.

3. $\alpha=\delta+1$ and $\beta=\eta+1$; $\gamma=\phi(\alpha,\eta+1)$, $\delta \underset{n}{\Longrightarrow} \mu$,

$\gamma[n+1]=\phi(\delta,\gamma[n])$, and $\gamma[n]=\omega_n(\delta,\phi\alpha\eta+1) \underset{n}{\Longrightarrow} \nu$.

(1) $\delta=\mu$.

(a) $\gamma[n]=\nu$; In the same way as 2. (1), (a), $\lambda=\gamma[n+1]$.

(b) $\gamma[n] \underset{n}{\longrightarrow} \nu$. By Lemma 3.6, $\phi(\delta,\phi\alpha\eta+1) \overset{\delta}{\underset{n}{\longrightarrow}} \phi\alpha\eta+1$. Then by Lemma 3.5, $\gamma[n]=\omega_n(\delta,\phi\alpha\eta+1) \overset{\delta}{\underset{n}{\longrightarrow}} \phi\alpha\eta+1 \overset{\delta}{\underset{n}{\longrightarrow}} \phi\alpha\eta$.

By Lemma 2.4(1), one of the following three cases occurs.

<1> $\nu \underset{n}{\longrightarrow} \phi\alpha\eta$. Since $\gamma[n] \overset{\delta}{\underset{n}{\longrightarrow}} \phi\alpha\eta$, $\gamma[n] \overset{\delta}{\underset{n}{\longrightarrow}} \nu$. Then by Proposition 3.4, $\gamma[n+1]=\phi(\delta,\gamma[n]) \overset{\delta}{\underset{n}{\longrightarrow}} \phi\delta\nu=\phi\mu\nu=\lambda$. Hence by Proposition 3.10, $\gamma \underset{n+1}{\longrightarrow} \lambda$.

<2> $\nu=\phi\alpha\eta$. Since $\lambda=\phi\mu\nu=\phi\alpha\eta=\nu$, $\gamma \underset{n}{\longrightarrow} \lambda$. Then by Proposition 3.10, $\gamma \underset{n+1}{\longrightarrow} \lambda$.

<3> $\phi\alpha\eta \underset{n}{\longrightarrow} \nu$. Since $r(\phi\alpha\eta)<r(\phi(\alpha,\eta+1))=r(\gamma)$, by induction hypothesis, $\phi\alpha\eta \underset{n+1}{\longrightarrow} \lambda$. Then since $\gamma \underset{n}{\longrightarrow} \phi\alpha\eta$, by Proposition 3.10, $\gamma \underset{n+1}{\longrightarrow} \lambda$.

(2) $\delta \underset{n}{\longrightarrow} \mu$. Quite the same as 2. (2).

4. α is limit and β is 0 or a successor. In either case, for some κ, $\gamma[n+1]=\phi(\alpha[n+1],\kappa)$. By Theorem 3.9,

$\alpha[n+1] \underset{n}{\longrightarrow} \alpha[n] \underset{n}{\Longrightarrow} \mu$ and $\gamma[n+1] \underset{n}{\longrightarrow} \gamma[n] \underset{n}{\Longrightarrow} \nu$.

By Proposition 2.10, $r(\gamma[n+1])<r(\gamma)$, so by induction hypothesis, $\gamma[n+1] \underset{n+1}{\longrightarrow} \lambda$. Therefore, $\gamma \underset{n+1}{\longrightarrow} \lambda$.

<u>Lemma 4.2.</u> If $\phi\alpha\beta$ is the normal form of $\gamma>1$, then $L(\beta+1)<L\gamma$.

<u>Proof.</u> Immediate from the definition of the length.

<u>Definition 4.3.</u> The <u>height</u> of γ, denoted by $h(\gamma)$, is defined by induction on $L\gamma$:

1. If $\gamma=0$, then $h(0)=0$.
2. If $\phi\alpha_1\beta_1+\cdots+\phi\alpha_k\beta_k$ is the normal form of γ, then $h(\gamma)=\max\{h(\alpha_i)+1,\ h(\beta_i);\ i=1,\cdots,k\}$.

<u>Lemma 4.4.</u> If $\gamma<\lambda$, then $h(\gamma)\leqq h(\lambda)$.

<u>Proof.</u> By induction on $L\gamma$ and on $L\lambda$.

The following lemma gives another proof of the fact that for every γ there exists $k<\omega$ such that $\gamma<\zeta_k$ (Cf. [6], Theorem 14.16).

<u>Lemma 4.5.</u> $h(\gamma)\leqq k$ if and only if $\gamma<\zeta_k$.

<u>Proof.</u> By induction on k and on $L\gamma$.

We may assume $k>0$ and γ is principal. Let $\phi\alpha\beta$ be the normal form of γ.

To prove $h(\gamma)\leqq k+1 \Leftrightarrow \gamma<\zeta_{k+1}$, assume that $h(\delta)\leqq k \Leftrightarrow \delta<\zeta_k$ for all δ, and that $h(\eta)\leqq k+1 \Leftrightarrow \eta<\zeta_{k+1}$ for all η such that $L\eta<L\gamma$. Since $L\alpha, L\beta<L\gamma$, by induction hypothesis,

$$h(\gamma)\leqq k+1 \Leftrightarrow h(\phi\alpha\beta)\leqq k+1 \Leftrightarrow h(\alpha)\leqq k \text{ and } h(\beta)\leqq k+1$$
$$\Leftrightarrow \alpha<\zeta_k \text{ and } \beta<\zeta_{k+1} \Leftrightarrow \gamma=\phi\alpha\beta<\phi(\zeta_k, 0)=\zeta_{k+1}.$$

By Proposition 1.1, it is clear that for every γ there are unique ordinals $\alpha_1,\cdots,\alpha_k$, $\beta_1,\cdots,\beta_k$ and unique numbers $n_1,\cdots,n_k$, m such that $\gamma=\phi\alpha_1\beta_1\cdot n_1+\cdots+\phi\alpha_k\beta_k\cdot n_k+m$, where $\beta_i<\phi\alpha_i\beta_i$ $(i=1,\cdots,k)$ and $\phi\alpha_1\beta_1>\cdots>\phi\alpha_k\beta_k>1$. Then the number $c(\gamma)$ is defined by induction on $L\gamma$;

$$c(\gamma)=\max\{c(\alpha_i)+1,\ c(\beta_i+1)+1,\ n_i,\ m\ ;\ i=1,\cdots,k\}.$$

<u>Proposition 4.6.</u> (Cf. Corollary 2.2.8 of [3])

If $h(\gamma)\leqq k$ and $c(\gamma)\leqq n$, then $\zeta_k \underset{n}{\longrightarrow} \gamma$.

<u>Proof.</u> By induction on k and on $L\gamma$.

If $k=0$ or $n=0$, then $\gamma=0$ and it is trivial.

To prove $h(\gamma)\leqq k+1$ and $c(\gamma)\leqq n+1$ imply $\zeta_{k+1} \underset{n+1}{\longrightarrow} \gamma$, assume that $h(\delta)\leqq k$ and $c(\delta)\leqq n$ imply $\zeta_k \underset{n}{\longrightarrow} \delta$ for all δ and n, and that $h(\eta)\leqq k+1$ and $c(\eta)\leqq n$ imply $\zeta_{k+1} \underset{n}{\longrightarrow} \eta$ for all η and n such that $L\eta<L\gamma$.

Suppose $\gamma>0$, $h(\gamma)\leqq k+1$, and $c(\gamma)\leqq n+1$. For some α, β, λ, and $m<\omega$, $\gamma=\phi\alpha\beta\cdot m+\lambda$ and $\beta, \lambda<\phi\alpha\beta$. Then obviously, $h(\alpha)\leqq k$, $h(\lambda)\leqq k+1$, $c(\alpha), c(\beta+1)\leqq n$, $m\leqq n+1$, $c(\lambda)\leqq n+1$, and $L\lambda<L\gamma$. Now $L(\beta+1)<L\gamma$ by Lemma 4.2, and $h(\beta+1)\leqq k+1$ by Lemma 4.4. Hence

by induction hypothesis, $\zeta_k \xrightarrow[n]{} \alpha$, $\zeta_{k+1} \xrightarrow[n]{} \beta+1$, and $\zeta_{k+1} \xrightarrow[n+1]{} \lambda$. By Proposition 4.1, $\zeta_{k+1} = \phi(\zeta_k, 0) \xrightarrow[n+1]{} \phi(\alpha, \beta+1)$, and by Lemma 3.6,

$$\phi(\alpha, \beta+1) \xrightarrow[n+1]{} \phi\alpha\beta\cdot(n+2) \xrightarrow[n+1]{} \phi\alpha\beta\cdot(m+1) \xrightarrow[n+1]{} \phi\alpha\beta.$$

Since $\phi\alpha\beta > \lambda$, $\phi\alpha\beta \xrightarrow[n+1]{} \lambda$ by Lemma 2.4(1). Hence by Lemma 2.5(1), $\phi\alpha\beta\cdot(m+1) \xrightarrow[n+1]{} \phi\alpha\beta\cdot m + \lambda = \gamma$. Therefore, $\zeta_{k+1} \xrightarrow[n+1]{} \gamma$ holds.

Now we identify the ordinals below Γ_0 with the corresponding numbers.

Theorem 4.7. There exists a primitive recursive function $g(\alpha, \beta)$ such that if $\alpha > \beta$ and $n = g(\alpha, \beta)$, then $\alpha \xrightarrow[n]{} \beta$.

Proof. Let $\alpha > \beta$ and define $g(\alpha, \beta) = \max\{c(\alpha), c(\beta)\}$. Obviously, the function $g(\alpha, \beta)$ is primitive recursive. Let $k = \max\{h(\alpha), h(\beta)\}$ and $n = g(\alpha, \beta)$. Then by Proposition 4.6, $\zeta_k \xrightarrow[n]{} \alpha$ and $\zeta_k \xrightarrow[n]{} \beta$. Therefore, by Lemma 2.4(1), $\alpha \xrightarrow[n]{} \beta$ holds.

References

[1] Kadota, N., and Aoyama, K., A note on Schmidt's built-up systems of fundamental sequences, to appear in RIMS, Kokyuroku, Kyoto Univ.

[2] Ketonen, J., and Solovay, R., Rapidly growing Ramsey functions, Ann. of Math. 113 (1981), 267-314.

[3] Kurata, R., Paris-Harrington principles, reflection principles, and transfinite induction up to ε_0, Ann. Pure Appl. Logic 31 (1986), 237-256.

[4] Kurata, R., and Shimoda, M., Some combinatorial principles equivalent to restrictions of transfinite induction up to Γ_0, manuscript.

[5] Schmidt, D., Built-up systems of fundamental sequences and hierarchies of number-theoretic functions, Arch. math. Logik 18 (1976), 47-53. Postscript, 18 (1977), 145-146.

[6] Schütte, K., Proof Theory, Springer, 1977.

The Continuum Hypothesis and the Theory of the Kleene Degrees

JUICHI SHINODA
THEODORE A. SLAMAN

§1. INTRODUCTION

A type 2 object on ω is a set of reals or more generally, function from $\omega^n \times (\omega^\omega)^m$ to ω or to ω^ω. Recursion relative to a type 2 object was first studied by Kleene [1] and [2]. Given type 2 objects A and B, A is said to be Kleene reducible to B, $A \leq_K B$, if A is recursive in B, ${}^2E$, and a real. This reducibility induces an equivalence relation on the type 2 objects. The equivalence class of A is called the Kleene degree of A. Let $\mathcal{K}$ be the partially ordered set consisting of the Kleene degrees of all type 2 objects with order $\leq_K$. $\mathcal{K}$ is an upper semi-lattice in which every countable set has an least upper bound.

In this paper, we will study the global structure of $\mathcal{K}$ under the assumption of the continuum hypothesis (CH). Seeing that the definition of $\mathcal{K}$ is naturally formulated in $\langle \omega, P(\omega), P(P(\omega)), \in, +, \times \rangle$, the first order theory of $\langle \mathcal{K}, \leq_K \rangle$ is canonically interpreted in the theory of $\langle \omega, P(\omega), P(P(\omega)), \in, +, \times \rangle$. That is, there is a recursive function $\psi \mapsto \tilde{\psi}$ which maps a first order sentence ψ in the language of $\mathcal{K}$ to a third order sentence $\tilde{\psi}$ so that

$$\langle \mathcal{K}, \leq_K \rangle \models \psi \iff \langle \omega, P(\omega), P(P(\omega)), \in, +, \times \rangle \models \tilde{\psi}.$$

We show that a reverse theorem is also true. Under the assumption of the continuum hypothesis, we can interpret the theory of $\langle \omega, P(\omega), P(P(\omega)), \in, +, \times \rangle$ in the first order theory of $\langle \mathcal{K}, \leq_K \rangle$.

THEOREM. *(CH) There is a recursive function $\varphi \mapsto \varphi^*$ which maps a sentence φ of third order arithmetic to a sentence φ^* in the first order language of $\mathcal{K}$ so that*

$$\langle \omega, P(\omega), P((\omega)), \in, +, \times \rangle \models \varphi \iff \langle \mathcal{K}, \leq_K \rangle \models \varphi^*.$$

In [7], we proved a similar result for the PTIME degrees of the recursive sets. There, we showed that the theory of $\langle \omega, +, \times \rangle$ is interpreted in the theory of the PTIME degrees of recursive sets of binary strings. We also showed that the theory of $\langle \omega, P(\omega), \in, +, \times \rangle$ is interpreted in the theory of the PTIME degrees of all sets of binary strings. The

---

This paper was prepared while Slaman visited Nagoya University with a fellowship from the Japan Society for the Promotion of Science. The first author was supported by Grant-in-Aid for Co-operative Research (No.61302010),The Ministry of Education, Science and Culture. The second author was supported by Presidential Young Investigator Award DMS-8451748 and N. S. F. research grant DMS-8601856.

coding scheme used to prove the theorem in this paper is a variation of that used in [7]. However, here we construct the coding parameters generically instead of by the priority argument used in [7].

We organize the paper as follows.

In §2, we give some basic definitions and notation.

In §3, we analyze the existence of exact pairs over countably closed ideals in $\mathcal{K}$ under the assumption of CH (Theorem 3.2). Following the work by Nerode-Shore [5] in the Turing degrees, this theorem is used to code ideals in $\mathcal{K}$ with $\aleph_1$ many generators.

In §4, we define a partial ordered set $P_{\mathbf{R}}$ which canonically represents the standard model of second order arithmetic. We also claim a technical theorem (Theorem 4.3) that ensures the existence of type 2 objects coding $P_{\mathbf{R}}$ inside $\mathcal{K}$. The proof of this theorem is given in §5–§7.

Our main theorem follows easily from Theorem 4.3. First, it is not difficult to show that the set of codes for $P_{\mathbf{R}}$ is definable in $\mathcal{K}$. By Theorem 4.3, it is not empty. Secondly, we can interpret second order quantifiers over the reals of a coding of $P_{\mathbf{R}}$ using exact pairs. Theorem 3.2 ensures that this interpretation is faithful.

In §5, we give a list of requirements sufficient to imply Theorem 4.3. We distinguish some of these as being global requirements.

In §6, we give a notion of forcing $\mathcal{P}$ used to generically construct the type 2 objects mentioned in Theorem 4.3. A condition has two parts: a Cohen type condition on the sets being constructed and a global constraint on the allowed extensions of the Cohen condition corresponding to global requirements given by §5. We show that each global requirement is forced in $\mathcal{P}$.

In §7, we prove that for each of the remaining requirements given in §5 is forced in $\mathcal{P}$. We also analyze exactly which dense sets must be met by the generic parameters to conclude that the requirements, which were forced, are actually true for these parameters. Then, we can take any parameters meeting these dense sets to prove our theorem. By the Continuum Hypothesis, there are only $\aleph_1$ many dense sets to meet, $\mathcal{P}$ is countably closed and so we obtain the required type 2 objects by an $\aleph_1$ length recursion.

Our proof of the theorem heavily depends on the continuum hypothesis. Therefore, it is an interesting problem whether this assumption is essential to the theorem.

§2. Preliminaries

Kleene Degrees. Suppose Y is a predicate. D. Normann [6] defines the functions which are E-recursive relative to Y. Given an integer e, let $\{e\}^Y$ denote the e-th E-recursive function relative to Y. Suppose f is a partial function from ω^ω to ω, Y is a type 2 object on ω, and a is a real. Let ${}^2E$ be the Kleene object of type 2. It is well-known that f is recursive in Y, ${}^2E$ and a if and only if there is an integer e such that $f(x) \simeq \{e\}^Y(x, a, \omega)$ for all $x \in \omega^\omega$. A Kleene functional is a functional of the form $\lambda Yx.\{e\}^Y(x, a, \omega)$, with real parameter a. The real coding $\langle e, a\rangle$ is the index of this functional. We use upper case Greek letters $\Phi, \Psi, \dots$ to indicate Kleene functionals. For the most part, Kleene functionals will be $\{0,1\}$-valued.

2.1. DEFINITION. Suppose A and B are two sets of reals.

(1) A is *Kleene reducible* to B, $A \leq_K B$, if there exists a Kleene functional Φ such that $A(x) = \Phi(B, x)$ for all $x \in \omega^\omega$.

(2) $A \equiv_K B$ if $A \leq_K B$ and $B \leq_K A$. The *Kleene degree* of A, $\deg_K(A)$, is the equivalence class of A under the relation $\equiv_K$.

(3) $\mathcal{K}$ is the partially ordered set consisting of all Kleene degrees with the order induced by $\leq_K$.

Given a type 2 object Y, let $L[\delta; x, Y]$ denote the sets constructed from x relative to Y in fewer than δ steps. δ is called (x, Y)-admissible if $L[\delta; x, Y]$ is a Y-admissible set. That is, the structure $\langle L[\delta; x, Y], \in, Y \cap L[\delta; x, Y] \rangle$ is a model of KP (Kripke-Platek set theory) phrased in the language of set theory with an additional predicate symbol for Y. The least ordinal which is (x, Y)-admissible and greater than ω is denoted by $\omega_1^{x,Y}$. Then, the set $L[\omega_1^{x,Y}; x, Y]$ is the least Y-admissible set containing x as an element. $L[\omega_1^{x,Y}; x, Y]$ is called the Y-admissible closure of x, and we use $\mathrm{cl}(x, Y)$ to denote the set $L[\omega_1^{x,Y}; x, Y] \cap \omega^\omega$. For a real z, z is an element of $\mathrm{cl}(x, Y)$ if and only if there is an e such that $\{e\}^Y(x) = z$.

Suppose a Kleene functional Φ with index $\langle e, a \rangle$ is given: $\Phi(Y, x) = \{e\}^Y(x, a, \omega)$. Then, only the values of Y on $\mathrm{cl}(\langle x, a \rangle, Y)$ are needed to compute the value $\Phi(Y, x)$. Indeed, let $\varphi(Y, x)$ denote the computation (computation tree) of $\Phi(Y, x)$, and $Y \restriction \varphi(Y, x)$ denote the set of reals which are queried to Y in $\varphi(Y, x)$. If $\Phi(Y, x)\downarrow$, then $\varphi(Y, x)$ is recursive in x, a, Y and ${}^2E$, and hence an element $L[\omega_1^{x,Y}; x, Y]$. When $\Phi(Y, x)\uparrow$, $\varphi(Y, x)$ is recursively enumerable in x, a, Y and ${}^2E$, and $\varphi(Y, x) \cap \omega^\omega$ is contained in $L[\omega_1^{x,Y}; x, Y]$. In either case, $Y \restriction \varphi(Y, x) \subseteq \mathrm{cl}(\langle x, a \rangle, Y)$. Further, note that if $\Phi(Y, x)$ is total, then the function $\varphi(Y, x)$ is recursive in a, Y and ${}^2E$.

Conversely to the above analysis of Kleene computations, let θ be a Σ_1 formula in the language of set theory with an additional predicate symbol and let a be a real. There is an e such that,

$$\left\{ y \;\middle|\; L[\omega_1^{y,a,Y}; a, y, Y] \models \theta \right\}$$

is equal to the domain of $\{e\}^Y(-, a)$. Further, the set theoretic formulation of the Gandy Selection Theorem [4] states that the function mapping y to the least witness in $L[\omega_1^{y,a,Y}; a, y, Y]$ to θ is uniformly Kleene recursive in Y.

Suppose Z is a type 2 object such that $Z(z) = Y(z)$ for all $z \in Y \restriction \varphi(Y, a)$, then it must hold that $\Phi(Z, a) = \Phi(Y, a)$. Also, for a partial type 2 object p such that $p(z)\downarrow = Y(z)$ on $Y \restriction \varphi(Y, a)$, we have $\Phi(p, a) = \Phi(Y, a)$.

The Continuum Hypothesis. Assume the continuum hypothesis (CH). Let $<$ be a well-ordering of the reals of length $\aleph_1$. We fix $<$ for the remainder of this paper.

For each α, let x_α be the α-th real in this ordering. We will often identify x_α with α itself. Hence, when $\alpha < \aleph_1$ and x is a real, $\alpha < x$ simply means that $x = x_\beta$ for some $\beta > \alpha$. Also, we will use the notation $M < x$ to indicate that all the reals in M are less than x. Thus, for example, $\mathrm{cl}(z, Y) < x$ means that x dominates all the reals in the Y-admissible closure of z.

Let W be a type 2 object such that, for each $\alpha < \aleph_1$, $W(x_\alpha)$ is an ω-sequence of reals. The first element of $W(x_\alpha)$ is a well ordering $\prec$ of ω of height α. The remaining elements of $W(x_\alpha)$ are the sequence of reals $\langle z_i \mid i \in \omega\rangle$ where, for each i, z_i is x_β and β is the ordinal height of i in $\prec$. The order $<$ is Kleene recursive in W. Further, if $\beta < \alpha$ then x_β is recursive in W, ${}^2E$ and x_α. We fix this type 2 object W throughout this paper. Also, to shorten the required notation, we will say that y is below x in W to indicate $y < x$. We also fix a pairing function mapping pairs of reals $\langle x, y\rangle$ to reals. We assume that $\langle x, y\rangle$ is always greater than either of x or y.

Forcing. Assume the continuum hypothesis. Suppose that I is an index set of cardinality at most $\aleph_1$ and $\{X_i \mid i \in I\}$ is a collection of sets of reals to be constructed. A *Cohen condition on* X_i is a $\{0,1\}$-valued function on a countable set of reals; a *Cohen condition on* $\{X_i \mid i \in I\}$ is a sequence of the form $\langle p(X_i) \mid i \in I_0\rangle$ with I_0 countable $\subseteq I$ such that each $p(X_i)$ is a Cohen condition on X_i. Suppose $p = \langle p(X_i) \mid i \in I_0\rangle$ is a Cohen condition on $\{X_i \mid i \in I\}$. We say that X_i is named in p when $i \in I_0$. Say that q is an extension of p if for every X_i named in p, X_i is also named in q and $q(X_i)$ is an extension of $p(X_i)$ as partial functions. A *condition on* $\{X_i \mid i \in I\}$ is a pair $\langle p, C\rangle$ consisting of a Cohen condition p and a set C of extensions of p. Then $\langle p_0, C_0\rangle$ is extended by $\langle p_1, C_1\rangle$ if p_1 is an element of C_0 and C_1 is a subset of C_0.

From our point of view, the Cohen condition p gives a countable initial segment of a construction of generic sets and the set C imposes a constraint on the possible continuations of this construction. Typically, we will specify a condition by giving a Cohen condition and its allowed extensions.

2.2. DEFINITION. (1) A Cohen condition p *strongly forces* $\Phi(\vec{X}, x) = m$ if every element of $\vec{X}$ is named in p, the domain of $p(\vec{X})$ $(\mathrm{dom}(p(\vec{X})))$ contains $\vec{X} \upharpoonright \varphi(\vec{X}, x)$, and the computation yields output m. Write $p \Vdash^* \Phi(\vec{X}, x) = m$. Similarly, p strongly forces the divergence $\Phi(\vec{X}, x)\uparrow$ if $\vec{X}$ are all named in p, $\vec{X} \upharpoonright \varphi(\vec{X}, x) \subseteq \mathrm{dom}(p(\vec{X}))$ and $\Phi(p(\vec{X}), x)\uparrow$.

We can determine whether $\Phi(\vec{X}, x)\downarrow$ by knowing the values of $\vec{X}$ on the countable set $\vec{X} \upharpoonright \varphi(\vec{X}, x)$. Strongly forcing the inequality $\Phi(\vec{X}, x) \neq \Psi(\vec{X}, x)$ is defined in the same way including the case that one of $\Phi(\vec{X}, x)$ or $\Psi(\vec{X}, x)$ diverges.

(2) Suppose $\mathcal{P}$ is a condition and $p \in \mathcal{P}$. *p forces* $\Phi(\vec{X}, x) = m$ in $\mathcal{P}$ if

$$(\forall q \leq_{\mathcal{P}} p)(\exists r \leq_{\mathcal{P}} q)(\exists y)\left[r \Vdash^{*} \Phi(\vec{X}, x) = y\right] \;\&$$
$$(\forall q \leq_{\mathcal{P}} p)\left[p \Vdash^{*} \Phi(\vec{X}, x) = y \implies y = m\right].$$

Write $p \Vdash_{\mathcal{P}} \Phi(\vec{X}, x) = m$.

(3) *p decides* $\Phi(\vec{X}, x)$ in $\mathcal{P}$ if either there is an m such that $p \Vdash_{\mathcal{P}} \Phi(\vec{X}, x) = m$ or there is no extension of p that forces a value for $\Phi(\vec{X}, x)$. Similarly, *p strongly decides* $\Phi(X, x)$ if p strongly forces some value for $\Phi(\vec{X}, x)$ or strongly forces divergence.

Note that the definition of strong forcing is independent of $\mathcal{P}$. Therefore, if p strongly forces $\Phi(\vec{X}, x) = i$, then p forces $\Phi(\vec{X}, x) = i$ in every $\mathcal{P}$ such that $p \in \mathcal{P}$.

Notation. Given a countable collection of type 2 objects $\{A_i \mid i \in I\}$, $\bigoplus_{i \in I} A_i$ denotes the recursive join of its elements.

Let A be a type 2 object and let x be a real. We use $A^{(x)}$ to denote the set $\{y \mid \langle x, y\rangle \in A\}$.

The operation $*$ will be used for various types of concatenation operations. If $p(X)$ is a condition on X, $p(X) * 0$ is the predicate which takes the value 0 at every point not in the domain of $p(X)$. If X is named in p, let $p * \{X(z) = m\}$ denote the condition which extends p and assigns the value m to X at z. Given an extension $q(X)$ of $p(X)$, $p * q(X)$ denotes the extension of p such that $(p * q(X))(Y) = p(Y)$ for every Y named in p other than X and $(p * q(X))(X) = q(X)$.

§3. Exact Pairs

Let $\mathcal{I}$ be an ideal in $\mathcal{K}$. $\mathcal{I}$ is *countably closed* if it is closed under countable join: for every countable subset $\mathcal{A}$ of $\mathcal{I}$, $\deg_{\mathcal{K}}(\bigoplus \mathcal{A})$ is an element of $\mathcal{I}$. In this paper, we will only consider countably closed ideals. So, when we refer to an ideal we will mean one that is countably closed.

3.1. Definition. (1) For $\mathcal{G}$ a subset of $\mathcal{K}$, $(\mathcal{G})$ denotes the ideal generated by $\mathcal{G}$.

(2) For A in $\mathcal{K}$, let (A) denote $\{X \mid X \leq_{\mathcal{K}} A\}$. Clearly, (A) is the ideal generated by $\{A\}$.

(3) Let $\mathcal{I}$ be an ideal in $\mathcal{K}$. An *exact pair* over $\mathcal{I}$ is a pair X, Y of elements of $\mathcal{K}$ such that

$$(X) \cap (Y) = \mathcal{I}.$$

One of the fundamental theorems about the Turing degrees is the Kleene-Post theorem [3] that the Turing degrees do not form a lattice. In [8], Spector gave an abstract version of the Kleene-Post argument to show that there is an exact pair over every countable ideal in the Turing degrees. Assuming the continuum hypothesis, we observe that Spector's argument applies to the Kleene degrees and that every ideal in $\mathcal{K}$ of cardinality at most $\aleph_1$ has an exact pair.

3.2. THEOREM. *(CH) For every ideal $\mathcal{I}$ in $\mathcal{K}$ of cardinality at most $\aleph_1$, there is an exact pair K_0 and K_1 over $\mathcal{I}$.*

PROOF: We will identify the reals with the ordinals less than $\aleph_1$. Therefore, K_0 and K_1 will be subsets of $\aleph_1$. Let $\langle A_\xi \mid \xi < \aleph_1 \rangle$ be an enumeration of the elements of $\mathcal{I}$, and $\langle \Phi_\alpha \mid \alpha < \aleph_1 \rangle$ be an enumeration of all $\{0,1\}$-valued Kleene functionals. Given two sets X, Y of reals, let $X =^* Y$ represent that $X(\xi) = Y(\xi)$ for all but countable many reals ξ. If $X =^* Y$, then we have $X \equiv_K Y$. We will construct K_0 and K_1 to satisfy the requirements

$$(T^0_\xi) \qquad K_0^{(\xi)} =^* A_\xi,$$

$$(T^1_\xi) \qquad K_1^{(\xi)} =^* A_\xi,$$

and

$$(R_{\alpha,\beta}) \qquad \Phi_\alpha(K_0) = \Phi_\beta(K_1) = Z \implies (\exists \xi)[Z \leq_K A_\xi].$$

It is clear that $\mathcal{I} \subseteq (K_0) \cap (K_1)$ follows from (T^0_ξ) and (T^1_ξ), and that $(K_0) \cap (K_1) \subseteq \mathcal{I}$ follows from $(R_{\alpha,\beta})$.

Let P be the set of Cohen conditions $\langle p_0, p_1 \rangle$ such that for $i \in \{0,1\}$, p_i is a Cohen condition on K_i. We work with conditions $\langle p, C \rangle$ where $p \in P$ and $C \subseteq P$. For each p in P and $\xi < \aleph_1$, let $\langle p, C_\xi \rangle$ be the condition defined by letting C_ξ be the set of extensions q of p such that for i in $\{0,1\}$ and $\eta < \xi$,

$$(3.3) \qquad (\forall x)\,[\langle \eta, x \rangle \in \mathrm{dom}(q_i) - \mathrm{dom}(p_i) \implies q_i(\langle \eta, x \rangle) = A_\eta(x)].$$

Let $\mathcal{P}$ be the set of conditions $\langle p, C_\xi \rangle$ as above. It is easy to see that for any p, and ξ and η less than $\aleph_1$, there exists a q in C_ξ such that $dom(q_0) = dom(q_1) \supseteq \eta$. Thus, any generic filter on $\mathcal{P}$ gives a pair of subsets K_0 and K_1 of $\aleph_1$. Further, because $\aleph_1$ is regular, $\mathcal{P}$ is countably closed. We simultaneously show that $\mathcal{P}$ forces that K_0 and K_1 form an exact pair over $\mathcal{A}$ and build a pair that is sufficiently generic for forcing to be the same as truth with regard to this statement. The countable closure of $\mathcal{P}$ allows us to proceed by a transfinite recursion of length $\aleph_1$.

Let $\xi = \langle \alpha, \beta \rangle$. Suppose $p^\eta = \langle \langle p_0^\eta, p_1^\eta \rangle, C_\eta \rangle$ $(\eta < \xi)$ are given such that if $\eta_1 < \eta_2$ then p^{η_1} is extended by p^{η_2}. Let $p_0^{<\xi} = \bigcup_{\eta<\xi} p_0^\eta$, let $p_1^{<\xi} = \bigcup_{\eta<\xi} p_1^\eta$, and let $p^{<\xi} =$

$\langle\langle p_0^{<\xi}, p_1^{<\xi}\rangle, C_\xi\rangle$. Then $p^{<\xi}$ is an element of P and extends all p^η ($\eta < \xi$). Define the condition $p^\xi = \langle\langle p_0^\xi, p_1^\xi\rangle, C_\xi\rangle$ to be a condition with the following features.

(1) $\mathrm{dom}(p_i^\xi) \geq \xi$ ($i = 0, 1$);

(2) $p^\xi \leq_{P_\xi} p^{<\xi}$;

(3) if there exist a real x and a Cohen condition $q = \langle q_0, q_1\rangle$ such that $q \leq_{P_\xi} p^{<\xi}$ and q strongly forces the inequality $\Phi_\alpha(K_0, x) \neq \Phi_\beta(K_1, x)$, then p^ξ is such a q.

For each i in $\{0,1\}$, let K_i be the union of all the p_i^ξ ($\xi < \aleph_1$). By (1), K_0 and K_1 are defined on all countable ordinals (*i.e.*, on all reals). By (2), for each ξ, K_0 and K_1 satisfy the requirements (T_ξ^0) and (T_ξ^1) respectively.

To show that K_0 and K_1 satisfy the requirement $(R_{\alpha,\beta})$, suppose $\Phi_\alpha(K_0) = \Phi_\beta(K_1) = Z$. Let $\xi = \langle\alpha, \beta\rangle$. There are two cases.

(i) There exist a real x and two extensions q_0 and q_0' of $p_0^{<\xi}$ which satisfy (3.3) with p_0 equal to $p_0^{<\xi}$ such that q_0 and q_0' strongly decide different values for $\Phi_\alpha(K_0, x)$.

(ii) Otherwise.

In the first case, take any extension q_1 of $p_1^{<\xi}$ which satisfies (3.3) relative to $p_1^{<\xi}$ and strongly decides $\Phi_\beta(K_1, x)$. One of q_0 or q_0' gives a value for $\Phi_\alpha(K_0, x)$ which is different from the value $\Phi_\beta(K_1, x)$ forced by q_1. Say that q_0 does so. Then $\langle q_0, q_1\rangle <_\xi p^{<\xi}$. By (3), p^ξ would be chosen so that

$$p^\xi \Vdash^* \Phi_\alpha(K_0, x) \neq \Phi_\beta(K_1, x).$$

Therefore, we would have $\Phi_\alpha(K_0) \neq \Phi_\beta(K_1)$, which is a contradiction.

For the second case, suppose that for every real x and every two extensions q_0 and q_0' of $p_0^{<\xi}$ satisfying (3.3) with p_0 equal to $p_0^{<\xi}$ and strongly deciding $\Phi_\alpha(K_0, x)$,

$$\Phi_\alpha(q_0, x) = \Phi_\alpha(q_0', x).$$

Let A be a type 2 object which extends $p_0^{<\xi}$ by coding $\langle A_\eta \mid \eta < \xi\rangle$ on the first ξ columns and being 0 elsewhere. Then A is Kleene reducible to $\bigoplus_{\eta<\xi} A_\eta$. By the failure of case 1 for every x,

$$\Phi_\alpha(K_0, x) = \Phi_\alpha(A, x).$$

Thus, Z is Kleene reducible to $\bigoplus_{\eta<\xi} A_\eta$. ◊

§4. Interpretations of Third Order Arithmetic

We begin by fixing a partially ordered set $P_{\mathbf{R}}$ with ordering $\leq_{\mathbf{R}}$, which codes the standard model of second order arithmetic. We present $P_{\mathbf{R}}$ in terms of generators and relations.

(i) The minimal elements of $P_{\mathbf{R}}$ form a set $\{a_i \mid i \in \omega\}$ which represents the set of nonnegative integers.

(ii) For each i and j, there is a unique pair $c_+(i,j)$ and $d_+(i,j)$ in $P_{\mathbf{R}}$ such that $a_i, a_j <_{\mathbf{R}} c_+(i,j)$ and $a_{i+j}, c_+(i,j) <_{\mathbf{R}} d_+(i,j)$.

(iii) For each i and j, there is a unique triple $c_\times(i,j)$, $d_\times(i,j)$ and $e_\times(i,j)$ in $P_{\mathbf{R}}$ such that $a_i, a_j <_{\mathbf{R}} c_\times(i,j)$; $a_{i\times j}, c_\times(i,j) <_{\mathbf{R}} d_\times(i,j)$ and $d_\times(i,j) <_{\mathbf{R}} e_\times(i,j)$.

(iv) For each subset X of ω, there is a unique element b_X in $P_{\mathbf{R}}$ such that for each integer i, $a_i <_{\mathbf{R}} b_X$ if and only if $i \in X$.

The only elements of $P_{\mathbf{R}}$ are the ones mentioned above. The only instances of $\leq_{\mathbf{R}}$ comparability are those needed to obtain a transitive relation satisfying the above clauses.

4.1. Definition. Given a partial ordered set $\mathcal{P} = \langle P, \preceq \rangle$, we define first order relations $Nat(a)$, $x + y = z$, $x \times y = z$, $Real(b)$ and $a \,\varepsilon\, b$ on $\mathcal{P}$ as follows.

$$Nat(a) \iff (\forall w)\neg[w \prec a];$$

$$x + y = z \iff Nat(x) \;\&\; Nat(y) \;\&\; Nat(z) \;\&\; (\exists c, d)\begin{bmatrix}(x, y \prec c) \;\&\; (z, c \prec d) \\ \&\; \neg(\exists w)(d \prec w)\end{bmatrix};$$

$$x \times y = z \iff Nat(x) \;\&\; Nat(y) \;\&\; Nat(z) \;\&\; (\exists c, d, e)\begin{bmatrix}(x, y \prec c) \;\&\; (z, c \prec d) \\ \&\; (d \prec e) \;\&\; \neg(\exists w)(e \prec w)\end{bmatrix};$$

$$Real(b) \iff \neg(\exists v)[b \prec v] \;\&\; (\forall w)[w \prec b \implies Nat(w)];$$
$$a \,\varepsilon\, b \iff Nat(a) \;\&\; Real(b) \;\&\; (a \prec b).$$

Note that in the case of $\mathcal{P} = P_{\mathbf{R}}$, the above interpretation of second order arithmetic gives a standard model.

4.2. Definition. Let $\mathcal{G}$ be a set of type 2 objects. $\mathcal{G}$ is *strongly independent* if for every X in $\mathcal{G}$ there is a type 2 object Y such that

$$(\forall Z \in \mathcal{G})[Z \neq X \iff Z \leq_K Y].$$

Note that strong independence is a first order property in $\mathcal{K}$ relative to $\mathcal{G}$.

4.3. Theorem. *(CH) Let W be the fixed type 2 object specified in §2. There exist type 2 objects A, B, C, U, L and a set $\mathcal{G} = \{G_\alpha \mid \alpha < \aleph_1\}$ of type 2 objects with the following properties.*

(1) *For each Y in $(\mathcal{G})$, either*
  (a) $(A) \cap (B \oplus Y \oplus U \oplus W) \not\subseteq (C \oplus Y \oplus U \oplus W)$ *or*
  (b) *there is a G_α such that $G_\alpha \leq_K C \oplus Y \oplus U \oplus W$.*

(2) *For each G_α, $(A) \cap (B \oplus G_\alpha \oplus U \oplus W) \subseteq (C \oplus G_\alpha \oplus U \oplus W)$.*

(3) *The set $\{C \oplus G_\alpha \oplus U \oplus W \mid \alpha < \aleph_1\}$ is strongly independent.*

(4) *For all α and β in $P_{\mathbf{R}}$, $G_\alpha \leq_K C \oplus G_\beta \oplus U \oplus W \oplus L$ if and only if $\alpha \leq_{\mathbf{R}} \beta$.*

We will present the proof of Theorem 4.3 in sections §5 to §7.

4.4. THEOREM. *(CH) There is an interpretation of the third order theory of arithmetic in the first order theory of $\mathcal{K}$. That is, there is a recursive function mapping $\varphi \mapsto \varphi^*$ such that for all third order sentences φ,*

$$\langle \omega, P(\omega), P(P(\omega)), \in, +, \times \rangle \models \varphi \iff \langle \mathcal{K}, \leq_K \rangle \models \varphi^*.$$

PROOF: Let A, B, C, U, L, W and $\mathcal{G}$ be given as in Theorem 4.3. By Theorem 3.2, there is an exact pair K_0 and K_1 over $(\mathcal{G})$. We will give an interpretation of the third order arithmetic in $\mathcal{K}$ which is defined with parameters A, B, C, U, L, W, K_0 and K_1. First we give an interpretation of $P_{\mathbf{R}}$ using these parameters. Then, we define the predicates Nat, $Real$, ε and the functions $+$, $\times$ as in 4.1. This gives a standard model of second order arithmetic. Following Nerode-Shore [5], the third order variables on the set of natural numbers are interpreted by the first order ones on $\mathcal{K}$ using exact pairs. This interpretation is described by a first order property of the parameters A, B, C, U, L, W, K_0 and K_1.

We begin by describing a specific coding of $P_{\mathbf{R}}$. In view of (1)–(3) of 4.3, the set $\mathcal{G} \oplus C \oplus U \oplus W = \{C \oplus G \oplus U \oplus W \mid G \in \mathcal{G}\}$ is defined by

(P1) $X \in \mathcal{G} \oplus C \oplus U \oplus W \iff$

$$X \text{ is minimal in } \left\{ C \oplus Z \oplus U \oplus W \;\middle|\; \begin{array}{c} Z \in (K_0) \cap (K_1) \ \& \\ (A) \cap (B \oplus Z \oplus U \oplus W) \subseteq (C \oplus Z \oplus U \oplus W) \end{array} \right\}.$$

Then, $\mathcal{G} \oplus C \oplus U \oplus W$ is definable in $\mathcal{K}$ from the parameters A, B, C, U, W and the exact pair K_0 and K_1. Now we define a partial ordering $\preceq$ on $\mathcal{G} \oplus C \oplus U \oplus W$ by

(P2) $C \oplus G_\alpha \oplus U \oplus W \preceq C \oplus G_\beta \oplus U \oplus W \iff C \oplus G_\alpha{}' | U \oplus W \leq_K C \oplus G_\beta \oplus U \oplus W \oplus L.$

By 4.3 (4), $\preceq$ is isomorphic to $\leq_{\mathbf{R}}$. Hence, if we define Nat, $+$, $\times$, $Real$, and ε on $\mathcal{G} \oplus C \oplus U \oplus W$ by the equations in 4.1, then we have specified a standard model of second order arithmetic.

Our next step is to quantify away the dependence on our particular sequence of parameters.

Let P^- be a finite set of first order sentences of arithmetic such that if M is a model of P^- and every countable subset of M has a least element, then M is a standard model of first order arithmetic. Since $\langle Nat, +, \times \rangle$ defined above is standard, we have

(P3) $$\langle Nat, +, \times \rangle \models P^-,$$

and

(P4) $$(\forall X)[(X) \cap Nat \neq \emptyset \implies (X) \cap Nat \text{ has a } \preceq\text{-least element}].$$

By clause (iv) in the definition of $P_{\mathbf{R}}$, every real has a corresponding element in *Real*. Therefore, for every coding of $P_{\mathbf{R}}$ in $\mathcal{K}$, $\mathcal{K}$ is a model of

$$\text{(P5)} \qquad (\forall X)(\exists Y)[Real(Y) \ \& \ (\forall Z \in Nat)(Z \leq_K X \iff Z \varepsilon Y)].$$

These first order properties characterize the codings of $P_{\mathbf{R}}$ in $\mathcal{K}$. In checking that a partial order P is isomorphic to $P_{\mathbf{R}}$, the only condition that is not first order is that every countable subset of Nat_P have a code in $P_{\mathbf{R}}$. This is expressed in $\mathcal{K}$ by (P5). Given a coding of P as above, its integers Nat_P form a strongly independent set. By the existence of countable least upper bounds in $\mathcal{K}$, every countable subset B of Nat_P has a least upper bound b in $\mathcal{K}$. By (P5), the existence of b ensures the existence of an element in the reals of P whose elements are exactly those in B.

Let Φ be a first order sentence in the parameters A, B, C, U, W, L, K_0 and K_1 expressing the conjunction of 4.3.(3), (P1)–(P4), and the equations of Definition 4.1. If A, B, C, U, W, L, K_0 and K_1 satisfy Φ, then the structure $\langle Nat, Real, \varepsilon, +, \times\rangle$ defined as above is a standard model of second order arithmetic.

We interpret the third order variables over Nat by first order ones ranging over $\mathcal{K}$ as follows. Suppose $\mathcal{Z}$ is a subset of *Real*. Using Theorem 3.2, we code $(\mathcal{Z})$ by a pair of type 2 objects Z_0, Z_1, an exact pair over $(\mathcal{Z})$. Then, we have

$$X \in \mathcal{Z} \iff Real(X) \ \& \ X \in (Z_0) \cap (Z_1).$$

The implication from right to left follows from the strong independence of $\mathcal{G} \oplus C \oplus U \oplus W$. Therefore, we can interpret a second order variable $\mathcal{X}$ over *Real* by and exact pair Z_0 and Z_1, with $X \in \mathcal{X}$ interpreted by $X \in (Z_0) \cap (Z_1)$.

Thus, we have given a definable collection of codes for the standard model of second order arithmetic together with a method to interpret second order quantifiers over the reals in the coded models. This is enough to conclude that a formula Φ is true in third order arithmetic if and only if there is a coding of second order arithmetic in which its interpretation holds. By Theorem 4.3, there is at least one coding of second order arithmetic. This gives the required interpretation of third order arithmetic in the first order theory of $\mathcal{K}$. $\diamond$

§5. Requirements

We are required to construct sets of reals A, B, C, U, L and a sequence of sets of reals $\mathcal{G} = \{G_\alpha \mid \alpha < \aleph_1\}$ which meet the conditions (1)–(4) of Theorem 4.3. Correspondingly, we divide the requirements into four parameterized families corresponding to the conditions (1)–(4).

Type I requirements. Suppose Φ is a Kleene functional and $\vec{G}$ is a countable sequence of elements of $\mathcal{G}$. Let Y denote $\Phi(\vec{G})$.

I(Φ, $\vec{G}$, global): Suppose this requirement is given index e. We require that there is a real a such that

$$A^{(e)}(x) = \Delta(B \oplus Y \oplus U \oplus W, x) \quad \text{for all } x \geq a,$$

where Δ is the $\{0,1\}$-valued Kleene functional computed as follows. $\Delta(B \oplus Y \oplus U \oplus W, x) = 1$ if and only if $x \in U$ and there are pos and neg, reals below x in W which code two disjoint countable sets, such that $pos \subseteq Y$, $neg \cap Y = \emptyset$ and $\langle x, 1, pos, neg\rangle \in B^{(e)}$.

Note, if this requirement is satisfied then $A^{(e)} =^* \Delta(B \oplus Y \oplus U \oplus W)$. That is, they agree except on a countable set.

I(Φ, $\vec{G}$, Ψ): Let e be the index of I(Φ, $\vec{G}$, global) given above. For each Ψ, we require one of the following. Either,

$$A^{(e)} \neq \Psi(C \oplus Y \oplus U \oplus W)$$

or for some G_α in $\vec{G}$,

$$G_\alpha \leq_K C \oplus Y \oplus U \oplus W.$$

It is easy to see that if all requirements of type **I(Φ, $\vec{G}$)** are satisfied then the condition (1) of Theorem 4.3 is satisfied.

Type II requirements.

II(Θ, Ω, α): Given Θ, Ω and α,

$$\Theta(A) = \Omega(B \oplus G_\alpha \oplus U \oplus W) = Z \implies Z \in (C \oplus G_\alpha \oplus U \oplus W).$$

Type III requirements.

III(Θ, α): Given Θ and α, we require

$$\Theta\Big(\bigoplus_{\xi \neq \alpha} C \oplus G_\xi \oplus U \oplus W\Big) \neq G_\alpha.$$

Type IV requirements.

IV(α, β, coding): Suppose e is the index of this requirement. If $\alpha \leq_{\mathbf{R}} \beta$, then we require that there is an b such that

$$G_\alpha(x) = \Lambda(C \oplus G_\beta \oplus U \oplus W \oplus L, \langle e, x\rangle) \quad \text{for all } x \geq b,$$

where Λ is the Kleene functional given by,

$$\Lambda(C \oplus G_\beta \oplus U \oplus W \oplus L, \langle e, x\rangle) = \begin{cases} L(\langle e, x\rangle), & \text{if } L(\langle e, x, 1\rangle) = 0 \\ G_\beta(x), & \text{otherwise.} \end{cases}$$

IV(Θ, α, β): Given Θ, if $\alpha \not\leq_{\mathbf{R}} \beta$, then

$$G_\alpha \neq \Theta(C \oplus G_\beta \oplus U \oplus W \oplus L).$$

If A, B, C, U, L and G_α $(\alpha < \aleph_1)$ are constructed to meet all of the above requirements then these sets satisfy the conditions of Theorem 4.3. We enumerate the requirements and attach a real number index to each requirement so that if e is the index of a requirement, say I(Φ, $\vec{G}$, global), then the index of Φ is less than e and all predicates in $\vec{G}$ are contained in $\{G_\alpha \mid \alpha < e\}$. We assume that the assignment of indices is made in the same way for other requirements.

The requirements of the form I(Φ, $\vec{G}$, global), II(Θ, Ω, α) or IV(α, β, coding) are called global requirements. For these requirements, we will globally impose constraints on the set of Cohen conditions. The remaining requirements are local in the sense that they are decided by simple Cohen conditions.

§6. Global Constraints

In this section, we specify the conditions used to force the global §5-requirements. We consider conditions on the predicates A, B, C, U, L and G_α $(\alpha < \aleph_1)$. As before, we will identify the reals with the countable ordinals.

Basic constraints. The first constraints are not associated with any requirement posed in §5, but rather establish the basic properties of the generic parameters. We define a preliminary collection of conditions $\mathcal{P}_0$. An element of $\mathcal{P}_0$ is determined by a pair $\langle p, |p|\rangle$: p is a Cohen condition and $|p|$ is a countable ordinal. We use $|p|$ as a parameter in the description of the allowed extensions of p.

6.1. Definition. Let $p \oplus W$ denote $\bigoplus\{p(X) * 0 \mid X \text{ is named in } p\} \oplus W$. $\mathcal{P}_0$ is the set of all conditions $\langle p, C\rangle$ defined by a pair $\langle p, |p|\rangle$ such that $|p|$ an countable ordinal, p is a Cohen condition on the predicates A, B, C, U, L the predicates G_α $(\alpha < \aleph_1)$, satisfying the following clauses.

(1) The predicates A, B, C, U and L are named in p. A countable ordinal is assigned to p, which we denote $|p|$. G_α is named in p if and only if $\alpha < |p|$. If X is named in p, then $\mathrm{dom}(p(X))$ is an initial segment of ω^ω, as ordered by $<$. All $p(X)$ other than $p(U)$ and $p(C)$ have the same domain, which we call $d(p)$. If $x \in d(p)$, then $\mathrm{cl}(x, p \oplus W) \subseteq d(p)$. Note, $p \oplus W$ is used as a predicate in evaluating this closure.

(2) *Restrictions on* A. All of the elements of $p(A)$ are of the form $\langle e, x\rangle$ for some $e < |p|$ and $x \in p(U)$.

(3) *Restrictions on* B. All of the elements of $p(B)$ are of the form $\langle e, \langle x, i, pos, neg\rangle\rangle$ where $x \in p(U)$, $i \in \{0, 1\}$, and pos and neg are reals lying below x in W and coding disjoint countable sets of reals.

(4) ***Restrictions on C.*** C is divided into two distinct parts, C_0 and C_1. All of the elements of $p(C_0)$ are of the form $\langle e, x, 1\rangle$ or $\langle e, x, 2, i\rangle$ where $e < x$, $x \in p(U)$ and $i \in \{0,1\}$. These elements are called *flags* for $\langle e, x\rangle$ recording whether $\langle e, x\rangle$ belongs to A in one of two ways. Further, for each x in $p(U)$ there can be at most one e such that there is a flag for $\langle e, x\rangle$ in $p(C)$. The elements of C_1 record whether $\langle e, x\rangle$ belongs to A in a way that is accessible to x's successor in U. That is $p(C_1)$ is defined on all points in $d(p)$ and also on all pairs $\langle x, y\rangle$ such that x is in $p(U)$ and y is less than x and in $d(p)$. For z in the domain of $p(C_1)$, z is in $p(C_1)$ if and only if z is of the form $\langle x, y\rangle$ with x in $p(U)$, y less than x and y in $p(A)$. We will impose conditions on U so that the information about whether y belongs to A is not embedded in the restriction of C_1 to the C_1-admissible closure of y.

(5) ***Restrictions on G_α.*** If $\alpha < |p|$, then $p(G_\alpha)$ is a subset of $p(U)$. For q to be an allowed Cohen extension of p, if α is not named in p but is named in q then then $q(G_\alpha)(x) = 0$ for all $x \in d(p)$.

(6) ***Restrictions on U.*** U is a set describing the potential elements of the other predicates. We require that $p(U) - d(p)$ be a singleton. If x is an element of $p(U)$ and z is less than x in W, then x is strictly larger than the $p \oplus W$-admissible closure of z together with the real coding the sequence $\langle y \in p(U) \mid y < x\rangle$, i.e.,

$$\mathrm{cl}(z, \langle y \in p(U) \mid y < x\rangle, p \oplus W) < x.$$

(Roughly speaking, the elements of $p(U)$ are very thinly distributed relative to the initial segments of the predicates that have been fixed by p and are strong closure points in W relative to these predicates.)

We have already described the way that conditions are ordered. $\langle p, C\rangle$ is extended by $\langle p_1, C_1\rangle$ if $p_1 \in C$ and $C_1 \subseteq C$. That is to say that p_1 is an allowed extension of p, in the sense of C and that the constraints imposed in C_1 include those of C.

Note that for each $\alpha < \aleph_1$, the set $\{p \in \mathcal{P}_0 \mid \alpha < |p| \ \& \ \alpha < d(p)\}$ is dense in $\mathcal{P}_0$. Also note that if $p \in \mathcal{P}_0$ and $\lambda < |p|$ then $\langle p(A), p(B), p(C), p(U), p(L), p(G_\alpha)\rangle_{\alpha<\lambda}$ is an element of $\mathcal{P}_0$.

Type I constraints. We allow further constraints for the requirements of the form I(Φ, $\vec{G}$, global). The eth of these ensures that the same set is recorded in $A^{(e)}$ as is recorded in $B \oplus \Phi(\vec{G}) \oplus U \oplus W$.

6.2. DEFINITION. An element of $\mathcal{P}_1$ is described by a Cohen condition p, a countable ordinal $|p|$ and a countable sequence of reals $\langle a_\alpha \mid \alpha < |p|\rangle$. In addition to the clauses specifying that $\langle p, |p|\rangle$ determines an element of $\mathcal{P}_0$, we require if I(Φ, $\vec{G}$, global) has index e less than $|p|$, then for all $x \geq a_e$, the following hold.

(1) Let Y denote $\Phi(\vec{G})$. If $\langle x, i, pos, neg\rangle \in p(B^{(e)})$, then p strongly decides whether $pos \subseteq Y$ and $neg \cap Y = \emptyset$. Further, if $\langle x, i, pos, neg\rangle \in p(B^{(e)})$ and p strongly forces $pos \subseteq Y \ \& \ neg \cap Y = \emptyset$, then $p(A^{(e)})(x) = i$.

(2) If $p(A^{(e)})(x) = 1$, then there is a quadruple $\langle x, 1, pos, neg\rangle$ in $p(B^{(e)})$ such that p strongly forces that $pos \subseteq Y$ and $neg \subseteq \bar{Y}$.

The first condition ensures, for almost all x, if a value is recorded for x in $B^{(e)} \oplus Y \oplus U \oplus W$, then the same value appears in $A^{(e)}$. The second condition ensures for almost all x, if no value is recorded in $B^{(e)} \oplus Y \oplus U \oplus W$, then $A^{(e)}$ takes the default value 0 at x. If the predicates A, B and G_α's are obtained as a limit of conditions in $\mathcal{P}_1$, then they satisfy the requirements of the form I(Φ, $\vec{G}$, global). Notice, the calculation relative to $B \oplus Y \oplus U \oplus W$ makes essential use of both B and Y. Thus, there remains a possibility to record information in their join so that it is not available to either constituent.

It is helpful to notice one of the consequences of the restriction on U imposed in $\mathcal{P}_0$. If e is less than $|p|$, then for p and q associated with conditions in $\mathcal{P}_1$, in order for q to be stronger than p, for each x in $d(q) - d(p)$ and in $q(U)$, q must force $cl(\langle y \in q(U) \mid y < x\rangle, C \oplus Y \oplus U \oplus W)$ to be contained in the reals below x. Thus, if x is an element of $q(U)$, then the value determined by q for $\Psi(C \oplus Y \oplus U \oplus W)$ on elements of $q(U)$ below x in W do not depend on the values of $q(G_\alpha)$ at x.

Type II constraints. Suppose II(Θ, Ω, α) is given. Given a condition p, we can extend p to force

$$\Theta(A) = \Omega(B \oplus G_\alpha \oplus U \oplus W) = Z \implies Z \in (C \oplus G_\alpha \oplus U \oplus W).$$

in one of two ways. The easier one is to find a z and an extension q of p such that q strongly forces the inequality $\Theta(A, z) \neq \Omega(B \oplus G_\alpha \oplus U \oplus W, z)$. If the easy method is not available, we must find a Kleene functional Σ and establish the equality $\Theta(A) = \Sigma(C \oplus G_\alpha \oplus U \oplus W)$. We allow conditions with the following refined constraints with the intention of forcing the disjunction of these two possibilities.

6.3. Definition. An element of $\mathcal{P}_2$ is defined from p, $|p|$, $\vec{a}$ as in $\mathcal{P}_1$, together with a sequence of reals $\vec{b} = \langle b_e \mid e < |p|\rangle$. In addition to the constraints mentioned in $\mathcal{P}_1$, if II(Θ, Ω, α) has index e less than $|p|$ then one of the following conditions must hold.

The first of these is that there is a $z \in d(p)$ and such that,

$$(6.4) \qquad p \Vdash^* \Theta(A, z) \neq \Omega(B \oplus G_\alpha \oplus U \oplus W, z).$$

The second of these is that there is an e_0 less than e such that the requirement II(Θ_{e_0}, Ω_{e_0}, α_{e_0}) is satisfied by achieving Equation 6.4 for the first time. Let x_1 be the greatest element of $p(U)$. To qualify under the second option, there must be a z in $d(p)$ which is greater than all of the elements of $p(U)$ other than x_1 such that

$$(6.5) \qquad p \Vdash^* \Theta_{e_0}(A, z) \neq \Omega_{e_0}(B \oplus G_{\alpha_{e_0}} \oplus U \oplus W, z).$$

The third qualifying option will allow the computation of $\Theta(A,z)$ to be simulated with a possibility of error in execution but obtaining the correct answer. Let b be the least element of $p(U)$ that is greater than b_e and for all e_0 less than e, greater than the least witness to Equation 6.5 if there is a witness in $d(p)$. If neither of the first two cases apply to p, we require, for all z greater than b and less than the greatest element x_1 of $p(U)$ and all x in $p(U)$ less than x_1, the following properties (1)–(3) hold.

(1) If $\langle e,x,1\rangle \in p(C)$, then for all $\xi < |p|$,

$$\langle e,x\rangle \in p(A) \iff x \in p(G_\xi).$$

(2) If $\langle e,x,2,i\rangle \in p(C)$, then

$$p(A)(\langle e,x\rangle) = i.$$

(3) Let x' be the successor of x in $p(U)$. If there is no flag in $p(C)$ for $\langle e,x\rangle$ and $\langle e,x\rangle \in p(A)$ then the values of $\Theta(A)$ strongly forced by p do not depend on the value of A at $\langle e,x\rangle$. By this we mean, for all z in the interval $x \le z < x'$, $p(A)\restriction x*0$ and $p(A)$ strongly force the same value for $\Theta(A,z)$. Here $p(A)\restriction x*0$ agrees with $p(A)$ up to but not including the argument $\langle e,x\rangle$ and is identically 0 at all arguments beyond that point. Note, by the sparseness of $p(U)$, the evaluation of $\Theta(A,z)$ does not depend on the value of A at arguments greater than or equal to x'.

Consider the effect of this condition on the generic sets. If the first clause obtains then p strongly forces $\Theta(A)$ to be unequal to $\Omega(B\oplus G_\alpha \oplus U \oplus W)$ and thereby satisfies the associated type II requirement. The second clause can apply to an e_0 below e at most once. By the regularity of $\aleph_1$, its effect is bounded by some b. Work above b and assume that the first clause does not apply.

The clauses (1)–(3) apply to the generic parameters for almost every z. Then, the evaluation of $\Theta(A)$ at an argument y above b can be simulated using C, G_α and U as follows. Let max be the supremum of the elements of U less than y. A query z to A below b is returned answer $A(z)$ using a real parameter coding $A\restriction b$. Otherwise, a query z to A is returned answer 0 if z is not of the form $\langle e^*,x\rangle$, for some e^* and $x\in U$. A query $z=\langle e^*,x\rangle$ of the appropriate form to A with x less than max is returned the value of $A(z)$ by reading off the correct answer from C_1, at $\langle x',z\rangle$ for x' the successor of x in U. Otherwise, we look for a flag in C. If either of the flags described in cases (1) and (2) appears in C, a query z to A is returned value $A(z)$ by appealing to C and G_α as determined by the flag for z in C. In case (3), where there is no flag in C, the query is returned an answer of 0. By the condition imposed in (3), this simulation will give the same answer as Θ on A.

Notation. We refer to an element of $\mathcal{P}_2$ by its defining parameters: p, $|p|$, $\vec{a}$ and $\vec{b}$. When the latter constituents are clearly determined, we will merely write p.

Type IV constraints. We include one further type of constraint in the description of a condition to force a global requirement of type IV.

6.6. DEFINITION. A condition in $\mathcal{P}$ is determined from a Cohen condition p, a countable ordinal $|p|$, a pair of countable sequences of reals $\vec{a}$ and $\vec{b}$ as in $\mathcal{P}_2$ together with a third sequence of reals $\langle c_e \mid e < |p| \rangle$. The additional constraint determined by the sequence $\vec{c}$ is that if IV(α, β, coding) is a requirement with index $e < |p|$ and $\alpha, \beta < |p|$, then for all $x \geq c_e$ such that $x \in d(p)$,

$$p(G_\alpha)(x) = \begin{cases} p(L)(\langle e, x \rangle), & \text{if } p(L)(\langle e, x, 1 \rangle) = 0 \\ p(G_\beta)(x), & \text{otherwise.} \end{cases}$$

In the rest of this paper, we will work with $\mathcal{P}$, and the forcing relation always means $\Vdash_{\mathcal{P}}$. Therefore, we will delete the subscript $\mathcal{P}$ systematically.

6.7. LEMMA. (1) *$\mathcal{P}$ is countably closed.*

(2) *For every $\alpha < \aleph_1$, the set $\{p \in \mathcal{P} \mid \alpha \leq |p| \ \& \ \alpha \leq d(p)\}$ is dense in $\mathcal{P}$.*

PROOF: (1). Suppose $\langle p_n \mid n \in \omega \rangle$ is a decreasing sequence in $\mathcal{P}$. We set $p = \bigcup_n p_n$, $d(p) = \bigcup_n d(p_n)$, $|p|$ equal to the supremum of $\{|p_n| \mid n \in \omega\}$ and $\vec{a}(p)$, $\vec{b}(p)$ and $\vec{c}(p)$ equal to the union of the sequences $\vec{a}(p_n)$, $\vec{b}(p_n)$ and $\vec{c}(p_n)$, respectively. The only clause in the definition of $\mathcal{P}_2$ that does not automatically hold of p by virtue of p's being the least extension of all of the p_n, is condition (6) of 6.1. This condition requires that the elements of U form a very sparse set with exactly one element in the complement of $d(p)$.

The sparseness condition on p is purely local and so follows from the same condition on the p_n. If there is an element of $p(U)$ that is not an element of $d(p)$, then p is a condition. Otherwise, we must add a new element to $p(U)$ to satisfy condition 6.1 (6). Define an extension q of p as follows.

(1.1) $|q| = |p|$.

(1.2) $d(q) = \bigcup_{z \in d(p)} \mathrm{cl}(z, \langle y \mid y \in \bigcup_n p_n(U) \rangle, p \oplus W)$.

(1.3) For all X other than U, $q(X)$ is compatible with $p(X) * 0$.

(1.4) Extend $p(U)$ to $q(U)$ by adding a big point x_1 so that $\mathrm{cl}(\langle z \mid z \in p(U) \rangle, p \oplus W) < x_1$ to ensure that $q(U)$ contains an element not contained in $d(q)$ and that this element satisfies the sparseness constraint on U. Extend $p(C_1)$ to code $q(A)$ as required in 6.1 (4).

Then, q is a common extension of p_n's in $\mathcal{P}$.

(2). Let p be a condition and let α be a countable ordinal. For the second claim of the lemma, we must show there is a condition q stronger than p extending both the domain of p and also p's sequences of reals to have length greater than α.

First we check that we can extend $d(p)$ to include the first α many reals in W. It is safe to assume that α is very large compared with $d(p)$. Define q on A, B, C, and all of the G_ξ named in p by extending $p(A)$, $p(B)$, $p(C)$, $p(G_\xi)$ and giving the default value 0 at all new points in the extended domains. Note that this is consistent with none of these points belonging to U. We can use the default value 0 for as many points as we wish; in particular, we can ensure that the common domain includes all of the elements of $p(U)$ and all of the first α reals in W. We can also ensure that this domain is sufficiently closed to satisfy 6.1 (1), namely x in $d(q)$ and $y \in cl(q \oplus W, x)$ implies that y is in $d(q)$. Set the values for L to satisfy the constraint given in 6.6. Extend $p(U)$ by adding x_{new} the least point that is not in $d(q)$. Extend $p(C_1)$ to code $p(A)$ as required. Finally, we extend the sequences $\vec{a}$, $\vec{b}$ and $\vec{c}$ to have length α by adding the constant value x_{new} at all of the newly filled places. Since x_{new} is large compared with the elements of $d(q)$ and the constraints imposed by the requirements associated with the newly filled places only refer to reals larger than x_{new}, there is no contradiction between the values of q on $d(q)$ and the constraints imposed by these requirements. The q so constructed is the desired extension of p. ◇

§7. Density: Proof of Theorem 4.3

We will first show that each of the requirements listed in §5 is forced in $\mathcal{P}$. Invoking CH, we will then show that there are $\aleph_1$ many dense sets such that any sequence of parameters that is generic with respect to these dense sets satisfies all of the requirements. Using the countable closure of $\mathcal{P}$, we can build a generic set of parameters meeting these dense sets by means of a construction of length $\aleph_1$. Thus, we obtain a set of type 2 objects $\{A, B, C, U, L, G_\alpha\}_{\alpha<\aleph_1}$ which satisfy all requirements of §5.

Φ-Splitting. Suppose $I(\Phi, \vec{G}, \text{global})$ is a type I global requirement with index e and $e < |p|$. Let Y denote $\Phi(\vec{G})$. Suppose p is an element of $\mathcal{P}$ such that $e < d(p)$, all the elements of $\vec{G}$ are named in p, and x is the element of $p(U) - d(p)$. By the sparse distribution of the elements of any $q(U)$ and hence of the elements any $q(G_\alpha)$, the values of $Y \restriction \text{cl}(x, p \oplus W)$ forced by the extensions of p are completely determined by p and the values of $\vec{G}$ at x.

By the constraint 6.1 (4), for any condition q extending p, there can be at most one flag in $q(C_0)$ relevant to some $\langle e, x\rangle$. Further, this flag can only apply to such a pair where e is below x in W. Since there are only finitely many types of flags, the set of possible values of C_0 on $cl(x, C \oplus Y \oplus U \oplus W)$ is recursive in $p(C)$ and x. Similarly, $p(C_1)$ on $cl(x, C \oplus Y \oplus U \oplus W)$ is decided by $p(A)$ below x. Thus, $cl(x, C \oplus Y \oplus U \oplus W)$ is determined by the values of C already fixed by $p(C)$, the values of U fixed by $p(U)$, Y and W.

Let $I(\Phi, \vec{G}, \Psi)$ be a local type I requirement such that the index of Ψ is less than $|p|$ and $d(p)$. Assume that p forces $\Psi(C \oplus Y \oplus U \oplus W)$ to be total; that is, for all q stronger than p and all x, q does not strongly force $\Psi(C \oplus Y \oplus U \oplus W, x)$ to diverge. Note, for any x, the evaluation of $\Psi(C \oplus Y \oplus U \oplus W, x)$ depends only on the values of

$C \oplus Y \oplus U \oplus W$ on the elements of $cl(x, C \oplus Y \oplus U \oplus W)$. Thus, q decides a value for this computation if it decides Y on $cl(x, C \oplus Y \oplus U \oplus W)$.

7.1. DEFINITION. Fix p and x as above.

(1) Y is *nonsplitting* at x over p if p decides $Y \restriction cl(x, C \oplus Y \oplus U \oplus W)$. That is, $Y \restriction cl(x, C \oplus Y \oplus U \oplus W)$ does not depend on the values of $\vec{G}$ at x. Otherwise, we say that Y is *splitting* at x over p.

(2) Y is *G_α-splitting* at x over p if

(a) For every $q \leq p$ such that $x \in d(q)$, $p * q(G_\alpha)$ decides $Y \restriction cl(x, C \oplus Y \oplus U \oplus W)$.

(b) If q and r are two extensions of p in $\mathcal{P}$ such that $q(G_\alpha)$ and $r(G\alpha)$ are incompatible at x, then q and r force incompatible values for $Y \restriction cl(x, C \oplus Y \oplus U \oplus W)$. (Note, in this condition we need not specify the version of $cl(x, C \oplus Y \oplus U \oplus W)$ as decided by q or r. This is immediate if the two closures are equal. If the two closures are not equal, then the reason for their difference must be an element they have in common where q and r force incompatible values for Y.)

(3) Y is *amorphous* at x over p if there is an extension q of p with $x \in d(q)$ such that for any $G_\xi \in \vec{G}$, $p * q(G_\xi)$ does not decide $Y \restriction cl(x, C \oplus Y \oplus U \oplus W)$.

7.2. LEMMA. *In the situation described above, suppose that x is the element of $p(U) - d(p)$. Then, one of the following three possibilities must hold.*

(1) *Y is nonsplitting at x over p.*

(2) *There is a G_α in $\vec{G}$ such that Y is G_α-splitting at x over p.*

(3) *Y is amorphous at x over p.*

PROOF: If Y is non-splitting at x over p, then the claim of the lemma is verified. Suppose that Y is splitting at x over p.

Recall that $p * \{G_\alpha(x) = i\}$ denotes the Cohen extension of p such that G_α has value i at x. If there is no $G_\alpha \in \vec{G}$ such that $p * \{G_\alpha(x) = 0\}$ decides $Y \restriction cl(x, C \oplus Y \oplus U \oplus W)$, then Y is amorphous at x over p. Again, this implies the claim of the lemma.

Suppose $p * \{G_\alpha(x) = 0\}$ decides $Y \restriction cl(x, C \oplus Y \oplus U \oplus W)$. Let Y_0 be the value of $Y \restriction cl(x, C \oplus Y \oplus U \oplus W)$ forced by $p * \{G_\alpha(x) = 0\}$. Now, we ask whether $p * \{G_\alpha(x) = 1\}$ decides $Y \restriction cl(x, C \oplus Y \oplus U \oplus W)$.

Suppose $p * \{G_\alpha(x) = 1\}$ decides $Y \restriction cl(x, C \oplus Y \oplus U \oplus W)$. Let Y_1 be the value of $Y \restriction cl(x, C \oplus Y \oplus U \oplus W)$ forced by $p * \{G_\alpha(x) = 1\}$. If $Y_0 = Y_1$, then Y would be nonsplitting at x over p. Thus, Y_0 and Y_1 must be incompatible, and it follows that Y is G_α-splitting at x over p.

Suppose $p * \{G_\alpha(x) = 1\}$ does not decide $Y \restriction cl(x, C \oplus Y \oplus U \oplus W)$. Then, there are two extensions q and r of $p * \{G_\alpha(x) = 1\}$ which force incompatible values for $Y \restriction cl(x, C \oplus Y \oplus U \oplus W)$. We may assume $q(G_\alpha) = r(G_\alpha)$ since the values of $G_\alpha(y)$ $(y \in U, x < y)$ do not affect the value of $Y \restriction cl(x, C \oplus Y \oplus U \oplus W)$. One of q and r, say q, forces $Y \restriction cl(x, C \oplus Y \oplus U \oplus W) \neq Y_0$. Let q' be the trivial extension of the Cohen condition obtained from $p * q(\vec{G})$ by changing the value of $G_\alpha(x)$ to 0. Then, q' must force $Y \upharpoonright cl(x, C \oplus Y \oplus U \oplus W) = Y_0$, and hence if $\xi \neq \alpha$, then $p * q(G_\xi)$ does not decide $Y \upharpoonright cl(x, C \oplus Y \oplus U \oplus W)$. Since q and r force different values for $Y \upharpoonright cl(x, C \oplus Y \oplus U \oplus W)$ and $q(G_\alpha) = r(G_\alpha)$, we also see that $p * q(G_\alpha)$ does not decide $Y \upharpoonright cl(x, C \oplus Y \oplus U \oplus W)$. Thus, Y is amorphous at x over p. ◇

Density for local type I requirements. Suppose $I(\Phi, \vec{G}, \text{global})$ has index e and $x \geq e$. By 6.2 and 6.7 (2), the set $\{p \in \mathcal{P} \mid p \Vdash A^{(e)}(x) = \Delta(B \oplus Y \oplus U \oplus W, x)\}$ is dense in $\mathcal{P}$. Therefore, to see that the conditions which force the local requirements of type I are dense in $\mathcal{P}$, it is sufficient to show the following lemma.

7.3. LEMMA. *Suppose Φ, Ψ and $\vec{G}$ are given and e is the index of the requirement* $I(\Phi, \vec{G}, \text{global})$. *Let Y denote $\Phi(\vec{G})$. Then, for every p, there is a $q \leq p$ such that either*

(1) $q \Vdash A^{(e)} \neq \Psi(C \oplus Y \oplus U \oplus W)$, *or*

(2) $(\exists G_\alpha \in \vec{G})[q \Vdash G_\alpha \leq_K C \oplus Y \oplus U \oplus W]$.

PROOF: Suppose p is given so that $e < \min(|p|, d(p))$ and $\vec{G}$ are named in p. We may assume that the index of Ψ is also less than $|p|$ and $d(p)$.

If there is a q extending p and a real z such that q strongly forces $\Psi(z, C \oplus Y \oplus U \oplus W)$ to diverge then the claim of the lemma is verified.

In the case that p forces $\Psi(C \oplus Y \oplus U \oplus W)$ to be total, we proceed by separately analyzing the three possibilities for the forcing relation on Y over p.

Case 1. There is a $q \leq p$ and $x \in q(U) - d(q)$ such that Y is nonsplitting at x over q.

Let q be such an extension of p. By definition of nonsplitting, x is the maximal element of $q(U)$ and q decides $Y \upharpoonright cl(x, C \oplus Y \oplus U \oplus W)$. This means that values of $\vec{G}$ at x do not affect the value of $Y \upharpoonright cl(x, C \oplus Y \oplus U \oplus W)$. Construct an extension r of q by extending only q's Cohen condition to have the following features.

(1.1) $d(r) = \text{cl}(x, q \oplus W)$.

(1.2) $\langle e; x, 1\rangle \in r(C_0)$. Thus, $C \oplus Y \oplus U \oplus W \upharpoonright cl(x, C \oplus Y \oplus U \oplus W)$ and therefore $\psi(C \oplus Y \oplus U \oplus W, x)$ have been decided.

(1.3) $r(A)(\langle e, x\rangle) = 1 - \Psi(r(C \oplus U \oplus Y \oplus W), x)$.

(1.4) For all $\alpha < |r|$, $r(G_\alpha)(x) = r(A)(\langle e, x\rangle)$.

(1.5) $r(L)(\langle e', x, 1\rangle) = 1$ for every index $e' \leq x$ of type IV global requirement.

(1.6) $\langle e, \langle x, i, \emptyset, \emptyset\rangle\rangle \in r(B)$, where i is the value of $r(A)$ at $\langle e, x\rangle$ and $\emptyset$ is a real coding the empty set.

(1.7) For every X named in r, $r(X)$ is compatible with $q(X) * 0$ at any point other than the points specified above.

(1.8) Obtain $r(U)$ by extending $q(U)$ by adding a single new element, sufficiently large to satisfy the restriction on U that $r(U)$ must contain an element bigger than $d(r)$ and $d(r)$ must be highly closed. Obtain $r(C_1)$ by coding $r(A) \upharpoonright x_1$ as required in Definition 6.1 (4).

The resulting condition r is an element of $\mathcal{P}$ by virtue of our flagging $\langle e, x\rangle$, coding A's value at $\langle e, x\rangle$ in all of the G_α and respecting the type IV coding constraints. By (1.3), r is an extension of p which satisfies the condition (1) of the lemma. So, we assume, hereafter, that for every $q \leq p$ and every $x \in q(U) - d(q)$, Y is splitting at x over q.

Case 2. There is a $q \leq p$ and $x \in q(U) - d(q)$ such that $e \leq x$ and Y is amorphous at x over q.

Take such q and x. We may assume $|p| = |q|$. Let r be an extension of q which decides $Y \restriction cl(x, C \oplus Y \oplus U \oplus W)$ amorphously. That is, $q * r(\vec{G})$ decides $Y \restriction cl(x, C \oplus Y \oplus U \oplus W)$ but for any $G_\xi \in \vec{G}$, $q * r(G_\xi)$ does not decide $Y \restriction cl(x, C \oplus Y \oplus U \oplus W)$. We may assume $|r| = |q|$. For each α less than $|q|$, let r_α be a condition extending p such that $r_\alpha(G_\alpha) = q(G_\alpha)$ but r_α decides $Y \restriction cl(x, C \oplus Y \oplus U \oplus W)$ differently from r. Let x_1 and x_2 be the least pair such that for all α, $\text{cl}(r, \max r_\alpha(U), r_\alpha \oplus W) < x_1$ and $\text{cl}(x_1, r, r_\alpha \oplus W) < x_2$. Define an extension s of r with the same real sequences as follows.

(2.1) $|s| = |q|$, $d(s) = \text{cl}(x_1, r \oplus W)$ and $\text{dom}(s(U)) = \text{cl}(x_2, r \oplus W)$.

(2.2) For every X named in s, $s(X)$ is compatible with $r(X)*0$ unless specified otherwise in (2.3)–(2.5). In particular, $s(C_0)$ and all of the $s(G_\alpha)$ are trivial extensions of the values given in r.

(2.3) x_1 and x_2 are the next two points of $s(U)$ after those in $r(U)$. $s(C_1)$ codes $s(A)$ as required in Definition 6.1 (4).

(2.4) $\langle e, x_1\rangle \in \text{dom}(s(A))$ and $s(A)(\langle e, x_1\rangle) = 1 - \Psi(s(C \oplus U \oplus Y \oplus W), x_1)$. Note that $\Psi(C \oplus Y \oplus U \oplus W, x_1)$ has been decided by the clauses (2.2)–(2.3) since the coding of A at $\langle e, x_1\rangle$ in C_1 occurs above x_2.

(2.5) $\langle e, \langle x, i, \emptyset, \emptyset\rangle\rangle \in s(B)$, where i is the value of $s(A)$ at $\langle e, x_1\rangle$.

Once we know that s is an element of $\mathcal{P}$, we see from (2.4) that s forces $A^{(e)}(x_1) \neq \Psi(C \oplus U \oplus Y \oplus W, x_1)$, and hence s satisfies the condition (1) of the lemma. Since we have taken special care to ensure that the enumerating function for $s(U)$ grow sufficiently quickly, we do not have the usual problem of verifying clause 6.1 (6). However, we have potentially put $\langle e, x_1\rangle$ into A without recording that fact by a flag in $r(C_0)$. We must verify the conditions of the type II constraints with index less than $|r|$ described in 6.3. These state that either we have established a witness to one of countably many inequalities or the values of $\Theta(A)$ do not depend on whether $\langle e, x_1\rangle$ belongs to A.

Suppose $\text{II}(\Theta, \Omega, \alpha)$ is a requirement with index less than $|s|$ such that α is less than $|s|$. If s strongly forces the inequality $\Theta(A) \neq \Omega(B \oplus G_\alpha)$, then s meets the constraint imposed by $\text{II}(\Theta, \Omega, \alpha)$ (see Definition 6.3).

For the type II constraints not satisfied as above, we want to prove that s satisfies the conditions (1)–(3) of 6.3. That is, we must show that if $\langle e, x_1\rangle \in s(A)$ then for each type II constraint with index below $|r|$, the values of $\Theta(A)$ for that requirement in the interval $[x_1, x_2)$ do not depend on whether $\langle e, x_1\rangle$ is an element of $s(A)$. For the sake of a contradiction, suppose this is not the case. Let e_0 be the least index for a type II constraint $\mathrm{II}(\Theta_{e_0}, \Omega_{e_0}, \alpha_0)$ for which there is a counter example. We will find a condition extending q that strongly forces $\Theta_{e_0}(A) \neq \Omega_{e_0}(B \oplus G_{\alpha_0} \oplus U)$.

Since e_0 is the index for a counter example as above, $\langle e, x_1 \rangle$ is in $s(A)$, and there is a real z in $[x_1, x_2)$ such that $\Theta_{e_0}(s(A,z))$ is unequal to $\Theta_{e_0}(r(A) * 0, z)$.

Let *pos* and *neg* be a pair of sets of reals, one of which is a singleton and the other empty, forced by both r and and r_{α_0} to lie in $Y \restriction cl(x, C \oplus Y \oplus U \oplus W)$, such that r forces *pos* to be contained in Y and *neg* to be contained in the complement of Y but r_{α_0} forces the negation of one of these inclusions. Now define an extension s' of q as follows.

(3.1) $d(s') = d(s)$ and $\mathrm{dom}(s'(U)) = \mathrm{dom}(s(U))$.

(3.2) $s'(G_{\alpha_0}) = s(G_{\alpha_0})$.

(3.3) $s'(B) \restriction \mathrm{cl}(x, q \oplus W) = r(B) \restriction \mathrm{cl}(x, q \oplus W)$; $\langle e, \langle x_1, 1, pos, neg \rangle\rangle \in s'(B)$ and otherwise $s'(B)$ is equal to 0 on the remainder of its domain. This explicitly sets the values in A and B compatibly with the type II constraints except possibily for the argument $\langle e, x_1 \rangle$.

(It is easy to see that the clauses (3.1)–(3.3) decide the values of $\Omega_{e_0}(B \oplus G_\alpha \oplus U \oplus W)$ on $[x_1, x_2)$.

(3.4) Chose $s'(A)$ to agree with $s(A)$ except possibly for argument $\langle e, x_1 \rangle$. Set the value of $s(A)(\langle e, x_1 \rangle)$ so as to force $\Theta_{e_0}(A, z)$ to be unequal to the value already decided for $\Omega_{e_0}(B \oplus G_\alpha \oplus U \oplus W, z)$.

(3.5) $s'(C_0)$ extends $r(C)$ to be defined on $d(t)$ so that, for every e' and z not in $d(r)$, $\langle e', z, 2, i \rangle \in s'(C_0)$, where $i = s'(A)(\langle e', z \rangle)$. $s'(C_1)$ is the extension of $r(C_1)$ that codes $s'(A)$ as required by Definition 6.1 (4).

(3.6) If $\xi < |s'|$ and $\xi \neq \alpha$, then

$$s'(G_\xi) = \begin{cases} s(G_\xi), & \text{if } s'(A)(\langle e, x_1 \rangle) = 1 \\ q(G_\xi) * \{r_\alpha(G_\xi) * 0\}, & \text{otherwise.} \end{cases}$$

(3.7) $s'(L)$ extends $q(L)$ and maintain to satisfy the constraint imposed in 6.6.

By the minimality of e_0, all of the type II constraints with index less than e_0 are satisfied by s'. By direct manipulation of s', the e_0th constraint will be satisfied by all conditions extending s' by means of the forced inequality. The remaining type II constraints are satisfied since the e_0th requirement was satisfied for the first time, as defined in Definition 6 3. We have to check the type I constraints described in Definition 6.2 to ensure that s' is an element of $\mathcal{P}$.

Suppose $s'(A)(\langle e, x_1 \rangle) = 1$. Then, $s'(\vec{G}) = r(\vec{G})$. Hence, s' strongly forces Y to satisfy the neighborhood condition $\langle pos, neg \rangle$. We have put $\langle e, \langle x_1, 1, pos, neg \rangle\rangle$ in $s'(B)$. Thus, since the value of $A(\langle e, x_1 \rangle)$ has been correctly recorded in $B \oplus Y \oplus U \oplus W$, s' satisfies the type I constraint with index e. The other type I constraints are automatically satisfied, as A and B take their default values on other than the eth column.

Now, suppose $s'(A)(\langle e, x_1 \rangle) = 0$. The type I constraints are all satisfied by default except for possibly the eth one. There we put a computation into B which, if it applied to $B \oplus Y \oplus U$ would produce an inequality between $A^{(e)}$ and the set computed by $B \oplus Y \oplus U$, on the eth coordinate.

However, for every $\xi \neq \alpha$, $s'(G_\xi)$ is an extension of $r_{\alpha 0}(G_\xi)$. Therefore, by the choice of r_α, s' strongly forces that Y does not satisfy $\langle pos, neg\rangle$. Thus, also in this case, s' satisfies the conditions of 6.2. Namely, the potential computation we recorded in B does not apply to $B \oplus Y \oplus U$.

The extension to satisfy a type II constraint with index less than e by forcing an inequality can only occur countably often since there are only countably many e_0 less than $|p|$. We can apply induction to obtain the desired condition provided that every extension of p has an extension that is either nonsplitting or amorphous.

Case 3. There is an extension of p, which we also call p, such that for any $q \leq p$ and $x \in q(U) - d(q)$, Y is neither nonsplitting nor amorphous at x over q .

In this case, we will show that there is a ξ such that p forces $C \oplus Y \oplus U \oplus W$ to compute G_ξ, thereby satisfying the type II requirement.

Let x_0 be the element of $p(U) - d(p)$. By Lemma 7.2, there is a $G_\alpha \in \vec{G}$ such that Y is G_α-splitting at x_0 over p. This α is the desired value for ξ.

We simultaneously define the reduction of G_α to $C \oplus Y \oplus U \oplus W$ and verify that it is forced by p to be correct.

Let q be an extension of p and z a real such that z is an element of $d(q)$. Compute the value $q(G_\alpha)(z)$ uniformly from p, z and $\langle q(G_\alpha)(y) \mid y < z\rangle$, $q(Y)$, relative to predicates for $q(C)$, $q(U)$, W and ${}^2E$, as follows.

If $z \notin q(U)$, then the computation gives answer 0. This value is correct since $q(G_\alpha) \subseteq q(U)$, by 6.1 (5). If $z \in d(p)$ then $q(G_\alpha)(z) = p(G_\alpha)(z)$. In these two cases, the computation is trivial. Further, the values it returns are forced to be correct by p without mention of the hypothesis of this case.

The interesting analysis occurs when z belongs to $q(U)$ but $G_\alpha(z)$ is not decided by p. Here we show that Y is consistently G_α-splitting.

Let us consider the case $z = x_0$. Let $cl(x_0, p, C \oplus Y \oplus U \oplus W)$ denote the closure of x_0 and p relative to the indicated predicates C and U as decided by q. Since Y is G_α-splitting at x_0 over p, for each $i \in \{0,1\}$, $p * \{G_\alpha(x_0) = i\}$ decides $Y \restriction cl(x_0, p, C \oplus Y \oplus U \oplus W)$ and the values forced are incompatible on some w that lies in this closure. Note that the closure does not depend on the particular value for Y as there are only two possibilities (by G_α-splitting) and both are recursive in p, relative to C and U. Further, since finding an incompatibility in the forcing for Y is an instance of finding a witness to a Σ_1 statement satisfied in $cl(x_0, p, C \oplus Y \oplus U \oplus W)$, there is a uniformly recursive function $w(-)$ such that $w(x_0, p, C \oplus U \oplus W)$ is a w as above.

Then, $q(G_\alpha)(x_0)$ is computed by

$$G_\alpha(x_0) = i \iff \left\{ \begin{array}{c} Y(w(x_0, p(G_\alpha), C \oplus U \oplus W)) \text{ is equal to the value} \\ \text{forced by } p(G_\alpha) * (G_\alpha(x_0) = i) \end{array} \right\}$$

It follows that $G_\alpha(x_0)$ is uniformly computable from x_0 and p, relative to $C \oplus Y \oplus U \oplus W$ and ${}^2E$.

We extend this argument to all $z \in d(q)$ such that $x_0 \leq z$ and $z \in q(U)$. Suppose $\langle q(G_\alpha)(y) \mid y < z\rangle$ is given. Let p' denote $p * \langle q(G_\alpha)(y) \mid y < z\rangle$, and p'_i denote $p' * \{G_\alpha(z) = i\}$ for $i \in \{0,1\}$. Now we claim that each p'_i decides $Y \upharpoonright cl(z, p, C \oplus Y \oplus U \oplus W)$

Suppose that the claim has been established. Since Case 1 does not happen, p'_0 and p'_1 must decide incompatible values for $Y \upharpoonright cl(z, p, C \oplus Y \oplus U \oplus W)$. As in the case $z = x_0$, we can apply the uniformly recursive function $w(-)$ to to z and $\langle q(G_\alpha)(y) \mid y < z\rangle$, relative to $q(C)$, $q(U)$, W and ${}^2E$, to obtain an instance of incompatibility. Then, $q(G_\alpha)(z)$ is computed by

$$G_\alpha(z) = i \iff \left\{ \begin{array}{r} Y(w(z, \langle q(G_\alpha)(y) \mid y < z\rangle, C \oplus U \oplus W)) \text{ is equal to the value} \\ \text{forced by } \langle q(G_\alpha)(y) \mid y < z\rangle * (G_\alpha(z) = i) \end{array} \right\}$$

Thus, the value of $q(G_\alpha)$ at z is uniformly computable from p, and $\langle q(G_\alpha)(y) \mid y < z\rangle$, relative to $q(Y)$, $q(U)$, W and ${}^2E$ as desired. Since functions defined by effective transfinite recursion are Kleene recursive, we have exhibited a Kleene computation of G_α as desired.

Now we return to the proof of the claim. Suppose that one of p'_0 and p'_1, say p'_0, does not decide $Y \upharpoonright cl(z, p, C \oplus Y \oplus U \oplus W)$. Then, there are two extensions of p'_0, r_1 and r_2 agreeing except for the G_ξ with $\xi \neq \alpha$, which strongly force incompatible values for Y on this set. By possibly taking a one point extension of $r_i(U)$, we may assume that z is sufficiently large to include the point of ambiguity in $cl(z)$. We may also assume that A, B and C_0 take the default value 0 beyond p in r_1 and r_2, since the values of A, B and C and do not affect the value for Y. Let r'_1 be the condition obtained from r_1 by changing the value of $G_\alpha(x_0)$ to $1 - q(G_\alpha(x_0))$. Then, r_1 and r'_1 must force incompatible values for $Y \upharpoonright cl(z, p, C \oplus Y \oplus U \oplus W)$ since Y is G_α-splitting at x_0 over p. It follows that $p'_0 * G_\xi(z)$ does not decide $Y \upharpoonright cl(z, p, C \oplus Y \oplus U \oplus W)$, for any $\xi \neq \alpha$. By the starting assumption in the verification of the claim, $p' * \{G_\alpha(z) = 0\}$ does not decide $Y \upharpoonright cl(z, p, C \oplus Y \oplus U \oplus W)$. Thus, Y is amorphous at z, which contradicts the assumption that Case 2 does not occur for any extension of p. ◊

Density for type II requirements.

7.4. Lemma. *Suppose Θ, Ω, and G_α are given. For every p, there exists a $q \leq p$ such that either*

(1) $q \Vdash \Theta(A) \neq \Omega(B \oplus G_\alpha)$, *or*
(2) $q \Vdash \Theta(A) \in (C \oplus G_\alpha \oplus U \oplus W)$.

Proof: The lemma is a direct consequence of the remarks following Definition 6.3. ◊

Density for type III requirements.

7.5. LEMMA. *Given Θ, α and p, there is a $q \leq p$ such that*

$$q \Vdash \Theta(\bigoplus_{\xi \neq \alpha} G_\xi \oplus U \oplus W) \neq G_\alpha.$$

PROOF: We may assume that G_α is named in p. Let x be the element of $p(U) - d(p)$. Set $G_\xi(x) = 0$ for all G_ξ named in p other than G_α. Then, $\Theta(\oplus_{\xi \neq \alpha} G_\xi \oplus U \oplus W, x)$ is decided based on the assumption that any G_ξ not named in p is empty. Now, set the value $G_\alpha(x)$ so that it is different from the value of $\Theta(\bigoplus_{\xi \neq \alpha} G_\xi \oplus U \oplus W, x)$. For A, B and C, set the value 0 on $\mathrm{cl}(x, p \oplus W) - d(p)$. Extend $p(L)$ on $\mathrm{cl}(x, p \oplus W)$ maintaining the global coding constraints of type IV. Finally, we extend $p(U)$ by adding a new big point after $\max p(U)$ and $p(C_1)$ to code A as required in 6.1 (4). By the constraint imposed in 6.1 (7), the predicates G_ξ not named by p will satisfy the implicit Cohen condition stating that they have no elements in $\mathrm{cl}(x, p \oplus W)$ Then, the extension so obtained forces the desired inequality at the point x. ◇

Density for type IV requirements. We imposed the constraints that the elements of $\mathcal{P}$ must force the global requirements of the form IV(α, β, coding). Therefore, to prove the density for the requirements of type IV, we need the following lemma.

7.6. LEMMA. *Suppose α and β are given so that $\alpha \not\leq_{\mathbf{R}} \beta$. Given p and Θ, there is a $q \leq p$ such that*

$$q \Vdash G_\alpha \neq \Theta(G_\beta \oplus U \oplus W \oplus L).$$

PROOF: We may assume that the index of Θ is less than $d(p)$ and that G_α and G_β are both named in p. Let x be the element of $p(U) - d(p)$ as in the previous proof. Set $G_\beta(x) = 0$. We want to set the values for L on $\mathrm{cl}(x, p \oplus W)$ without deciding the value of $G_\alpha(x)$. This is possible by setting the values for L at the points $\langle e, x \rangle$ and $\langle e, x, 1 \rangle$ appropriately, where e is the index of a global requirement of type IV such that $e \leq x$.

We give an example. Suppose $\xi < |p|$ and $\alpha \leq_{\mathbf{R}} \xi$. Let e be the index of the requirement IV(α, ξ, coding). Then, we set $L(\langle e, x, 1 \rangle) = 1$ to satisfy the requirement since otherwise we must set $L(\langle e, x \rangle) = G_\alpha(x)$ and the value of $G_\alpha(x)$ must depend on the value for $L(\langle e, x \rangle)$. In this case, further, suppose $\beta \leq_{\mathbf{R}} \xi$ and the requirement IV(β, ξ, coding) has index e'. Then, we set $L(\langle e', x, 1 \rangle) = 0$ and $L(\langle e', x \rangle) = 0$ to satisfy the requirement since otherwise we must have $G_\xi(x) = G_\beta(x) = 0$ and thus $G_\alpha(x)$ must take the value 0.

Other cases can be treated in the same manner, and we leave it to the reader.

Now, we have decided $\Theta(G_\beta \oplus U \oplus W, x)$ without deciding $G_\alpha(x)$. Therefore, we can set the value $G_\alpha(x)$ different from the value $\Theta(G_\beta \oplus U \oplus W, x)$. We set values $G_\xi(x)$ for $\xi \neq \alpha, \beta$ to satisfy the conditions of 6.6. Set the default value 0 for A, B and C on $\mathrm{cl}(x, p \oplus W) - d(p)$. Finally, extend U by adding a new big point and extend C_1 to code A with the appropriate delay. Then, we have an extension of p which strongly forces $G_\alpha(x) \neq \Theta(G_\beta \oplus U \oplus W \oplus L, x)$. ◇

Proof of Theorem 4.3. Let $\langle R_e \mid e < \aleph_1 \rangle$ be the enumeration of all requirements given in §5. For each e, let $\mathcal{D}_e$ be the set of elements of $\mathcal{P}$ which forces the requirement R_e. We have shown that $\mathcal{D}_e$'s are all dense in $\mathcal{P}$. Therefore, we can find a decreasing sequence $\langle p_e \mid e < \aleph_1 \rangle$ such that for every e, $p_e \in \mathcal{D}_e$, $e \leq |p_e|$ and $e \leq d(p_e)$. For each predicate X, let $X = \bigcup_{e<\aleph_1} p_e(X)$. Then, the sets thus obtained satisfy the conditions of Theorem 4.3. ◊

References

[1] S. C. Kleene, *Recursive functionals and quantifiers of finite types I*, Trans. Amer. Math. Soc. **91** (1959), 1–52.

[2] ———, *Recursive functionals and quantifiers of finite types II*, Trans. Amer. Math. Soc. **108** (1963), 106–142.

[3] S. C. Kleene and E. L. Post, *The upper semi-lattice of degrees of recursive unsolvability*, Ann. of Math. (2) **59** (1954), 379–407.

[4] Gandy, R.O., *Generalized recursive functionals of finite type and hierarchies of functionals*, Ann. Fac. Sci. Univ. Clermont-Ferrand **35** (1967), 5–24.

[5] A. Nerode and R. A. Shore, *Reducibility orderings: theories, definability and automorphisms*, Ann. Math. Logic **18** (1980), 61-89.

[6] D. Normann, *Set recursion*, in "Generalized Recursion Theory II," North-Holland, Amsterdam, 1978, pp. 303–320.

[7] J. Shinoda and T. A. Slaman, *On the theory of the PTIME degrees of recursive sets*, (to appear).

[8] C. Spector, *On degrees of recursive unsolvability*, Ann. of Math. (2) **64** (1956), 581–592.

Department of Mathematics; Nagoya University; Nagoya 464; JAPAN
Department of Mathematics; The University of Chicago; Chicago, IL 60637; U. S. A.

Σ_1-Collection and the Finite Injury Priority Method

THEODORE A. SLAMAN
W. HUGH WOODIN

ABSTRACT. We show that there is an intermediate recursively enumerable Turing degree in every model of $P^- + B\Sigma_1$. The proof is not uniform, depending on whether $I\Sigma_1$ holds. There is a model of $P^- + B\Sigma_1$ in which there is a least recursively enumerable degree strictly above the recursive degree. Thus, the Sacks Splitting Theorem cannot be proven in $P^- + B\Sigma_1$.

§1. INTRODUCTION

In recursion theory, the fundamental technical tool is the organization of an effective construction using the priority method. A theorem may be classified as part of recursion theory solely on the basis of the appearance of the priority method in its proof. In [3], Groszek and Slaman have given a classification of the standard priority methods based their combinatorial complexity. The finite injury method constitutes the first two levels in their hierarchy, Π_1 and Σ_2.

Typically, a finite injury construction is used to define a collection of computable subsets of the integers. The construction is organized in terms of stages, which are just the non-negative integers viewed in the context of a definition by recursion. During each stage, the extension of the recursive definitions of the sets under construction is determined by a collection of strategies. A strategy is a procedure that imposes enough constraints on the construction to ensure that the resulting sets satisfy an associated requirement. We view strategies as simple machines with states and rules for changing states. A state corresponds to a particular set of constraints. The strategy changes state during a stage when the construction fulfills its transition conditions. The strategies are given a priority ranking so that if there is a contradiction between constraints, the strategy of higher priority takes precedence.

The Π_1-method was independently introduced by Friedberg [2] and Mučnik [6]. It is characterized by two critical features. First, for each strategy, there is a fixed finite bound on the number of times that it can change state. Second, when a strategy changes state all strategies of lower priority are returned to their initial state.

The Σ_2-method was introduced by Sacks [10]. It is distinguished from the Π_1-method in that there is no inherent bound on the number of times that a strategy can change state. Using the Σ_2-method, either all of the strategies change state only finitely often or there is one strategy that changes state infinitely often. In Sacks original arguments,

The first author was partially supported by a grant from the Japan Society for the Promotion of Science during the preparation of this paper. Both authors were partially supported by N.S.F. grants and by Presidential Young Investigator Awards.

the second case lead to a contradiction. In a general application, we must argue that the theorem is true in either case.

In this paper, we examine some of the theorems of recursion theory in the context of fragments of arithmetic to see to what extent the apparent complexity in their proofs is necessary. We regard an independence result, saying that φ cannot be proven in the theory T, as meaning the proof of φ is more complicated than any proof in T. This is appropriate in the setting of arithmetic, where theories are strengthened by adding a syntactically more complicated scheme for collection or induction.

Let P^- be the finitely axiomitized theory summarizing the algebraic properties of successor, addition, multiplication and exponentiation on the non-negative integers. Our inclusion of exponentiation is somewhat eccentric. However, exponential algorithms are so common in recursion theory that leaving the exponential function out of our base theory would also leave out the nature of the subject.

Let $B\Sigma_k$ and $I\Sigma_k$ be the schemes of collection and induction for the Σ_k definable sets, respectively. We work over the base theory $P^- + I\Sigma_0$. Let N denote the standard model of arithmetic or the standard part of a nonstandard model of the base theory. The necessary material on fragments of arithmetic is reviewed in section §2.

In [**3**], Groszek and Slaman show that the abstract success of the Π_1-method is equivalent to the scheme $I\Sigma_1$. Similarly, they show that the success of the Σ_2-method is equivalent to $I\Sigma_2$. We analyze the two critical applications of these methods, the Friedberg-Mučnik solution to Post's problem and the Sacks splitting theorem, in the theory $P^- + B\Sigma_1$.

Post's problem was proposed in [**9**, Post]. He asked whether there is a recursively enumerable set that is neither recursive nor Turing complete among the recursively enumerable sets. Such a set is said to have intermediate degree. Friedberg and Mučnik introduced the Π_1-method to construct a pair of recursively enumerable sets of incomparable Turing degree. As was first proven by Simpson [**11**] and follows from the Groszek-Slaman analysis, the Friedberg-Mučnik construction can be shown to produce a solution to Post's problem in $P^- + I\Sigma_1$.

In section §3, we show that $P^- + B\Sigma_1$ proves that there is an intermediate degree. Our proof is not uniform. We show that in every model M of $P^- + B\Sigma_1$, there is an intermediate degree: either, M satisfies $I\Sigma_1$ and we apply Simpson's result, or M contains a Σ_1-cut, which we show has intermediate degree. In section §5, we pose the question of whether the full Friedberg-Mučnik theorem can be proven in $P^- + B\Sigma_1$.

Sacks introduced the Σ_2-method to prove the Splitting Theorem: every nonrecursive recursively enumerable degree is the join of two smaller recursively enumerable degrees.

In section §4, we show that the splitting theorem cannot be proven in $P^- + B\Sigma_1$. We work with a cut M in an ω-saturated model of Peano arithmetic. M is the downward closure of the finite iterations of the exponential function applied to a nonstandard element. By a standard argument, M is a model of $P^- + B\Sigma_1$. In M, N has least Turing degree among the recursively enumerable sets that are not recursive.

In some sense, our result is best possible as Mytilinaios [7] has shown that the splitting theorem can be proven in $P^- + I\Sigma_1$.

M provides a good model for the complete breakdown of the priority method. The priority method is based on the use of strategies to decide the convergence or divergence of computations using finite conditions. Thus, the standard priority constructions produce amenable (see Definition 2.7) predicates. In M, every amenable recursively enumerable predicate is recursive.

There is a gap between our results and the theoretical limits imposed by Groszek-Slaman. In section §5, we pose some questions aimed at closing the gap. Namely, are there recursion theoretic theorems proven using the Π_1 and Σ_2 methods that cannot be proven in $P^- + B\Sigma_1$ and $P^- + B\Sigma_2$, respectively?

§2. Fragments of Arithmetic

Fragments of Peano Arithmetic. Following standard usage, say that a formula is Σ_0 or Π_0 if all of its quantifiers are bounded. A formula is Σ_{k+1} (equivalently, in Σ_{k+1}), if it is a string of existential quantifiers followed by a formula in Π_k. A formula is Π_{k+1}, if it is the negation of a Σ_{k+1} formula. A set X is Σ_n with parameters in a model M if there is a Σ_n formula φ and parameters $\vec{p}$ in M such that for all x in M, x is an element of X if and only if $\varphi(x, \vec{p})$ is satisfied in M.

2.1. Definition. (1) By P^- we mean the subtheory of Peano Arithmetic with successor, addition, multiplication and exponentiation containing no instances of the induction scheme. It consists of the universal closures of the following axioms.

$$x' \neq 0$$
$$(x' = y') \implies (x = y)$$
$$x \neq 0 \implies 0' \leq x$$
$$x < y \iff (\exists t)[x + t' = y]$$
$$x < y \lor x = y \lor x > y$$
$$x + y = y + x; \quad x \times y = y \times x$$
$$x + (y + z) = (x + y) + z; \quad x \times (y \times z) = (x \times y) \times z$$
$$x + 0 = x; \quad x \times 0 = 0; \quad x^0 = 1$$
$$x + y' = (x + y)'; \quad x \times y' = (x \times y) + x; \quad x^{y'} = x^y \times x$$
$$x \times (y + z) = (x \times y) + (x \times z)$$
$$x + y = x + z \implies y = z$$

Here, we use the abbreviations: $x \leq y$ for $(\exists t)[x + t = y]$ and $x < y$ for $x \leq y$ & $x \neq y$.

(2) If Γ is a set of formulas, $B\Gamma$ is the *bounding scheme* for elements of Γ. It consists of the universal closures of the formulas

$$(\forall x)\left[(\forall y < x)(\exists w)\varphi(y,w) \implies (\exists b)(\forall y < x)(\exists w < b)\varphi(y,w)\right] \tag{2.2}$$

for each φ in Γ. Formula 2.2 states that the search for the witnesses $w(y)$ for a bounded set $\{\varphi(y,-) \mid y < x\}$ of instances of φ, is itself bounded.

(3) A related and usually stronger scheme is $I\Gamma$, the *induction scheme* for elements of Γ. It consists of the universal closures of the formulas

$$\Big(\varphi(0) \,\&\, (\forall x)[\varphi(x) \implies \varphi(x+1)]\Big) \implies (\forall x)\varphi(x)$$

for φ in Γ.

(4) *Peano Arithmetic* (PA) is defined by

$$PA = P^- \cup \bigcup_{k\in\omega} I\Sigma_k.$$

2.3. **Definition.** Suppose that M is a model of $P^- + I\Sigma_0$. A *cut* in M is a nonempty subset of M that is closed under the successor function and closed downward. I is a *proper cut* in M if, in addition, the complement of I in M is not empty.

In a model of P^-, there is a proper cut with a Σ_k definition if and only if $I\Sigma_k$ fails to hold. Thus, in a model of PA, there is no definable proper cut.

The basic result in the area of fragments of arithmetic is due to Kirby and Paris, 1977.

2.4. **Theorem** [5]. *For all k, the following implications hold in the presence of $P^- + I\Sigma_0$.*

$$I\Sigma_{k+1} \implies B\Sigma_{k+1} \implies I\Sigma_k$$
$$I\Sigma_k \iff I\Pi_k$$

Further, the only true implications are the ones indicated.

Nonstandard Finite Sets. We take $P^- + I\Sigma_0$ as our base theory. In this theory, one can prove the basic facts of number theory. We can use the exponential function to represent sets of elements by single elements. If c is a number and F is a set, say that c *codes* F if

$$m \in F \iff \text{The } m\text{th digit in the binary representation of } c \text{ is } 1.$$

2.5. Definition. Suppose that M is a model of $P^- + I\Sigma_0$. A set F is *M-finite* if and only if it has a code in M.

We will treat the M-finite sets exactly like elements of M. Since the coding and decoding functions are recursive and we treat the recursive functions as primitive objects, there is no visible difference between an M-finite set and its code. Similarly, we will treat M as if it were closed under M-finite sequences. We will denote the sequence with elements $m_1, \ldots, m_k$ by $\langle m_1, \ldots, m_k \rangle$.

2.6. Notation. If M is a model of $P^- + I\Sigma_0$ and m is an element of M, let $<m$ denote the set of x in M such that x is less than m.

2.7. Definition. If M is a model of $P^- + I\Sigma_0$ and Z is contained in M, then Z is an *amenable* subset of M if for every m in M, the intersection of Z with $<m$ is M-finite.

Recursion theoretic definitions. For the basics of recursion theory, we follow the standard presentation. For example, see Soare [**12**].

2.8. Definition. A subset W of M is *recursively enumerable* in M if W is Σ_1 with parameters in M.

A recursively enumerable set W has a natural enumeration given in terms of the witnesses to the existential condition defining W. Let $W[s]$ be the collection of elements of W that are less than s and enumerated in W by witnesses less than s. Clearly, W is the union of the sets $W[s]$. Further, in a model M of $P^- + I\Sigma_0$, each set $W[s]$ is M-finite.

2.9. Definition. Suppose that M is a model of $P^- + I\Sigma_0$.

(1) A *Turing functional* Φ is a recursively enumerable set of quadruples $\langle x, y, pos, neg \rangle$ where x and y are elements of M and pos and neg are M-finite sets. We refer to the elements of Φ as *computations*. Using finite sequences to code syntax, let $\langle \Phi_m \mid m \in M \rangle$ be a uniformly recursive listing of all the recursive functionals.

(2) Suppose that X is a subset of M. $\Phi(x, X) = y$ if and only if there is a computation $\langle x, y, pos, neg \rangle$ in Φ, so that $pos \subseteq X$ and $neg \cap X = \vee$. In this case, we say that the computation $\langle x, y, pos, neg \rangle$ *applies* to X. Similarly, we say that $\langle x, y, pos, neg \rangle$ is a computation from X, or relative to X, or an X-computation. Note, we allow the possibility that $\Phi(-, X)$ does not define a function; it may not be single valued.

(3) If X and Y are subsets of M then X is *pointwise recursive* in Y, $X \leq_{pT} Y$, if there is a Turing functional Φ such that, for all x in M, $X(x) = \Phi(x, Y)$.

(4) X is *recursive* in Y, $X \leq_T Y$, if the set

$$\left\{ \langle pos, neg \rangle \;\middle|\; \begin{array}{c} pos \text{ is an } M\text{-finite subset of } X \text{ and} \\ neg \text{ is an } M\text{-finite set of the complement of } X. \end{array} \right\}$$

is pointwise recursive in Y. The elements $\langle pos, neg \rangle$ are called *neighborhood conditions* on X.

A function is recursive if it is recursive in the empty predicate $\emptyset$.

The important difference between the two notions of relative recursiveness is that the latter is transitive by its very definition. The transitivity of the pointwise reducibility below Y depends on whether M-finitely many Y-computations can be amalgamated to produce a single M-finite computation. Groszek-Slaman [4], show that there is a model of $P^- + I\Sigma_1$ in which the pointwise reduction does not give a transitive relation on the recursively enumerable sets. Of course, the two reducibilities coincide in the standard model.

If M is a model of $P^- + I\Sigma_0$, the pairing function can be used to show that there is a universal recursively enumerable set. Let $\emptyset'$ denote such a set. That is, every recursively enumerable set in M is recursive in $\emptyset'$.

The Pigeon Hole Principle. The pigeon hole principle states that if n is less that m, then there is no injective function with domain $<m$ and range $<n$. In nonstandard models, only limited versions of this principle can be proven. The following proposition is well known.

2.10. PROPOSITION. *Suppose that M is a model of $P^- + B\Sigma_1$. If m and n are elements of M with m greater than n and f is a recursive function from $<m$ into $<n$, then f is not injective.*

PROOF: Recall, we are working in models with the exponentiation function. Suppose that f, m and n are a counter example to the proposition. We will produce a counter example to $I\Sigma_0$ in M, thereby contradicting $B\Sigma_1$ (see Theorem 2.4).

By $B\Sigma_1$, let w be an upper bound on the witnesses needed to verify the values of f on $<m$. Applying $I\Sigma_0$, where we bound all quantifiers by w, f is an M-finite set. Let I be the set of x less than or equal to n such that there is no M-finite injective function f_x with domain $<y$ and range contained in $<x$ with y strictly greater than x. Clearly, 0 is an element of I. If $x+1$ is not an element of I, then neither is x: given an injective function f_{x+1} from $<y+1$ into $<x+1$ we produce f_x with domain $<y$ by setting

$$f_x(z) = \begin{cases} f_{x+1}(z), & \text{if } f_{x+1}(z) \neq x; \\ f_{x+1}(y), & \text{otherwise.} \end{cases}$$

Thus, I is a cut in M. It is Σ_0, as we can bound all of the quantifiers in its definition by the set of codes for M-finite subsets of $<m$, an M-finite set. It is a proper cut as n is an element of its complement. ◇

2.11. COROLLARY. *Suppose that M is a model of $P^- + B\Sigma_1$ and I is a proper cut in M. There is no injective recursive function with domain M and range contained in M.*

Saturation. As in Mytilinaios-Slaman [8], we will use an ω-saturated model of PA to deduce independence results for various statements about the recursively enumerable Turing degrees. For an explanation of saturation and a proof that ω-saturated models exist, consult Chang-Keisler [1].

2.12. DEFINITION. Suppose that M is a model of $P^- + B\Sigma_1$. A subset r of N is in the *standard system* of M if there is an M-finite set R such that, for all integers i

$$i \in R \iff i \in r. \tag{2.13}$$

2.14. PROPOSITION. *There are models M and P such that*

(1) *M is a cut in P.*
(2) *P is a model of PA and M is a model of $P^- + B\Sigma_1$.*
(3) *The standard system of M is equal to 2^N, the set of all subsets of N.*
(4) *N is a Σ_1 cut in M.*

PROOF: Let P be an ω-saturated model of PA. For each r as above, there is a type extending the set of formulas indicated in Formula 2.13. By ω-saturation, P must realize all of these types and so the standard system of P is equal to 2^N.

Let a be a nonstandard element of P and let M be the set of y that are less than some finite iterate of the function $x \mapsto 2^x$ applied to a. The code for a P-finite subset of a is a number less than 2^a. As $M \cap {<}2^a$ is equal to $P \cap {<}2^a$, a subset of ${<}a$ is M-finite if and only if it is P-finite. Any subset of N in the standard system of P can be extended to a P-finite subset of ${<}a$. Thus, the standard system of M is equal to 2^N.

M is a model of $P^- + B\Sigma_1$ because it is a proper cut in a model of PA. An unbounded enumeration in M of an M-finite set F would give a definition of M in P: m would be an element M if and only if there is an element of F that is not enumerated by any n less than or equal to m. This is impossible, as P is a model of PA.

Finally, N is Σ_1 in M because n is an element of N if and only if there is an M-finite sequence $\vec{s}$ of length n, with first element equal to a and satisfying the recursion relation $\vec{s}(x+1) = 2^{\vec{s}(x)}$. ◇

§3. POST'S PROBLEM

In [9], Post posed the question that has become known as Post's Problem: Is there a recursively enumerable set W such that $\emptyset <_T W <_T \emptyset'$? The affirmative solution to Post's problem by Friedberg [2] and Mučnik [6] heralded the introduction of the Π_1-priority method. In this section, we show the affirmative solution to Post's problem can be shown without using the full proof theoretic strength of the Friedberg-Mučnik finite injury method, which is $I\Sigma_1$ as measured in [3, Groszek-Slaman].

3.1. THEOREM. *In every model of $P^- + B\Sigma_1$ there is a recursively enumerable set W such that $\emptyset <_T W <_T \emptyset'$.*

PROOF: Suppose that M is a model of $P^- + B\Sigma_1$. The proof divides into two cases depending on whether M is a model of $I\Sigma_1$.

First, suppose that M is a model of $I\Sigma_1$. In this case, Simpson [11] has shown that the Friedberg-Mučnik construction, executed in M, provides a recursively enumerable set of intermediate degree in M. In fact, Groszek and Slaman [3] have shown that any theorem with proof using the combinatorial pattern of the Friedberg-Mučnik argument can be formalized in $P^- + I\Sigma_1$. We refer the reader to [3] for the proofs of these results.

Next, suppose that M is a model of $P^- + B\Sigma_1$ in which $I\Sigma_1$ fails. All subsequent discussion takes place in M. By the failure of $I\Sigma_1$, let I be a proper Σ_1 cut. I is a recursively enumerable set. By $B\Sigma_1$, I is not recursive. We will show that I is a solution to Post's problem by showing that it is not complete in M. In fact, we will show that there is a recursively enumerable set A such that A is not pointwise computable from I.

In Definition 2.9, we defined a computation from a predicate X to be a quadruple $\langle x, y, pos, neg\rangle$. The computation says that if pos is contained in X and neg is contained in the complement of X then y is the answer at argument x. Now suppose that X is closed downward. A set pos is contained in X if and only if its greatest element is in X; similarly, neg is contained in the complement of X if and only if its least element is in the complement of X. Thus, we can canonically replace an arbitrary computation by one in which pos and neg are singletons. As I is closed downward, we will assume that all computations relative to I have this special form. Let $\langle x, y, p, n\rangle$ denote the computation with positive neighborhood condition $\{p\}$ and negative neighborhood condition $\{n\}$.

Let f be an increasing recursive function with domain I and range cofinal in M. Typically, $f(i)$ is the supremum of the witnesses to the effect that $<i$ is contained in I.

We recursively enumerate A as follows. We regard A as a disjoint recursive union of sets A_m, $A = \{\langle m, x\rangle \mid x \in A_m\}$. We use A_m to ensure that $\Phi_m(I)$ is not equal to A. A is enumerated in I many steps. During the ith step, we enumerate at most one element into each A_m based on witnesses below $f(i)$. Let $A_m[f(i)]$ be the set of elements enumerated in A_m during the previous steps. Let $\Phi_m[f(i)]$ be the set of computations of the form $\langle x, y, p, n\rangle$ enumerated into Φ_m by witnesses less than or equal to $f(i)$. Say that $\langle x, y, p, n\rangle$ applies to I during step i if $\langle x, y, p, n\rangle$ is in $\Phi_m[f(i)]$, x is less than or equal to $f(i)$, p is less than or equal to i and n is greater than i.

If there is a computation $\langle\langle m, x\rangle, 0, p, n\rangle$ that applies to I during step i, let n_{max} be maximal such that there are x and p with $\langle\langle m, x\rangle, 0, p, n_{max}\rangle$ applying to I during stage i. If there is such a computation $\langle\langle m, x\rangle, 0, p, n_{max}\rangle$ where x is an element of $A_m[f(i)]$, then it appears that $\Phi_m(I)$ is not equal to A at $\langle m, x\rangle$. In this case, we do not enumerate any new element in A_m during the ith step. Otherwise, we enumerate the least x such that there is a p such that $\langle\langle m, x\rangle, 0, p, n_{max}\rangle$ applies to I during step i.

If there is no computation $\langle\langle m, x\rangle, 0, p, n\rangle$ that applies to I during step i, we do not enumerate any element into A_m during that step.

We claim that A is not pointwise recursive in I. For a contradiction, suppose that $\Phi_m(I)$ is equal to A. Consider the function $e : x \mapsto e(x)$ that maps an element of M to the step when it is enumerated into A_m. Since M is a model of $P^- + B\Sigma_1$ and the range of e is bounded, Corollary 2.11 implies that there is an element x_0 of M that is not in the domain of e. Thus, $\langle m, x_0\rangle$ is not an element of A and $\Phi_m(\langle m, x_0\rangle, I)$ must equal 0.

Let $\langle\langle m, x_0\rangle, 0, p_0, n_0\rangle$ be a computation in Φ_m that applies to I. Since the range of f is cofinal in M, there is an i such that this computation applies to I during every step greater than or equal to the ith one. But during the ith step, we ensured that there was an x in A_m and a computation $\langle x, 0, p, n_{max}\rangle$ in Φ_m such that $p \in I$, by virtue of being less than or equal to i, and such that $n_{max} \geq n_0$. Since $\Phi_m(\langle m, x\rangle, I)$ is equal to $A(\langle m, x\rangle)$, the computation $\langle x, 0, p, n_{max}\rangle$ cannot apply to I. As p was known to be an element of I during step i, the only way for the computation to fail to apply to I is for n_{max} to be an element of I. As I is a cut, this implies that n_0 is also an element of I. This contradicts the assumption that $\langle\langle m, x_0\rangle, 0, p_0, n_0\rangle$ applies to I. ◇

§4. The Sacks Splitting Theorem

In this section, we show that the standard theorems proven using the Sacks Preservation Strategy cannot be proven in the theory $P^- + B\Sigma_1$.

4.1. Theorem. *Suppose that M is the model of $P^- + B\Sigma_1$ constructed in Proposition 2.14. The standard system of M is 2^N; N is a Σ_1 cut in M; and M is a cut in a model of Peano arithmetic. In M, N is the $\leq_T$-least recursively enumerable set that is not recursive.*

4.2. Corollary. *The following theorems of* [**10**, Sacks] *cannot be proven in $P^- + B\Sigma_1$.*

(1) *The Splitting Theorem. For every recursively enumerable set W that is not recursive, there are recursively enumerable sets A and B such that $W \leq_T A \oplus B$ and $A, B <_T W$.* $\oplus$ denotes recursive join.

(2) *For every recursively enumerable set W such that $0 <_T W <_T 0'$, there is a recursively enumerable set A such that A and W are $\leq_T$-incomparable.*

We prove Theorem 4.1 by a pair of lemmas. Let M be fixed as in Proposition 2.14. Let f be a strictly increasing recursive function mapping N cofinally into M.

Working in M, we show in Lemma 4.3 that every amenable recursively enumerable predicate is recursive. In Lemma 4.5, we show if W is a recursively enumerable predicate that is not amenable then $N \leq_T W$. Since M is a model of $P^- + B\Sigma_1$, it cannot have a proper Δ_1 cut. Thus, N is a recursively enumerable set that is not recursive and is below any other recursively enumerable set that is not recursive.

4.3. Lemma. *Suppose that W is a recursively enumerable amenable subset of M. W is recursive.*

Proof: Let φ be a Δ_0 predicate and $\vec{p}$ a sequence of parameters from M such that for all x

$$x \in W \iff (\exists y)\varphi(x, y, \vec{p}).$$

Since W is amenable, every initial segment of W is M-finite. By $B\Sigma_1$, the enumeration of any M-finite set is bounded. Let s be a function from N to N such that, for all integers i

$$(4.4) \qquad (\forall x < f(i)) \left[x \in W \iff \left(\exists y < f(s(i))\right) \varphi(x, y, \vec{p})\right].$$

In essence, the function s provides a Skölem function for $\varphi(\vec{p})$.

By assumption, the standard system of M is 2^N. Regarding a function from N to N as a set of integers coding pairs, let S be an M-finite set with standard part equal to s. We recursively compute W as follows. Given a pair $\langle pos, neg\rangle$, we are required to decide whether it is a neighborhood condition on W. We first compute the least i such that $f(i)$ is greater than all of the elements of *pos* and *neg*. Next, we compute $W\left[f(S(i))\right]$. If *pos* is a subset of $W\left[f(S(i))\right]$ and $neg \cap W\left[f(S(i))\right]$ is empty, then we answer *yes*; otherwise, answer *no*.

Given an argument $\langle pos, neg\rangle$, since f maps N cofinally in M, there will be a least integer i as above. By Formula 4.4 and the fact that S is a nonstandard extension of s, $W \cap {<}f(i)$ is equal to $W\left[f(S(i))\right] \cap {<}f(i)$. The pair $\langle pos, neg\rangle$ is a neighborhood condition on W if and only if it is one on $W\left[f(S(i))\right]$. Thus, the procedure described above will correctly compute W. ◇

4.5. LEMMA. *Suppose that W is a recursively enumerable subset of M that is not amenable. N is recursive in W.*

PROOF: Let m be an element of M such that $W \cap {<}m$ is not M-finite. Consider the recursive function h from N into 2^m given by

$$(4.6) \qquad h(i) = W\left[f(i)\right] \cap {<}m.$$

Since M is a cut in a model P of Peano arithmetic, h has an extension to an M-finite set H, corresponding to interpreting Equation 4.6 in P for nonstandard arguments i. H is a map from its domain d, an element of M, into 2^m. Since the range of f is cofinal in M, the intersection of W with ${<}m$ is equal to the union of $\{H(i) \mid i \in N\}$. Since $W \cap {<}m$ is not M-finite, for every x in the domain of H, either x is an integer or there is an element of $H(x)$ that is not an element of $W \cap {<}m$. Thus, we have the condition

$$(4.7) \qquad i \in N \iff H(i) \subset W.$$

Arguing as in the previous section, to compute N from W, it is enough to be able to identify the neighborhood conditions $\langle n, p\rangle$, where n in nonstandard and p is standard. That is, it is enough to show that N is pointwise recursive in W. This is clearly the case by Formula 4.7. ◇

§5. Open Problems

The Π_1-priority method was introduced by Friedberg and Mučnik to solve Post's problem. Formulated as an abstract scheme over $P^- + I\Sigma_0$, Groszek-Slaman show that the success of Π_1-method is equivalent to $I\Sigma_1$. However, we have shown in Theorem 3.1 that the most simple application of this method can be proven in the weaker theory $P^- + B\Sigma_1$.

5.1. Question. Is there a theorem of recursion theory proven using the Π_1-priority method that cannot be proven in $P^- + B\Sigma_1$? Can the Friedberg-Mučnik Theorem stating that there are recursively enumerable sets that are $\leq_T$ incomparable be proven in $P^- + B\Sigma_1$?

The Σ_2-priority method was introduced by Sacks [**10**] to prove the Splitting Theorem. Groszek-Slaman show that the success of the Σ_2-method is equivalent to $I\Sigma_2$. Our independence result states that the Splitting Theorem cannot be proven in $P^- + B\Sigma_1$ and Mytilinaios [7] has shown that it can be proven in $P^- + I\Sigma_1$.

5.2. Question. Is there a theorem of recursion theory proven using the Σ_2-priority method that cannot be proven in $P^- + B\Sigma_2$? Using the Σ_2-method, Harrington has shown that the creative sets form an orbit in the lattice of the recursively enumerable sets modulo the finite sets, see [**12**, Soare]. Can Harrington's theorem be proven in $P^- + B\Sigma_2$?

References

1. Chang, C. C. and Keisler, H. J., "Model Theory," (Studies in Logic and Foundations of Mathematics, vol. 73), North Holland, Amsterdam, 1973.
2. Friedberg, R. M., *Two recursively enumerable sets of incomparable degrees of unsolvability*, Proc. Natl. Acad. Sci. **43** (1957), 236-238.
3. Groszek, M. J. and Slaman, T. A., *Foundations of the priority method I: finite and infinite injury*, (to appear).
4. Groszek, M. J. and Slaman, T. A., *On Turing reducibility*, (to appear).
5. Kirby, L. A. S. and Paris, J. B., *Σ_n-collection schemas in arithmetic*, in "Logic Colloquium '77," North Holland, Amsterdam, 1978, pp. 199-209.
6. Mučnik, A. A., *On the unsolvability of the problem of reducibility in the theory of algorithms*, Dokl. Akad. Nauk SSSR, N. S. **108** (1956), 194-197, (Russian).
7. Mytilinaios, M. E., *Finite injury and Σ_1-induction*, Jour. Sym. Log. (to appear).
8. Mytilinaios, M. E. and Slaman, T. A., *Σ_2-collection and the infinite injury priority method*, Jour. Sym. Log. (to appear).
9. Post, E. L., *Recursively enumerable sets of positive integers and their decision problems*, Bull. Amer Math. Soc. 50 (1944), 284-316.
10. Sacks, G. E., *On the degrees less than* 0′, Ann. of Math. (2) **80** (1963), 211-231.
11. Simpson, S. G., private correspondence (1983).
12. Soare, R. I., "Recursively Enumerable Sets and Degrees," Springer-Verlag, Berlin, 1987.

Department of Mathematics; The University of Chicago; Chicago, Illinois, 60637, U. S. A.

Department of Mathematics; The California Institute of Technology; Pasadena, California, 91125 U. S. A.

COMPUTATIONAL COMPLEXITY OF LANGUAGES COUNTING RANDOM ORACLES

Tomoyuki Yamakami

Department of Mathematics, Rikkyo University
Nishi-ikebukuro, Tokyo, 171, Japan

ABSTRACT. The nature of the recursive real $\mu(U_B(x))$ in Structural Complexity Theory is discussed, in which μ is Lebesgue measure of random oracles and $U_B(x)$ denotes the set of every oracle Z such that the polynomial-time computable oracle-dependent language B^Z accepts x. To formalize this argument we introduced the class $L\mathcal{S}$ of any language Lm(B) expressing a lower bound of its measure of random oracles on an oracle-dependent language B^Z in a relativized class $\mathcal{S}^Z$, and we investigated the nature of this class. The class $C\mathcal{S}$ of languages was further introduced to examine the circumstances of $L\mathcal{S}$, each language L of which indicates a lower bound of the cardinality of elements in L by exponentially-many searchings. Originating from the P-hierarchy, the related LP- and CP-hierarchies was obtainable and their structures were also studied. Our best result for the lower and upper bounds of the computational complexity of LP is that PP $\le_m^p$ LP$\subseteq$PSPACE. More from an angle of relativization, a natural relativization of each $L\mathcal{S}$ and $C\mathcal{S}$ is treated, and the probability-one separations of the first levels of the LP- and CP-hierarchies are shown.

§1. *INTRODUCTION*. Recent investigations of the polynomial-time hierarchy in Structural Complexity Theory, simply denoted by the *P-hierarchy*, seem to have elucidated in the following aspects.

One direction for enriching the discussion of the P-hierarchy is to fractionalize the hierarchy and to explore some special properties classifying the intractabilities among languages. Meyer and Stockmeyer's early studies deepened the elucidation of its structure by beneficial concepts (see e.g.[12,14,15]).

Another fruitful argument is on the relativizations. Since Baker, Gill, and Solovay[3] showed, in 1975, that there exist two recursive languages being positive and negative answers on the relativized P=?NP question, the relativization of classes plays an important role in improving the original problems of the P-hierarchy.

As an elongation of the arguments on the complexity of computing functions by fixed-depth Boolean circuits, in 1985 Yao[17] have determined the oracle separation of the whole relativized P-hierarchy. To contribute to solving the collapse problem of the P-hierarchy, however, we should construct somewhat simpler oracles, e.g., sparse or tally set, to separate the relativized P-hierarchy owing to Long and Selman's investigations [12].

On the other hand, Bennett and Gill[5] proved, in 1981, the probability-one separation of the relativized P and NP with the concept of Lebesgue measure of random oracles.

In the stream of random oracle discussions on the relativized P-hierarchy, here we wish to take up a natural question of the measure of random oracles in the following setting. Let μ be Lebesgue measure of random oracles. We denote by B^Z a polynomial-time computable oracle-dependent language and define $U_B(x)$ as the set of oracles Z satisfying $x \in B^Z$. Note that, since the value $\mu(U_B(x))$ can be computed by some Turing machine, it must be a recursive real number on the unit interval. Then our question is: *What kind of recursive real number does $\mu(U_B(x))$ express* ?

To deal with this question as the computational complexity of languages, we need to introduce a language $Lm(B)$ intuitively asserting a lower bound of the value $\mu(U_B(x))$ for each string x. Strictly speaking, Lm(B) consists of every element $\langle x, z_k, z_n \rangle$ which expresses the relation $\mu(U_B(x)) > n/k$, where z_k is the $(k+1)$-th string by the natural order on Σ^*. Hence we adjust the question to the following form: *Can the language Lm(B) be computable in either polynomial time or polynomial space* ?

To study this in more general aspects, for a relativized class $\mathcal{B}^Z$ we further introduce a useful class $L\mathcal{B}$ of languages Lm(B) induced from oracle-dependent languages B^Z in $\mathcal{B}^Z$. Due to this concept we can formulate the above question as follows. Assume $\mathcal{B}^Z$ is in a class in the relativized P-hierarchy.

$$(1)\ P \supseteq ?L\mathcal{B} \qquad (2)\ PH \supseteq ?L\mathcal{B} \qquad (3)\ PSPACE \supseteq ?L\mathcal{B}.$$

Section 3 illustrates basic properties of the *LP-hierarchy* which is induced from the relativized P-hierarchy and thus can be found lower and upper bounds of its computational complexity. Main result in this section is that $PP \leq_m^P LP \subseteq LPH \subseteq EXPSPACE$. Hence we get the result $P \not\supseteq LP$ as an answer to the first question $P \supseteq ?L\mathcal{B}$ on the assumption that $P \neq NP$.

Section 4 introduces another class $C\mathcal{B}$ from a class $\mathcal{B}$ of languages to argue the circumstances of the LP-hierarchy. Any language in $C\mathcal{B}$ informally expresses the number of accepted elements by exponentially-many checkings among elements of a language in $\mathcal{B}$. The concept of the *CP-hierarchy* is also obtained. The consequence that PSPACE is an upper bound of the complexity of LP is demonstrated in this section.

In Section 5 we further study natural relativizations of the LP- and CP-hierarchies and give some positive and negative answers for the collapse questions of the *relativized LP- and CP-hierarchies*. As an answer to the second question in the relativized form, we here show the existence of an oracle D satisfying $PH^D \neq LP^D$. Also shown are some more applicable results relating to the probability-one separation of the first level classes in the relativized LP- and CP-hierarchies.

Problems still remaining unsolved in this paper are listed in the last section.

§2. *PRELIMINARIES*. We wish to make precise some of the notions and notations which we shall use here.

Let Σ be the set $\{0,1\}$ throughout this paper and suppose every element in Σ^* is enumerated by the natural ordering as $\lambda<0<1<00<\cdots$, where λ is the empty string. To the readability, we set z_i to be the $(i+1)$-th element in this ordering starting from $z_0=\lambda$. More strictly z_n is defined by:

$$z_n=x \text{ if and only if } 1x=\text{the binary representation of } n+1.$$

μ denotes Lebesgue measure on the power set $\mathcal{P}(\Sigma^*)$ identifying the Cantor-space Σ^ω (see e.g.[1,5]). Note that μ is a measure on the unit interval.

Let $\leq_m^P$ be the polynomial-time many-one reducibility between two languages (see e.g.[1,2,11]). Here can be used the relation $\mathcal{A}\leq_m^P\mathcal{B}$ even for two classes $\mathcal{A}$ and $\mathcal{B}$ as the abbreviation of: $\forall L\in\mathcal{A}\exists K\in\mathcal{B}(L\leq_m^P K)$. Furthermore $\mathcal{A}\equiv_m^P\mathcal{B}$ expresses both $\mathcal{A}\leq_m^P\mathcal{B}$ and $\mathcal{B}\leq_m^P\mathcal{A}$, and $\mathcal{A}<_m^P\mathcal{B}$ expresses both $\mathcal{A}\leq_m^P\mathcal{B}$ and $\mathcal{A}\not\equiv_m^P\mathcal{B}$.

Assume the pairing $\langle x_1,x_2,\cdots,x_k\rangle$ of k-tuples satisfies, for some polynomial r, $|x_i|<|\langle x_1,x_2,\cdots,x_k\rangle|\leq r(|x_1|+|x_2|+\cdots+|x_k|)$, where $1\leq i\leq k$.

We denote by $K\oplus L$ the natural join of two languages K and L, that is,

$$K\oplus L=\{0x|x\in K\}\cup\{1x|x\in L\}.$$

For each non-negative integer m and n strings $x_1,x_2,\cdots,x_n$, let the notation $(\#k\leq m)(R(x_1,x_2,\cdots,x_n,k))$ denote the cardinality of the set

$$\{k\in\omega|k\leq m,R(x_1,x_2,\cdots,x_n,k)\},$$

where R is a relation with $n+1$ argument-places.

The characteristic function χ_A for a set A is defined by

$$\chi_A(y)=\begin{cases}1 & \text{if } y\in A,\\ 0 & \text{otherwise.}\end{cases}$$

We assume that any oracle-dependent language L^Z treated in the present paper is obtained by a natural relativization of the basic language L. Let A be a language and k be a positive integer. $NP^{A[k]}$ denotes the family of languages accepted by nondeterministic oracle Turing machines in polynomial time with k oracle queries to A in each computation path.

The reader's familiarity with the P-hierarchy is assumed here. In particular, remark that we adopt the following definition of the class EXPSPACE differing from e.g.[12]: $\mathrm{EXPSPACE}=\bigcup_{k\geq 1}\mathrm{DSPACE}(2^{n^k})$. Additionally, PSPACE^X indicates the set of languages witnessed by deterministic oracle machines with polynomial-space bounded input/work and oracle tapes, and with the oracle X.

§3. *LANGUAGES COUNTING RANDOM ORACLES*. To investigate the behavior of random oracles in oracle-dependent languages, we wish to introduce the useful class $\mathcal{LB}$ which is based on Lebesgue measure of random oracles on each input.

Now let us begin with the definition.

***DEFINITION* 3.1.** Let B^Z be an oracle-dependent language with an oracle Z and $\mathcal{B}^Z$ be a class of oracle-dependent languages with Z. Define $Lm(B)$ and $\mathcal{LB}$ by:

$$\mathrm{Lm(B)} = \{\langle x, z_k, z_n\rangle \in \Sigma^* \mid \mu(\{Z\subseteq\Sigma^* \mid x\in B^Z\}) > n/k;\ k,n\in\omega\},\ \text{and}$$

$$\mathcal{LB} = \{\mathrm{Lm(B)} \subseteq \Sigma^* \mid B^Z \in \mathcal{B}^Z\}.$$

Each element in a language Lm(B) induced from B^Z intuitively indicates how many oracles are there which B^Z accepts x.

Now let $U_B(x)$ denote the set of oracles: $\{Z\subseteq\Sigma^* \mid x\in B^Z\}$. For any given input $x\in\Sigma^*$, the value of the measure of $U_B(x)$ is clearly able to be computed by some Turing machine. Therefore $\mu(U_B(x))$ always expresses a recursive real number on the unit interval.

The main questions we raise here are described as follows. Suppose $\mathcal{B}^Z$ is an oracle-dependent class in the relativized P-hierarchy.

(1) $P \supseteq ?\mathcal{LB}$ (2) $PH \supseteq ?\mathcal{LB}$ (3) $PSPACE \supseteq ?\mathcal{LB}$.

To aim at considering the computational complexity of these languages counting random oracles, relating to the P-hierarchy, we need to make the following definition.

***DEFINITION* 3.2.** $\{L\Delta_k^p, L\Sigma_k^p, L\Pi_k^p \mid k\in\omega\}$ is called the *LP-hierarchy*.

This definition simply shows that $L\Sigma_k^p \cup L\Pi_k^p \subseteq L\Delta_{k+1}^p \subseteq L\Sigma_{k+1}^p \cap L\Pi_{k+1}^p$ holds for every integer $k\in\omega$, however, it is still unknown whether these inclusions are really proper or not.

Before discussing lower and upper complexity bounds of the LP-hierarchy, let us first see some basic nature of the hierarchy.

Provided that $\mathcal{A}^X = \mathcal{B}^X$ holds for any oracle X not directly depending on the peculiarity of X, clearly $\mathcal{LA} = \mathcal{LB}$ holds. Now denote by NP_B^X the class of languages recognizable by some nondeterministic polynomial-time bounded Turing machine in which the number of all query strings in the computation tree is below some polynomial (see [6,12]). Let us define, for all $k>0$,

$$\Sigma_{B,1}^{p,X} = NP_B^X \text{ and } \Sigma_{B,k+1}^{p,X} = NP_B^{\Sigma_k^{p,X}}.$$

***PROPOSITION* 3.3.**

(1) *$P=NP$ implies $LP=LNP_B$.*

(2) *$L\Delta_k^p = L\Sigma_{B,k}^p$ if $k>1$.*

Proof. (1) This follows by certifying the fact in [6], that P=NP implies $P^Z = NP_B^Z$ for all Z.

(2) Similar to the proof of (1), we can see $\Delta_k^{p,Z} = \Sigma_{B,k}^{p,Z}$ holds for any Z if $k>1$. □

***PROPOSITION* 3.4.** *$L\Delta_k^p = L\Sigma_k^p$ if and only if $L\Delta_k^p = L\Pi_k^p$, where $k\in\omega$.*

Proof. The if-part will be only demonstrated here. We assume that $L\Delta_k^p = L\Pi_k^p$. For any language $A^Z \in \Sigma_k^{p,Z}$, the complement $\overline{A}^Z$ is in $\Pi_k^{p,Z}$. Since $U_A(x)$ and $U_{\overline{A}}(x)$ are distinct and $U_A(x) \cup U_{\overline{A}}(x) = \mathcal{P}(\Sigma^*)$, then $\mu(U_A(x)) + \mu(U_{\overline{A}}(x)) = 1$ can be concluded. Under the hypothesis there exists a language $B^Z \in \Delta_k^{p,Z}$ such that $\mu(U_B(x)) = \mu(U_{\bar{A}}(x))$. Therefore $\mu(U_A(x)) = \mu(U_{\bar{B}}(x))$ for any string x, and $Lm(A) = Lm(\bar{B})$. The fact $\bar{B}^Z \in \Delta_k^{p,Z}$ leads to $Lm(A) \in L\Delta_k^p$. □

This proposition guarantees that $L\Delta_k^p = L\Sigma_k^p$ implies $L\Sigma_k^p = L\Pi_k^p$. On the other hand, we have no proof that the downward separation holds also in the LP-hierarchy. Unless LP=LNP ensures LP=LPH, unfortunately it may be quite difficult to investigate the collapse problem of the LP-hierarchy.

Additionally, let us see the important concept of the reducibility of classes in the LP-hierarchy.

PROPOSITION **3.5.** *Suppose that $\mathcal{B}^Z$ is a class in the relativized P-hierarchy and both L and K are any two languages.*

(1) $\mathcal{B} \leq_m^p L\mathcal{B}$.

(2) $L \oplus K \leq_m^p L\mathcal{B}$ *if* $L, K \leq_m^p L\mathcal{B}$.

(3) *Each* $L\Sigma_k^p$ *has* $\leq_m^p$*-complete languages, where* $k \geq 0$.

Proof. (1) For any language $A \in \mathcal{B}$, it suffices to define A^Z as A itself. Then a result $A \leq_m^p Lm(A)$ is obviously derived.

(2) Now take languages $K^Z, L^Z \in \mathcal{B}^Z$ and define a polynomial-time transducer f as follows: $f(u) = \langle kx, y, z \rangle$ if $u = kv$, $k \in \{0,1\}$ and $v = \langle x, y, z \rangle$ are satisfied, otherwise $f(u) = 0$. If we write $(L \oplus K)^Z$ instead of $L^Z \oplus K^Z$, then $(L \oplus K)^Z \in \mathcal{B}^Z$ and hence $Lm(L) \oplus Lm(K) \leq_m^p Lm(L \oplus K)$ by f.

(3) Take any $\leq_m^p$-complete language A^Z in $\Sigma_k^{p,Z}$ which each reducing transducer has no reliance on the nature of Z. Accordingly, $Lm(A)$ becomes $\leq_m^p$-complete in $L\Sigma_k^p$. □

The latter part of this section is devoted to investigate upper and lower bounds of the complexity of the classes in the LP-hierarchy.

Relating to an upper bound of the complexity of the LP-hierarchy, we are able to construct exponential-space bounded Turing machines which have the power of witnessing each element in LPSPACE. Hence LPSPACE ⊆ EXPSPACE holds. This is an upper bound of the LP-hierarchy.

THEOREM **3.6.** *LP ⊆ LNP ⊆ LPH ⊆ LPSPACE ⊆ EXPSPACE.*

Next we show a lower complexity bound of the LP-hierarchy by using the probabilistic Turing machines. The concept of the probabilistic Turing machine was first argued by Gill[9]. To see the lower bound, we introduce the class PROB that intuitively seems to be an extension of the class PP.

Let $\mathcal{M}^P$ be the set of any probabilistic Turing machine which always halts in some polynomial time. Without loss of generality, it can be further supposed that any nondeterministic branch of each configuration in a probabilistic Turing machine is two (we may allow the same instructions) and any computation path terminates in just $p(|x|)$-many steps for some fixed polynomial p. Owing to this set, a class PROB is induced as follows.

DEFINITION **3.7.** Let M be a probabilistic Turing machine. Define:

$Pt(M) = \{\langle x, z_k, z_n \rangle \in \Sigma^* \mid Pr(M(x)=1) > n/k;\ k, n \in \omega\}$, and

$PROB = \{Pt(M) \subseteq \Sigma^* \mid M \in \mathcal{M}^P\}$.

LEMMA 3.8. *$PP \le_m^p PROB$.*

Proof. Suppose A is any element in PP. Then there exists a polynomial-time probabilistic Turing machine M satisfying the condition that its error probability is bounded below 1/2 and recognizing A as $A=\{x|Pr(M(x)=1)>1/2\}$. It is enough to choose a polynomial-time transducer f as $f(x)=\langle x,z_2,z_1\rangle$. M can be easily seen to be in $\mathcal{M}^P$ and thus $A\le_m^P Pt(M)$ holds. □

The following is the key lemma in this section.

LEMMA 3.9. *PROB=LP*

Proof Sketch. We now show an outline of a proof of PROB=LP. Let M be any probabilistic machine in $\mathcal{M}^P$. For a given input x, if the i-th configuration of M has two different instructions, let $C_0(i)$ and $C_1(i)$ be these two instructions. Otherwise, let $C_0(i)$ denote the unique instruction.

Since each computation path $C=\langle C_{i_0}(0),C_{i_1}(1),\cdots,C_{i_{p(|x|)}}(p(|x|))\rangle$ on an input x is identified with the set $B_C=\{z_{i_0},z_{i_1},\cdots,z_{i_{p(|x|)}}\}$, C is simulated by some deterministic machine M_0 with B_C as the oracle. Thus we can get that Pt(M) =Lm(L(M_0)) and Pt(M)∈LP.

Conversely, let Lm(A) be in LP and A^Z be recognizable by a polynomial-time deterministic oracle Turing machine M. To show Lm(A)∈PROB, it is enough to construct a probabilistic machine M_0 simulating M. So M_0 is defined as follows. M_0 simulates M while M does not enter any query state. Whenever M queries a string y, M_0 stores y and tosses a coin to simulate an answer from the oracle. Obviously $M_0\in\mathcal{M}^P$ and Lm(L(M))=Pt(M_0). Thus Lm(A)=Pt(M_0).

Therefore PROB=LP is proved. □

The above two lemmas derive the next desired consequence which shows a lower bound of the complexity of the LP-hierarchy.

THEOREM 3.10. *$PP\le_m^p L\mathcal{B}$, where $L\mathcal{B}$ is in the LP-hierarchy.*

Remark that it is unknown whether this reducibility is proper (i.e. $PP<_m^P L\mathcal{B}$). Section 5 will give a relativized answer to this question with probability one.

The next corollary partially answers the first and second questions which we raised at the beginning of this section.

COROLLARY 3.11.

(1) *P⊇LP implies P=PP (and then P=NP).*

(2) *PH⊇LP implies PH⊇PP.*

From this result, it can be concluded that P⊉$L\mathcal{B}$ holds for any class $L\mathcal{B}$ in the LP-hierarchy on the assumption of P≠NP. We wish to mention that, relating to the second question PH⊇?LP, one negative answer to the relativized case can be demonstrated in Section 5.

To see the complexity of LP, we lastly note the relation between LP and the class #P discussed by Valiant[16]. Remark that a function $g:\Sigma^*\to\omega$ is in #P if and only if there exist a probabilistic machine $M\in\mathcal{M}^P$ and a function f: $\Sigma^*\to\omega$ such that, for any input x, $f(x)$ indicates the number of all terminal configurations in every computation path of M(x), and the equation $Pr(M(x)=1)=g(x)/f(x)$ holds.

***PROPOSITION* 3.12.** *For any* $g\in\#P$, $A_g\in NP^{LP[2]}$, *where* A_g *be* $\{\langle x,z_n\rangle \mid g(x)=n\}$.

Proof Sketch. Let us take a machine M in $\mathcal{M}^P$ and a function f defining the given function g. To show $A_g\in NP^{LP[2]}$, we must design the following nondeterministic oracle Turing machine M_0 with Pt(M) as the oracle since PROB=LP. On input $\langle x,z_{g(x)}\rangle$, M_0 guesses a string $z_{f(x)}$ and queries both $s_0=\langle x,z_{g(x)},z_{f(x)}\rangle$ and $s_1=\langle x,z_{g(x)+1},z_{f(x)}\rangle$ to Pt(M). If $s_0\in Pt(M)$ and $s_1\notin Pt(M)$, M_0 accepts $\langle x,z_{g(x)}\rangle$. It is clear M_0 witnesses A_g relative to Pt(M). Hence $A_g\in NP^{Pt(M)[2]}\subseteq NP^{LP[2]}$. □

§4. *LANGUAGES CHECKING EXPONENTIALLY-MANY STRINGS*. The previous section showed that EXPSPACE is an upper bound of the complexity of the LP-hierarchy. This section will investigate some other circumstances of the LP-hierarchy and see the third question PSPACE≥?L$\mathcal{B}$.

Our basic idea is to introduce the special families of languages below PSPACE not to depend on the random oracles, which have a close connection with L$\mathcal{B}$. Founded on the idea, a meaningful language is needed to introduce, which expresses the number of accepted strings by checking exponentially-many strings. The following is the formal definition.

***DEFINITION* 4.1.** Let f be a function on ω, $B\subseteq\Sigma^*$ and let $\mathcal{B}$ be a class of languages. Define $L(f,B)$ and $C\mathcal{B}$ by:

$$L(f,B)=\{\langle x,z_m\rangle\in\Sigma^* \mid (\#k\le 2^{f(|x|)})(\langle x,z_k\rangle\in B)>m;\ m\in\omega\},\text{ and}$$

$$C\mathcal{B}=\{L(p,B)\subseteq\Sigma^* \mid p\text{ is a polynomial, }B\in\mathcal{B}\}.$$

$L(p,B)$ expresses a lower bound of the cardinality of $\{\langle x,z_k\rangle\in B \mid k\le 2^{p(|x|)}\}$ for $B\in\mathcal{B}$. It seems to behave as a census function for the given language B.

Similar to the LP-hierarchy, the CP-hierarchy can be introduced from the P-hierarchy.

***DEFINITION* 4.2.** $\{C\Delta_k^P, C\Sigma_k^P, C\Pi_k^P \mid k\in\omega\}$ is called the *CP-hierarchy*.

To begin with, we shall observe some relationships of these families of languages checking exponentially-many strings to their basic class of languages. It should be noted that the definiton directly shows $C\mathcal{A}=C\mathcal{B}$ if $\mathcal{A}=\mathcal{B}$ holds. Hence CP≠CNP concludes P≠NP. Also in the CP-hierarchy, $C\Sigma_k^P\cup C\Pi_k^P\subseteq C\Delta_{k+1}^P\subseteq C\Sigma_{k+1}^P\cap C\Pi_{k+1}^P$ holds for all $k\in\omega$.

***DEFINITION* 4.3.** $C\mathcal{B}$ is *condensable* if and only if

$\forall K\in C\mathcal{B}\ \forall p$:polynomial $\exists B\in\mathcal{B}$ such that $(\forall x)[\langle x,z_{2^{p(|x|)}+1}\rangle\notin K]$ implies $K=L(p,B)$.

***PROPOSITION* 4.4.** *Assume* $\mathcal{B}$ *is a class in the P-hierarchy.*

(1) $\mathcal{B}\le_m^p C\mathcal{B}$.

(2) $C\mathcal{B}\subseteq C(C\mathcal{B})$.

(3) *Suppose* $C\Delta_k^p$ *is condensable. Then* $C\Delta_k^p=C\Sigma_k^p$ *if and only if* $C\Delta_k^p=C\Pi_k^p$.

Proof. (1) For any language $A\in\mathcal{B}$, define $B=\{\langle x,z_0\rangle \mid x\in A\}$ and $f(u)=\langle u,z_0\rangle$. Then this comes to the conclusion that $A\le_m^P L(0,B)$ by f.

(2) This follows from the fact that, for any polynomials p,q and a language $A\subseteq\Sigma^*$, $L(p,A)=L(q,L(p,A))$ if $(\forall n\in\omega)[q(n)\ge p(n)]$.

(3) The only-if part will be proved here. It is supposed that $A\in\Pi_k^P$ and

p is any polynomial. Since $C\Delta_k^P = C\Sigma_k^P$, there exist a language $B \in \Delta_k^P$ and a polynomial q so that, for all x, $(\#k \le 2^{p(|x|)})(\langle x, z_k\rangle \in \bar{A}) = (\#k \le 2^{q(|x|)})(\langle x, z_k\rangle \in B)$ holds. This concludes $\langle x, z_{2^{p(|x|)}+1}\rangle \notin L(q,B)$.

From the condensability of $L(q,B)$, $L(q,B) = L(p,C)$ follows for some $C \in \Delta_k^P$. Therefore we can obtain $L(p,A) = L(p,\bar{C})$ and $L(p,\bar{C}) \in C\Delta_k^P$. □

Recall that a set A is sparse if and only if there exists a polynomial p such that $(\#k \le 2^{n+1})(|z_k| \le n$ and $z_k \in A) \le p(|x|)$ for all $n \in \omega$. This sparseness plays a singular role in the languages checking exponentially-many strings. Here we show some typical cases.

PROPOSITION 4.5. *Let A be a language and p be a polynomial.*

(1) *$L(p,A) \in \Sigma_k^p$ if A is sparse in Σ_k^p.*

(2) *There exists a sparse set B such that $L(p,A) = L(p,B)$ if $L(p,A)$ is sparse.*

(3) *$C\Sigma_k^p$ has no $\le_m^p$-complete set which is sparse unless $\Sigma_k^p \equiv_m^p C\Sigma_k^p$, where $k \in \omega$.*

Proof. (1) The sparseness of A shows the existence of a polynomial upper bound q such that $(\forall x)[(\#k \le 2^{p(|x|)})(\langle x, z_k\rangle \in A) \le q(|x|)]$. This leads to the set $L(p,A)$ being recognized by $q(|x|)$-many nondeterministic searchings. Thus $L(p,A)$ is in Σ_k^P.

(2) Define B as the set $\{\langle x, z_k\rangle \in A \mid k \le 2^{p(|x|)}\}$.

(3) This comes from (1) and (2). It is left to the reader. □

As to the P-hierarchy, here suffices it to consider that the class $C\Sigma_k^P/\mathrm{Str}$ is equivalent to Σ_k^P, which is a subset of $C\Sigma_k^P/\mathrm{Poly}$, where *Str* is a set of all functions h from ω into Σ^* such that there is a polynomial p satisfying $h(n) = z_{p(n)}$ for any integer $n \in \omega$ (see e.g. [12]).

PROPOSITION 4.6. *$C\Sigma_k^p/Str = \Sigma_k^p$, where $k \in \omega$.*

Proof Sketch. For any language $A \in \Sigma_k^P$, let B be a set $\{\langle x, z_0\rangle \mid x \in A\}$ and h be a function as $h(n) = z_0$ for all $n \in \omega$. So B is still in Σ_k^P. Now we set E to be $\{\langle x, z_m\rangle \mid (\#k \le 1)(\langle x, z_k\rangle \in B) > m\}$. Then $A = \{x \mid \langle x, h(|x|)\rangle \in E\}$ and $A \in C\Sigma_k^P/\mathrm{Str}$. Thus $\Sigma_k^P \subseteq C\Sigma_k^P/\mathrm{Str}$.

Conversely, suppose A is in $C\Sigma_k^P/\mathrm{Str}$, then there exist a language $B \in \Sigma_k^P$ and two polynomials p and q satisfying $A = \{x \mid (\#k \le 2^{p(|x|)})(\langle x, z_k\rangle \in B) > q(|x|)\}$. Therefore A is recognizable by $q(|x|)+1$-many nondeterministic searchings on B. Hence $A \in \Sigma_k^P$ and $C\Sigma_k^P/\mathrm{Str} \subseteq \Sigma_k^P$. □

By $C\Sigma_k^P/\mathrm{Str} \subseteq C\Sigma_k^P/\mathrm{Poly}$, the above proposition is able to infer $\Sigma_k^P \subseteq C\Sigma_k^P/\mathrm{Poly}$, however, whether $C\Sigma_k^P \subseteq C\Sigma_k^P/\mathrm{Poly}$ is still open.

The CP-hierarchy has a clearer upper bound than the LP-hierarchy does. Since exponentially-many checkings of accepting strings are deterministically traced by a polynomial-space machine, hence obviously we conclude as:

PROPOSITION 4.7. *CPSPACE $\equiv_m^p$ PSPACE.*

COROLLARY 4.8. *CP $\subseteq$ CNP $\subseteq$ CPH $\subseteq$ CPSPACE $\subseteq$ PSPACE.*

Therefore the CP-hierarchy is informaly said to be located between PP and PSPACE. Let us next prove the main theorem in this section concerning the relation of L𝓑 and C𝓑.

***THEOREM* 4.9.**

(1) *If 𝓑 is a class in the P-hierarchy then* $C\mathcal{B}\leq^p_m L\mathcal{B}$.

(2) $CP\equiv^p_m LP$.

Proof Sketch. (1) Take B in 𝓑 and a polynomial p. Now let us construct a polynomial-time oracle Turing machine M_0 so that, for an oracle Z, M_0 queries the strings $\{z_0,z_1,\cdots,z_{p(|x|)+1}\}$ and defines t_i by the answer of the oracle Z, that is, $t_i=\chi_Z(z_i)$ for $0\leq i\leq p(|x|)+1$. Next compute a string $s_x=t_{j+1}\cdots t_{p(|x|)+1}$, where $j=\inf\{i\leq p(|x|)+1 \mid t_i=1\}$; and if $s_x\leq z_{2^{p(|x|)}}$, then check whether $\langle x,s_x\rangle\in B$ holds. It is obvious that $L(M_0)$ is in 𝓑 and $Lm(L(M_0))\in L\mathcal{B}$.

Using this machine M_0 let $B_x=\{Z\mid(\forall k>p(|x|)+1)(z_k\notin Z)\}$ and $N'_x=\{Z\in B_x\mid M_0(x)=1\}$. Note that $\mu(\{Z\mid M_0^Z(x)=1\})=|N'_x|/2^{p(|x|)+2}$. If we let $K_x=\{z_k\mid k\leq 2^{p(|x|)},\langle x,z_k\rangle\in B\}$ then the equation $|N'_x|=|K_x|=(\#k\leq 2^{p(|x|)})(\langle x,z_k\rangle\in B)$ is obtained since N'_x is isomorphic to K_x.

Thus $(\#k\leq 2^{p(|x|)})(\langle x,z_k\rangle\in B)=2^{p(|x|)+2}\cdot\mu(\{Z\mid M_0^Z(x)=1\})$. We next define a polynomial-time transducer h as follows: if both $u=\langle x,z_m\rangle$ and $r=2^{p(|x|)+2}$ are satisfied, then let $h(u)=\langle x,z_r,z_m\rangle$, otherwise $h(u)=0$.

Therefore $L(p,B)\leq^p_m Lm(L(M_0))$ by h.

(2) Let us take a language $L^Z\in P^Z$. Then there exists a polynomial-time deterministic oracle machine M satisfying $L^Z=L(M^Z)$. Without loss of generality M queries just $p(|x|)$-many strings of length $\leq p(|x|)-3$ for some polynomial p. Assume $M^Z(x)$ queries the strings $\{u_0,u_1,\cdots,u_{p(|x|)}\}$, where $u_0<u_1<\cdots<u_{p(|x|)}$. Let us define $s_{x,Z}=s_0t_0s_1t_1\cdots s_{p(|x|)-1}t_{p(|x|)-1}$, where each s_i is $0^{p(|x|)-|u_i|-3}1u_i$ and t_i is defined by $t_i=\chi_Z(u_i)$. Therefore $s_{x,Z}$ uniquely corresponds to the computation of $M^Z(x)$.

Now construct the following polynomial-time deterministic Turing machine M_0. On input $\langle x,s_{x,Z}\rangle$, M_0 simulates the computation of $M^Z(x)$ in a way that, if $M^Z(x)$ queries a string u_i, M_0 checks whether $t_i=1$ or not. Put C_x as

$$C_x=\{s\in\Sigma^*\mid s=s_0t_0\cdots s_{p(|x|)-1}t_{p(|x|)-1},|s_i|=p(|x|)-1,|t_i|=1,s_0<\cdots<s_{p(|x|)-1}\}.$$

When U_x is $\{s\in C_x\mid M_0(\langle x,s\rangle)=1\}$, the equation $\mu(\{Z\mid M^Z(x)=1\})=|U_x|/2^{p(|x|)}$ is concluded. From $|U_x|=(\#k\leq 2^{p(|x|)^2})(\langle x,z_k\rangle\in L(M_0))$, we can infer

$$(\#k\leq 2^{p(|x|)^2})(\langle x,z_k\rangle\in L(M_0))=2^{p(|x|)}\cdot\mu(\{Z\mid M^Z(x)=1\}).$$

To reach the desired assertion it is enough to define a polynomial-time transducer f as follows: $f(u)=\langle x,z_r\rangle$ if both $u=\langle x,z_m,z_n\rangle$ and $r=[n2^{p(|x|)}/m]$ hold, otherwise $f(u)=0$, where [ ] denotes Gauss' symbol. Hence we get that $Lm(L(M))\leq^p_m L(p^2,A)$ by this transducer f. □

***COROLLARY* 4.10.** $CP<^p_m CNP$ *implies* $LP\subsetneq LNP$.

Theorem 4.9 partially gives us an affirmative answer to the third question, that is, PSPACE⊇?L𝓑 for any class L𝓑 in the LP-hierarchy. The following is the best result known to us on the upper complexity bounds of only LP.

THEOREM **4.11.** *PSPACE* $\supseteq$ *LP*.

It remains still unsolved whether even PSPACE $\supseteq$? LNP holds or not.

§5. *ORACLES*. We shall turn our viewpoint on a feature of relativized languages $L\mathcal{B}^Z$ and $C\mathcal{B}^Z$. Oracle treatments give us pertinent information on the properties in the LP- and CP-hierarchies similar to the P-hierarchy. This section investigates one natural relativization of $L\mathcal{B}$ and $C\mathcal{B}$.

DEFINITION **5.1.** Let B^X be an oracle-dependent language with an oracle X, and $\mathcal{B}^X$ a class of oracle-dependent languages with X. Let f be a function on ω and Y an oracle.

$$Lm(B)^Y = \{\langle x, z_k, z_n\rangle \in \Sigma^* \mid \mu(\{Z \mid x \in B^{Z \oplus Y}\}) > n/k;\ k, n \in \omega\}.$$

$$L\mathcal{B}^Y = \{Lm(B)^Y \subseteq \Sigma^* \mid B^Y \in \mathcal{B}^Y\}.$$

$$L(f,B)^Y = \{\langle x, z_m\rangle \in \Sigma^* \mid (\#k \leq 2^{f(|x|)})(\langle x, z_k\rangle \in B^Y) > m;\ m \in \omega\}.$$

$$C\mathcal{B}^Y = \{L(p,B)^Y \subseteq \Sigma^* \mid p \text{ is a polynomial, } B^Y \in \mathcal{B}^Y\}.$$

Note that $L\mathcal{B}^{\emptyset} = L\mathcal{B}$ and $C\mathcal{B}^{\emptyset} = C\mathcal{B}$.

PROPOSITION **5.2.**

(1) $CNP^{\Sigma_k^p} = C\Sigma_{k+1}^p$ *and* $LNP^{\Sigma_k^p} \subseteq L\Sigma_{k+1}^p$ *for any integer* $k \in \omega$.

(2) $A \leq_m^p LP^A$ *for any oracle A.*

(3) $L\mathcal{B} \leq_m^p L(L\mathcal{B}))$ *for any relativized class* $\mathcal{B}^Z$ *in the relativized P-hierarchy.*

(4) $LPP \leq_m^p L(LP)$.

Proof. (1) By definition.

(2) If $f(x) = \langle x, z_2, z_1\rangle$ and $B^Z = \{x \mid 1x \in Z\} \in P^Z$ are taken, $A \leq_m^p Lm(B)^A$ holds by the transducer f for any oracle A.

(3) Define f as: $f(u) = \langle u, z_2, z_1\rangle$ if $u = \langle x, y, v\rangle$ and $y > z_0$, otherwise $f(u) = 0$. For any language $A^Z \in \mathcal{B}^Z$, let us consider two sets $Lm(A)$ and $Lm(A)^Z$. Therefore a relation $Lm(A) \leq_m^p Lm(Lm(A))$ can be derived from f.

(4) This is a direct consequence of $PP^Y \leq_m^p LP^Y$ for any oracle Y. Note that this is shown, not depending on the property of Y, according to Theorem 3.10. □

On account of the difference of relativizations between $L\mathcal{B}^Y$ and $C\mathcal{B}^Y$, it is not known whether $LNP^{\Sigma_k^p} = L\Sigma_{k+1}^p$ for all $k > 0$ or not.

Baker, Gill, and Solovay's techniques of constructions of special oracles carry on the relativized LP- and CP-hierarchies (see [3]).

PROPOSITION **5.3.** *There exist recursive languages A, B, and C such that*

(1) $CP^A = CNP^A$,

(2) $LP^B \equiv_m^p LNP^B \equiv_m^p P^B$ *and*

(3) $CP^C \neq C\Pi_1^{p,C}$.

Proof Sketch. (1) Take any oracle A so that $P^A = NP^A$ then $CP^A = CNP^A$ holds.

(2) It is enough to construct B satisfying the relationship that $LNP^B \subseteq P^B$. Suppose NP^Z is enumerated as $L(M_0), L(M_1), \cdots$, where each M_i is a polynomial-time nondeterministic Turing machine. Then let B be a set

$\{\langle z_j, x, z_m, z_n, 0^k\rangle \mid \mu(\{X \mid M_j^{X\oplus B}$ accepts x in $\le k$-steps$\}) > n/m\}$.

For any element $L\in LNP^B$, L is obviously recognized by a deterministic machine with the help of B.

(3) We here assume that some fixed polynomial r satisfies, for all strings x and y, $|\langle x,y\rangle| \le r(|x|+|y|)$. Now define C in the following way. Suppose CP^Y is enumerated as $L(p_0,B_0)^Y, L(p_1,B_1)^Y, \cdots$, and each B_i^Y is recognized by a deterministic machine M_i^Y bounded by some polynomial q_i.

(i) Let $n_{-1}=-1$ and $C_{-1}=\emptyset$.

(ii) Let n_k be $\inf\{j\in\omega \mid q_k(r(j+p_k(j)+1))<2^j, n_{k-1}+1<j\}$ and C_k be defined as follows. Put E_k be $\{y \mid |y|=n_k\}$. If $\exists m\le 2^{p_k(n_k)}\exists D\subseteq E_k(M_k^{A_{k-1}\cup D}(\langle 0^{n_k}, z_m\rangle)=1)$ holds, choose the least m and the set D so that the cardinality of D depending on the number m is the least. Now let C_k be $C_{k-1}\cup D$. Otherwise let C_k be $C_{k-1}\cup E_k$.

Lastly let C be $\bigcup_{k\in\omega} C_k$. The verification of $CP^C \ne C\Pi_1^{p,C}$ is left to the reader. □

Now let us recall the second question, that is, $PH \supseteq ? L\mathcal{B}$ for any class $L\mathcal{B}$ in the LP-hierarchy. Although Corollary 3.10 has given a partial solution of this question, we here demonstrate a negative answer to the question in the relativized form in the following theorem.

***THEOREM* 5.4.** *There exists an oracle D satisfying $PH^D \not\supseteq L\mathcal{B}^D$ for any class $L\mathcal{B}^Z$ in the relativized LP-hierarchy.*

Proof. Let $L^Y=\{1^n \mid \exists m\le 2^{n-1}$ s.t. $(\#k\le 2^{n+1})(|z_k|=n$ and $z_k\in Y)=2m+1\}$. Note that there exists a language D satisfying $L^D\notin PH^D$ from the result gained by Yao [17].

To obtain the desired result, it suffices to show that $PH^D \not\supseteq CP^D$ for this D. Suppose $PH^D \supseteq CP^D$. We define $A^Y=\{\langle 1^n, z_k\rangle \mid |z_k|=n, z_k\in Y\}$ and $h(n)=n$ for all $n\in\omega$. Since $L(h,A)^D\in CP^D$, $L(h,A)^D\in\Sigma_k^{p,D}$ is derivable for some $k\in\omega$. A simple consideration shows that L^D is recognized by some polynomial-time nondeterministic oracle machine relative to $L(h,A)^D$. This expresses that L^D is in $NP^{L(h,A)^D}$. Then $L^D\in NP^{L(h,A)^D}\subseteq NP^{\Sigma_k^{p,D}}=\Sigma_{k+1}^{p,D}\subseteq PH^D$. Hence $L^D\in PH^D$ follows.

However this result $L^D\in PH^D$ contradicts to the previous fact $L^D\notin PH^D$. □

Next we shall discuss the probability-one separation of the lower classes in the relativized LP- and CP-hierarchies. This stronger separation, first investigated by Bennett and Gill[5], gives an important suggestion to the collapse questions of both the relativized LP- and CP-hierarchies.

***LEMMA* 5.5.** *There exist two languages $L^X, K^X\in\Pi_1^{p,X}$ such that*

(1) *$L^X \le_m^p CNP^X$, $L^X\notin CNP^X$ and $L^X\in C\Pi_1^{p,X}$ with probability one, and*

(2) *$K^X \le_m^p LNP^X$, $K^X\notin LNP^X$ and $K^X\in L\Pi_1^{p,X}$ with probability one.*

Proof Sketch. (1) Let $\eta_Z(u)$ be

$$\eta_Z(u)=\begin{cases}\lambda & \text{if } u=\lambda,\\ \chi_Z(u10)\chi_Z(u10^2)\cdots\chi_Z(u10^{|u|}) & \text{otherwise.}\end{cases}$$

Let F^Z be $\{\langle x,z_0\rangle \mid (\forall k\le 2^{|x|+1})(\eta_Z(z_k)\ne x)\}$. A consideration of $L(0,F)^Z$

clearly infers that it is in $C\Pi_1^{p,Z} \cap \Pi_1^{p,Z}$. Now define C_n^0 and C_n^1 as

$C_n^0 = \{Z \mid (\forall u \leq z_{2^{n+1}})(\eta_Z(u) \neq 1^n)\}$ and $C_n^1 = \{Z \mid \eta_Z(1^n) \neq 1^n, (\exists! u \leq z_{2^{n+1}})(\eta_Z(u) = 1^n)\}$.

Note that $\mu(C_n^0) = \mu(C_n^1) = (1-1/2^n)^{2^n} \downarrow 1/e$ as $n \to \infty$.

Suppose every element in CP^Z is enumerated as $L(p_0, B_0)^Z, L(p_1, B_1)^Z, \cdots$. Furthermore we let two sets $C_n^{0,i}$ and $C_n^{1,i}$ corresponding to the i-th language $L(p_i, B_i)^Z$ be

$C_n^{0,i} = \{Z \in C_n^0 \mid (\#k \leq 2^{p_i(n)})(\langle x, z_k \rangle \in B_i^Z) > 0\}$ and $C_n^{1,i} = \{Z \in C_n^1 \mid (\#k \leq 2^{p_i(n)})(\langle x, z_k \rangle \in B_i^Z) > 0\}$.

It can be seen that there exists a sequence $\{\tau_n\} \downarrow 0$, by similar evaluations of the measures of $C_n^{1,i}$ and $C_n^{0,i}$ to [7], such that $\mu(C_n^{1,i}) \geq (1-\tau_n)\mu(C_n^{0,i})$.

Thus the error probability $\varepsilon_i = \mu(\{Z \mid L(0,F)^Z \neq L(p_i, B_i)^Z\})$ has a lower bound of $1/e$ since $\varepsilon_i \geq \mu(C_n^1) \cdot \mu(C_n^{1,i}) + \mu(C_n^0) \cdot [1 - \mu(C_n^{0,i})] \geq 1 - \tau_n \cdot \mu(C_n^{0,i}) \geq 1/e$. This least bound $1/e$ guarantees the theorem holds.

(2) A similar argument can be gone through by defining $F^{Y \oplus Z}$ as F^Z of (1). □

THEOREM 5.6.

(1) *$CP^X \neq C\Pi_1^{p,X} \neq CNP^X$ with probability one.*

(2) *$LP^X \neq L\Pi_1^{p,X} \neq LNP^X$ with probability one.*

Finally we shall show a lower complexity bound of the relativized CP-hierarchy with probability one. This result is due to [5].

THEOREM 5.7. *$PP^X <_m^p CP^X$ with probability one.*

Proof. Let ODD^Y be $\{x \mid x \neq z_0, (\exists k \leq 2^{|x|}-1)(|z_{2k+1}| = |x|$ and $z_{2k+1} \in Y)\}$. By Bennett and Gill's result of $ODD^Y \notin PP^Y$ with probability one, if $ODD^Y \leq_m^P CP^Y$ is shown to hold then the theorem is proved. Since ODD^Y is determined by exponentially-many checkings of every string z_{2k+1}, on input x, whether both $|x| = |z_{2k+1}|$ and $z_{2k+1} \in Y$ hold or not, thus ODD^Y is in CP^Y. □

COROLLARY 5.8. *$P^X <_m^p CP^X$, $NP^X <_m^p CNP^X$, co-$NP^X <_m^p C\Pi_1^{p,X}$ and $PP^X <_m^p CPP^X$ with probability one.*

COROLLARY 5.9. *For any class $L\mathcal{B}^Z$ in the relativized LP-hierarchy, $PP^X <_m^p L\mathcal{B}^X$ with probability one.*

§6. ***REMAINING PROBLEMS.***

We have observed on the three questions related to the measure of random oracles, which is raised in the beginning of the present paper, however, a little part of the questions was only demonstrated.

Let us give here the presentation of open problems left on both the LP- and CP-hierarchies other than the above questions. The reader may already find some of the unsolved questions mentioned in the previous sections. One of the bigger questions is whether these hierarchies collapse or not. The difficulty in settling this question essentially depends on that of the P=?NP question. As big a question as the collapse problem of the LP- and CP-hierarchies, whether $LP \equiv_m^P LNP$ is also a fundamental question. To make clear the difference between the equality and the equivalence of reducibility brings us useful information on the structure of the hierarchies.

Additionally, we shall give some others in the non-relativized form.

PROBLEMS 6.1.

(1) *Does* $L\Delta_k^p \equiv_m^p L\Sigma_k^p$ *imply* $L\Delta_k^p = L\Sigma_k^p$?

(2) *Does* $PSPACE \leq_m^p LNP$ *hold* ?

(3) *Does* $C(\Sigma_k^p/Poly) \equiv_m^p C\Sigma_k^p/Poly$ *hold* ?

(4) *Does* $NP^{CP} = co\text{-}NP^{CP}$ *hold* ?

In connection with the relativization of the LP- and CP-hierarchies, similar questions in the relativized form are also presented. We shall list some others here.

PROBLEMS 6.2.

(1) *Is there an oracle A such that* $L\Delta_k^{p,A} \neq L\Sigma_k^{p,A}$ *and* $L\Delta_k^{p,A} \equiv_m^p L\Sigma_k^{p,A}$?

(2) *Is there an oracle B such that* $\Delta_k^{p,B} \neq \Sigma_k^{p,B}$ *and* $C\Delta_k^{p,B} = C\Sigma_k^{p,B}$?

(3) *Does* $\Sigma_k^{p,X} <_m^p C\Sigma_k^{p,X} <_m^p L\Sigma_k^{p,X}$ *hold with probability one* ?

(4) *Does* $L\Delta_k^{p,X} <_m^p L\Sigma_k^{p,X}$ *hold with probability one* ?

(5) *Is* $L\Delta_k^p = L\Sigma_k^p$ *equivalent to* $L\Delta_k^{p,X} = L\Sigma_k^{p,X}$ *for any tally language X* ?

I express my gratitude to Professors Hisao Tanaka and Kojiro Kobayashi, both of whom gave me many stimulating suggestions about the LP- and CP-hierarchies. I was also encouraged throughout the preparation of this paper by fruitful conversations with Professor Takakazu Simauti.

REFERENCES

[1] K.AMBOS-SPIES, Randomness, relativizations, and polynomial reducibility, *Structure in Complexity Theory* in *Lecture Notes in Computer Science*, 223(1986), 23-34.

[2] K.AMBOS-SPIES, Minimal pairs for polynomial time reducibilities, *Computer Theory and Logic* in *Lecture Notes in Computer Science*, 270 (1987), 1-13.

[3] T.BAKER, J.GILL, and R.SOLOVAY, Relativizations of the P=?NP question, *SIAM Journal on Computing*, 4(1975), 431-442.

[4] T.P.BAKER and A.L.SELMAN, A second step toward the polynomial hierarchy, *Theoretical Computer Science*, 8(1979), 177-187.

[5] C.H.BENNETT and J.GILL, Relative to a random oracle A, $P^A \neq NP^A \neq co\text{-}NP^A$ with probability 1, *SIAM Journal on Computing*, 10(1981), 96-113.

[6] R.V.BOOK, T.J.LONG, and A.L.SELMAN, Quantitative relativizations of complexity classes, *SIAM Journal on Computing*, 13(1984), 461-487.

[7] J.CAI, Probability one separation of the Boolean hierarchy, *STACS 87* in *Lecture Notes in Computer Science*, 247(1987), 148-158.

[8] M.FURST, J.B.SAXE, and M.SIPSER, Parity, circuites, and the polynomial-time hierarchy, *Mathematical Systems Theory*, 17(1984), 13-27.

[9] J.GILL, Computational complexity of probabilistic Turing machines, *SIAM Journal on Computing*, 6(1977), 675-695.

[10] J.HARTMANIS, Some observations about NP complete sets, *Fundamentals of Computation Theory* in *Lecture Notes in Computer Science*, 278(1987), 185-196.

[11] R.E.LADNER, On the structure of polynomial time reducibility, *Journal of the Association for Computing Machinery*, 22(1975), 153-171.

[12] U.SCHÖNING, *Complexity and Structure* in *Lecture Notes in Computer Science*, 211(1986).

[13] M.SIPSER, Borel sets and circuit complexity, *Proceedings of the 15th Annual ACM Symposium on Theory of Computing*, 1983, 61-69.

[14] L.J.STOCKMEYER, The polynomial-time hierarchy, *Theoretical Computer Science*, 3(1977), 1-22.

[15] L.STOCKMEYER, Classifying the computational complexity of problems, *The Journal of Symbolic Logic*, 52(1987), 1-43.

[16] L.G.VALIANT, The complexity of computing the permanent, *Theoretical Computer Science*, 8(1979), 189-202.

[17] A.C.YAO, Separating the polynomial-time hierarchy by oracles, *Proceedings of the 26th Annual IEEE Symposium on Foundations of Computer Science*, 1985, 1-10.

Tomoyuki Yamakami
Department of Mathematics
Rikkyo University
Nishi-ikebukuro, Tokyo
171 Japan

Infinitesimal calculus interpreted in infinitary logic

Mariko Yasugi
Faculty of Science, Kyoto Sangyo University
Kita-ku, Kyoto, Japan 603

Dedicated to Professor Shôji Maehara on his sixtieth birthday

Introduction.

Infinitesimal calculus has been theorized as a part of nonstandard analysis, which is the consequence of an application of model theory. There have been attempts to reexamine nonstandard mathematics (from calculus to set theory) in syntactical lights. (A few of them are listed in the references.)

The author wished to single out the essence of the metatheory of nonstandard (infinitesimal) calculus as a "trick of the language." For this purpose, we set up a formal system *IR* of infinitary logic which is just adequate to develope elementary calculus. Let *A* be the axiom set of reals and let *C* be the axiom that specifies the domain to (standard) reals. A (possibly infinitary) sequent $\Gamma \to \Delta$ is a theorem of elementary calculus if and only if *A*, *C*, $\Gamma \to \Delta$ is a theorem of *IR*. The collection of such sequents will be called *SR*, the theory of standard reals.

Let *D* be the set of theorems of *SR* consisting of finitary closed formulas. A sequent $\Gamma \to \Delta$ is said to be a theorem of *GR* (general theory of reals) if *D*, $\Gamma \to \Delta$ is provable in *IR*.

For any A a statement of elementary calculus, let A^i be the version of A expressed in term of infinites and infinitesimals.

Then we can show that A is a theorem of *SR* if and only if A^i is a theorem of *GR*. This fact can be established by means of several proof-theoretical metatheorems (linkage principles) which hold for the systems *SR* and *GR*. They are the following.

1°.	(SPC)	Specification of the domain in *SR*
2°.	(CML)	Completeness of *SR*
3°.	(SBT)	Subtheory property of *GR*
4°.	(CMP)	Compactness property of *GR*
5°.	(TRF)	Transfer principle between *SR* and *GR*

It is characteristic of our treatment that existences of nonstandard objects are not assumed in *GR* ; they are only consistent with *GR*. This corresponds to Kakuda's finite satisfiability in [Kk]. This reference was assuring to the author.

Internality of a set (a formula) is defined to be first order definability in *GR*.

The work is supported in part by a Grand-in-Aid for Scientific Research (No.62540171).

Externality of the set of standard natural numbers, for example, can be easily proved as an application of [CMP].

§1. Elementary infinitary logic.

Definition 1.1. We first define the syntax of a portion of infinitary logic, which will be called "elementary infinitary logic" and will be abbreviated to *EIL*.

1) Language and formation rules. Let κ_0, κ_1, κ_2, κ_3 and κ_4 be some ordinals, where κ_0 is sufficiently large and κ_4 will be regulated to a certain extent. Our language consists of the following symbols.

1.1) Individual constants ordered in κ_1-type.

1.2) Function constants of finite arities ordered in κ_2-type.

1.3) Predicate constants of finite arities ordered in κ_3-type ; = is included in particular.

1.4) Free variables : $a_0, a_1, \cdots, a_\lambda, \cdots, \lambda < \kappa_0$. Bound variables : $x_0, x_1, \cdots, x_\lambda, \cdots, \lambda < \kappa_0$.

1.5) Logical symbols : $\neg, \wedge, \vee, \vdash, \forall, \exists$, where $\wedge$ and $\vee$ are of any arities below κ_4, while $\forall$ and $\exists$ are of finite property.

2) Terms and formulas are defined as usual except for $\wedge$ and $\vee$. If $\{A_\iota\}_{\iota<\mu}$ is a sequence of formulas where $\mu < \kappa_4$, then $\wedge\, [\iota < \mu]\, A_\iota$ and $\vee\, [\iota < \mu]$ are formulas. (We may use the familiar expression such as $\overset{\frown}{c}$ F(c).) = is allowed for objects of individual type.

3) Let κ be the cardinality of the set of all formulas in our language. $\Gamma \to \Delta$ is a sequent if Γ and Δ are respectively sequences of formulas of length $< \kappa^+$.

The familiar notions concerning our language will be assumed.

Note. We may and shall deal with many sorted systems ; here a single sorted system is presented for the sake of simplicity.

Definition 1.2. System *EIL*. Our system *EIL* (elementary infinitary logic) is a simplified version of the homogeneous system in Definition 22.1 of [Tk].

1) The rules of inference of *EIL* are the following.

1.1) (Weak) structural rule of inference :

$$\frac{\Gamma \to \Delta}{\Gamma' \to \Delta'}$$

where every formula occurring in Γ occurs in Γ' ; similarly for Δ and Δ'.

Subsequently $\gamma < \kappa^+$ will be assumed.

1.2) $\neg$

left:
$$\frac{\Gamma \to \Delta,\ \{A_\lambda\}_{\lambda<\gamma}}{\{\neg A_\lambda\}_{\lambda<\gamma},\ \Gamma \to \Delta}$$

right:
$$\frac{\{A_\lambda\}_{\lambda<\gamma},\ \Gamma \to \Delta}{\Gamma \to \Delta,\ \{\neg A_\lambda\}_{\lambda<\gamma}}$$

1.3) $\wedge$ Suppose $\beta_\lambda < \kappa_4$ for all $\lambda < \gamma$.

left:
$$\frac{\{A_{\lambda,\mu}\}_{\lambda<\beta_{\lambda},\lambda<\gamma},\ \Gamma \to \Delta}{\{\wedge\,[\mu < \beta_\lambda]\, A_{\lambda,\mu}\}_{\lambda<\gamma},\ \Gamma \to \Delta}$$

right: $$\frac{\Gamma \to \Delta, \ \{A_{\lambda,\mu_\lambda}\}_{\lambda<\gamma} \text{ for all } \{\mu_\lambda\}_{\lambda<\gamma}; \mu_\lambda < \beta_\lambda, \lambda < \gamma}{\Gamma \to \Delta, \{\bigwedge[\mu < \beta_\lambda] A_{\lambda,\mu}\}_{\lambda<\gamma}}$$

1.4) $\bigvee$ Dual to $\bigwedge$.

1.5) $\vdash$ Similarly to $\neg$.

1.6) $\forall$

left: $$\frac{\{A_\lambda(t_\lambda)\}_{\lambda<\gamma}, \ \Gamma \to \Delta}{\{\forall x_\lambda A_\lambda(x_\lambda)\}_{\lambda<\gamma}, \ \Gamma \to \Delta},$$

where each t_λ is an arbitrary term.

right: $$\frac{\Gamma \to \Delta, \ \{A_\lambda(a_\lambda)\}}{\Gamma \to \Delta, \{\forall x_\lambda A_\lambda(x_\lambda)\}_{\lambda<\gamma}},$$

where the eigenvariables a_λ do not occur in the lower sequent and they are all mutually distinct.

1.7) $\exists$ Dual to $\forall$.

1.8) Cut rule:

$$\frac{\{\Gamma \to \Delta, A_\lambda\}_{\lambda<\gamma} \qquad \{A_\lambda\}_{\lambda<\gamma}, \Pi \to \Lambda}{\Gamma, \Pi \to \Delta, \Lambda}$$

2) A proof of *EIL* is defined as usual with initial sequents of the form $D \to D$.

Since our system is a portion of the general infinitary logic with homogeneous quantifiers, the facts proved for the general case hold here also.

Proposition 1.1. 1) The consistency and the completeness (together with the cut elimination theorem) hold for *EIL*. (See Propositions 22.16 and Theorem 22.17 in [Tk].)

2) The dualities respectively between $\forall$ and $\exists$ and between $\bigwedge$ and $\bigvee$ are valid (provable in *EIL*).

Theorem 1. (The compactness theorem for the first order predicate calculus ; see Problem 22.26 of [Tk].) Let $\Gamma \to \Delta$ be a sequent of *EIL* consisting of first order formulas (formulas of the finite part of the language). If it is *EIL* -provable, then there exist finite subsets Γ_0 and Δ_0 respectively of Γ and Δ for which $\Gamma_0 \to \Delta_0$ is first order provable. That is, $\Gamma_0 \to \Delta_0$ is a theorem of first order predicate calculus. For the proof of the theorem, see [Tk].

§2. Standard theory of real analysis.

We shall henceforth assume the soundness of elementary theory of standard reals.

Definition 2.1. Language of standard reals, *LR*. *LR* is the language in Definition 1.1 specified (and modified) as follows.

1) The ordinals $\kappa_0 \sim \kappa_4$ are determined by c (the cardinality of the continuum). It is sufficient to assume $\kappa_0 \sim \kappa_3$ to be c, and κ_4 to be $c + 1$.

2) There are three sorts, N, Q and R, representing respectively the set of natural numbers, the set of rationals and the set of reals. The same letters as these will be used for the corresponding sets.

3) Individual constants are prepared for each sort ; those of N and Q are ordered in ω-type , while those of R in c-type.

4) Variables are prepared for each sort (ordered in *c*-type). Although free variables and bound variables are assumed to be distinguished, we may not do so in practice.

5) Among the constant symbols are included 0, 1, +, $\cdot$, =, <, ≦, −, $^{-1}$, +, max, min, | |, and various familiar functions which appear in the calculus. The sorts of these will be distinguished if necessary. We may write 0_Q for instance to denote the 0 of sort Q. Objects of *LR* which are supposed to represent mappings will be said to be function-terms.

6) Special function constants ϕ_1, ϕ_2, ψ_1 and ψ_2, which represent the mappings respectively from N to Q, Q to R, Q to N and R to N, will be assumed.

7) Auxiliary symbols λ and Σ (for term-forming) will be used.

Definition 2.2. Formation rules.

1) Term-formation. Term (representing respectively numbers and functions) are defined as usual. If f is a (numerical) term and x is a variable, then λxf is a function term. If F represents a binary function and n is of sort N, then Σ(n ; F, x) is a numerical term.

2) Formulas. Formulas are defined as in Definition 1.1. X=Y is admitted only if X and Y are of the same sort.

3) The logical system which underlies standard real analysis is the *EIL* in the Definition 1.2. applied to the language *LR*. It will be called *IR*.

Note. Let us agree on the following notational conventions.

k, *l*, m, n, ⋯ will denote terms or variables of sort N;

p, q, r, s, ε, δ, ⋯ will denote those of sort Q;

a, b, c, x, y, z, ⋯ will denote those of sort R.

f, g, h, ⋯ will denote function symbols.

Definition 2.3. Axioms of real calculus.

(Group E) Equality axioms of each sort.

(Group N) Axioms on N : the axioms on the arithmetic of natural numbers (expressed in first order *LR*-sentences) as well as mathematical inductions.

(Group Q) Axioms on the arithmetic of Q.

(Group N−Q) Relations between N and Q. In particular, ϕ_1 is the embedding of N into Q with respect to functions and relations of N, and (ψ_1, ψ_2) is its inverse.

(Group R) axioms on R.

(Group Q−R) Relations between Q and R. In particular, ϕ_2 is the embedding of Q into R with respect to functions and relations.

We shall identify the images of ϕ_1 and ϕ_2 respectively with the coimages.

(Group Σ, λ) $\Sigma(0; F, x) = F(0, x)$,

$$\Sigma(n+1; F, x) = \Sigma(n; F, x) + F(n+1, x).$$

$$(\lambda x f(x))(t) = f(t).$$

(Group F, P) Axioms on individual functions and predicates : the values of functions and relations as well as their properties expressed in the first order language.

(Group CMR) Completeness of standard reals. For example, let { s_n } be a sequence of reals and let p and q be rational constants. Then

$$\forall n\,(q < s_n < p) \rightarrow \exists x \forall y\,(y \leqq x \wedge x = \limsup s_n \wedge y = \liminf s_n).$$

(Group C) Specification of individual domains.

$$\forall l \bigvee_{n} l = n,$$

where n ranges over all the constants of sort N ; that is, those representing standard natural numbers.

$$\forall p \bigvee_{r} p = r,$$

where r ranges over all the constants representing rationals.

$$\forall x \bigvee_{c} x = c,$$

where c ranges over all the constants representing standard reals.

The set of axioms except (Group C) will be denoted by *A* and (Group C) will be denoted by *C*. Notice that each axiom in *A* is a first order sentence.

Definition 2.4. Theory *SR*. Let $\Gamma \rightarrow \Delta$ be a sequent of *LR*. It will be said to be a theorem of *SR*, the theory of standard reals, if *A*, *C*, $\Gamma \rightarrow \Delta$ is provable in *IR*.

Theorem 2 (Specification principle). The quantifiers can be replaced by $\wedge$ and $\vee$ in *SR*; namely (for example),

$$\forall x F(x) \leftrightarrow \bigwedge_{c} F(c)$$

and

$$\exists x F(x) \leftrightarrow \bigvee_{c} F(c)$$

are theorems of *SR*.

The proof is trivial since (Group C) immediately implies these equivalences.

Theorem 3 (Completeness of *SR*). For every sentence of *LR*, it is true in standard theory of real calculus if and only if it is a theorem of *SR*, and hence for every sentence A either A or $\neg$A is a theorem of *SR*.

Proof. The consistency of *SR* is guaranteed by the soundness of the axioms (which is assumed) and that of the basic logic *IR*. For an atomic sentence A, we may assume exactly one of A and $\neg$A is included in the axioms. By virtue of Theorem 2 above, quantifiers can be replaced by $\wedge$ and $\vee$. Consider, for example, $\wedge[\lambda < \beta]\, A_\lambda$. By the induction hypothesis, either A_λ is a theorem for every λ or $\neg A_\lambda$ is a theorem for some λ. If the former is the case, $\wedge[\lambda < \beta]\, A_\lambda$ is derivable in *IR*. Otherwise *A*, *C*, $A_\lambda \rightarrow$ is provable, and hence $\neg\wedge[\lambda < \beta]\, A_\lambda$ is a theorem.

§3. General theory of real calculus.

Definition 3.1. The general theory of real calculus, *GR*, is defined in the same way as *SR* except that here the axiom set *D* consists all the closed, first order theorems of *SR*.

Corollary. 1) *D* contains all the first order axioms of *SR*. It can contain consequences of the axiom set *C* also as long as they are of first order.

2) (Subtheory property) *GR* is a subtheory of *SR*.

As an immediate consequence of Theorem 1, we have the Theorem 4 (Compactness of *GR*). If $\Gamma \to \Delta$ is a theorem of *GR* consisting of first order formulas, then there are finite sequences of formulas D_0, Γ_0 and Δ_0, which are subsets of respectively D, Γ, and Δ, such that $D_0, \Gamma_0 \to \Delta_0$ is first order provable. If Γ and Δ are finite, in particular, then, for some finite D_0, $D_0, \Gamma \to \Delta$ is first order provable.

We also obtain the following facts at our disposal.

Theorem 5. 1) Every theorem of *SR* is consistent with *GR*.

2) Axioms in *C* (of *SR*) are not theorems of *GR*.

3) (Completeness of the quantifier-free part of *GR*) Let A be a closed, quantifier-free formula of *GR*. A is true (relative to the axiom set *D*) if and only if A is a theorem of *GR*. For any such A, exactly one of A and $\neg$A is a theorem of *GR*.

4) Suppose A_λ, $\lambda < \mu$, are quantifier-free. $\bigwedge [\lambda < \mu] A_\lambda$ is a theorem of *GR* if and only if $A_{\lambda_1} \wedge \cdots \wedge A_{\lambda_n}$ is for every $\lambda_1, \cdots, \lambda_n < \mu$.

5) (Transfer principle) For a first order sentence A, A is a theorem of *SR* if and only if it is a theorem of *GR*, and hence exactly one of A and $\neg$A is a theorem of *GR*.

6) If $\Gamma \to \Delta$ is a theorem of *SR*, where Γ and Δ consist of sentences of first order, then $\Gamma \to \Delta$ is a theorem of *GR*.

Proof. 2) Suppose otherwise : $D \to \forall x \bigvee_c x = c$ is provable in *IR*. Then $D \to \bigvee_c a = c$ also, and hence $D \to \{ a = c \}_c$ also. By Theorem 4, $D_0 \to a = c_1, \cdots, a = c_k$ for some finite D_0 a subset of *D* and some $c_1, \cdots, c_k$ is first order provable. From this follows $D_0 \to \forall x(x = c_1 \vee \cdots \vee x = c_k)$. Since D_0 is a theorem of *SR*, $\forall x(x = c_1 \vee \cdots \vee x = c_k)$ must be also, which will yield a contradiction.

3) For every closed, atomic A, exactly one of A and $\neg$A is an axiom in *A*, and hence either is a theorem of *GR*. For the compound A, apply the induction hypotheses.

6) By 5) and Theorem 3.

Note. 1) Conversions of quantifiers into $\bigwedge$ and $\bigvee$ (see Theorem 2) are not necessarily possible in *GR*.

2) The converse of the 1) in Theorem 5 is not necessarily true.

Theorem 6. Let A(l) be of first order. If $\forall k \exists l > k A(l)$ is consistent with *GR*, then $\exists l(A(l) \wedge \bigwedge_n (l > n))$ is also.

Proof. Suppose the latter were inconsistent with *GR*. Then

$$D, \exists l(A(l) \wedge \bigwedge_n (l > n)) \to$$

is *IR*-provable, and hence so is

$$D, A(l), \{ l > n \}_n \to$$

with l a free variable. By the compactness theorem (Theorem 4), it follows

$$D_0, A(l),\ l > m \to$$

is first order provable for a fixed natural number m and for D_0 a finite subset of *D*. From this successively hold

$$D_0,\ \exists l(l > n \wedge A(l)) \to$$

and

D_0, $\forall k \exists l (l > k \wedge A(l)) \rightarrow$;

that is , $\forall k \exists l > kA(l)$ is inconsistent with D in first order logic, and is hence with GR, contrary to the hypothesis.

Note. The same holds for sorts Q and R.

Theorem 7. Let A be any subset of C (reals) and let ϕ be a any binary first order formula. Suppose, for each finite set $a_1, \cdots a_n$ of elements of A, there is a b in C such that $\phi(a_1, b) \wedge \phi(a_2, b) \wedge \cdots \wedge \phi(a_n, b)$ is a theorem of SR. (ϕ is said to be finitely concurrent in A.) Then $\exists y \bigwedge [a \in A]\phi(a, y)$ is consistent with GR.

Proof. Assume finite concurrence of ϕ and suppose $\exists y[a \in A]\phi(a, y)$ were inconsistent with GR ; that is, $\neg \exists y \bigwedge [a \in A]\phi(a, y)$ were GR-provable. Then

$D \rightarrow \{ \neg\phi(a, y) \}_{a \in A}$

would be IR-provable with y a free variable. Then by the compactness theorem

$D_0 \rightarrow \neg\phi(a_1, y), \cdots, \neg\phi(a_n, y)$

would be first order provable for a finite D_0 and some $a_1, \cdots, a_n$ in A, or

$D_0 \rightarrow \neg(\phi(a_1, y) \wedge \phi(a_2, y) \wedge \cdots \wedge \phi(a_n, y))$

for any b, contradicting the assumption.

Remark. 5) of Theorem 5 corresponds to the transfer principle, while Theorem 7 corresponds to the enlargement (or weak concurrence) principle in the usual theory of non-standard analysis. The latter is an immediate consequence of the compact theorem for GR (Theorem 4), and in our approach it will be used in the form of the compactness theorem. In fact this is the essence of the syntactical interpretation of nonstandard analysis, and hence we shall list it as one of the principles linking the standard theory and nonstandard one.

[Linkage Principles]

[SPC : specification principle]	Theorem 2
[CML : completeness of SR]	Theorem 3
[SBT : subtheory property]	2) of Corollary to Definition 3.1
[CMP : compactness of GR]	Theorem 4
[TRF : transfer principle]	5) of Theorem 5

These are fundamental principles in the foundations of the infinitesimal calculus and will be freely used.

Definition 3.2. We shall agree on the following notational conventions. s and t will denote any terms, and r (>0) will denote rational constants.

$s \approx t$ (s and t are infinitely close.) : $\bigwedge [r > 0] | s - t | < r$

$i(s)$ (s is infinitesimal.) : $s \approx 0$

$ni(s)$ (s is nonstandard, infinitesimal.) : $i(s) \wedge \neg s = 0$

$\infty(s)$ (s is infinite.) : $\bigwedge_r (| s | > r)$

$\infty(+ ; s)$ (s is positive infinite.) : $\bigwedge [r > 0](s > r)$

$\infty(- ; s)$ (s is negative infinite.) : $\bigwedge [r > 0](s > -r)$

fnt(s) (s is finite): $\bigwedge [r > 0](|s| \leqq r)$

nfnt(s) (s is nonstandard finite.): fun(s) $\wedge \bigvee_c \neg s = c$

st(x) (x is standard of sort C): $\bigvee_c x = c$

(The same definitions for sorts N and Q)

nst(x) (x is nonstandard.): ¬st(x)

Corollary. Every s which satisfies one of ni(s), ∞(s), ∞(+ ; s), ∞(− ; s), nfnt(s) is nonstandard.

Proposition 3.1. 1) The existence of a nonstandard object of each kind defined is *GR*-consistent.

2) $\forall x(\text{fut}(x) \leftrightarrow \bigvee_c (x \approx c))$ is *GR*-consistent.

Proof. 1) We shall work on a few cases as examples.

∃x ni(x). Suppose this were inconsistent with *GR*. That is,

$D, \exists x\, \text{ni}(x) \rightarrow$

is *IR*-provable. Then, so is

$D, \text{ni}(x) \rightarrow$

with x a free variable, that is,

$D, \neg x = 0, \{|x| < r\}_{r>0} \rightarrow$.

By [CMP],

$D_0, \neg x = 0, |x| < p \rightarrow$

for a $p > 0$ a rational constant and D_0 a finite subset of *D*. Take any q such that $0 < q < p$.

$D_0, \neg q = 0, |q| < p \rightarrow$

follows by substitution, and hence *D* and $\neg q = 0 \wedge |q| < p$ would be inconsistent. But $\neg q = 0 \wedge |q| < p$ can be regarded as an axiom in *D*, yielding a contradiction.

∃y(nfnt(y) ∧ ¬ni(y)). This will follow from the provability of $i(x) \rightarrow c + x \approx c$ for any $c > 0$ a real constant and the consistency of ∃x ni(x) as follows.

From $i(x) \rightarrow c + x \approx c$ follow successively in *GR*:

$\text{ni}(x) \rightarrow c + x \approx c \wedge (\neg c + x = c)$,

$\exists x\, \text{ni}(x) \rightarrow \exists y(y \approx c \wedge \neg y = c)$,

$y \approx c \wedge \neg y = c \rightarrow (\neg \bigwedge y = d) \wedge y \approx c \wedge (\neg y \approx 0)$

where d ranges over all real constants,

$\exists x\, \text{ni}(x) \rightarrow \exists y(\text{nfnt}(y) \wedge \neg \text{ni}(y))$.

We have just proved.

(∗) $D, \exists x\, \text{ni}(x) \rightarrow \exists y(\text{nfnt}(y) \wedge \neg \text{ni}(y))$.

in *IR*.

Suppose ∃y(nfnt(y) ∧ ¬ni(y)) were inconsistent; that is,

$D, \exists y(\text{nfnt}(y) \wedge \neg \text{ni}(y)) \rightarrow$

were provable. From this and (∗) above,

$$D, \exists x\, \mathrm{ni}(x) \to,$$

contradicting the consistency of $\exists x\, \mathrm{ni}(x)$. So, $\exists y(\mathrm{nfnt}(y) \wedge \neg \mathrm{ni}(y))$ is consistent.

2) $\forall x(\exists r > 0(|x| \leqq r) \vdash \bigvee_c x = c)$ is a theorem of *SR*, and by virtue of [SPC] this can be equivalently expressed as

$$\forall x(\bigvee[r > 0](|x| \leqq r) \vdash \bigvee_c \bigwedge[r > 0](|x - c| < r)$$

in *SR*, and hence it is consistent with *GR* (1) of Theorem 5).

This is $\forall x(\mathrm{fnt}(x) \vdash \bigvee_c x \approx c)$. $x \approx c \to \mathrm{fnt}(x)$ is *GR*-provable, and hence is consistent.

Note, The consistency of $\mathrm{fnt}(x) \to \bigvee_c x \approx c$ follows without the completeness of standard reals.

Proposition 3.2. 1) $a \approx b \leftrightarrow a = b$ in *GR* for any real constants a and b.

2) $\approx$ is an equivalence relation in *GR* with respect to the arithmetic of reals.

3) $\infty(x) \leftrightarrow \mathrm{ni}(1/x)$ and $0 < r \leqq |x| \leftrightarrow \mathrm{fnt}(1/x)$ in *GR*.

4) Natural numbers are non-negative in *GR*

Proof. 1) $a = b \to a \approx b$ is a theorem of *GR*. If a and b are identical, then $a = b$ is a theorem of *GR*, and hence so is $a \approx b \to a = b$. Otherwise $a \approx b \to$ is a theorem of *GR*, and hence so is $a \approx b \to a = b$.

2) Consider, as an example, $x \approx 0, \mathrm{fnt}(y) \to xy \approx 0$. For any $\varepsilon, r > 0$,

$$|x| < \varepsilon/r, |y| < r \to |xy| < \varepsilon$$

for $r > 0$. So

$$[\delta > 0]|x| < \delta, \quad [r > 0]|y| < r \to \bigwedge[\varepsilon > 0]|xy| < \varepsilon$$

is a theorem of *GR*.

3) and 4) are immediate.

Definition 3.3. Let $\phi(x)$ be any unary formula. ϕ is said to be internal if, for a first order $A(x)$,

$$\forall x(A(x) \leftrightarrow \phi(x))$$

is *GR*-provable. ϕ is external otherwise.

The A as above will be called a defining formula for ϕ.

Corollary. If $\phi(x)$ is of first order, then it is trivially internal.

Proposition 3.3. 1) The "set of standard natural numbers" is external. That is, there is no defining formula for $\bigvee_n x = n$. The same holds for rationals and reals.

2) The "set of all natural numbers" is internal ; the same holds for rationals and reals.

3) The set of infinitesimals is external.

Proof. 1) Suppose $\forall x(A(x) \vdash\dashv \bigvee_n x = n)$ in *GR* for a first order $A(x)$. Then

$$A(x) \to \bigvee_n x = n$$

with x free and

$$\bigvee_m m = n \to A(m)$$

for each constant m. From the first sequent follows

$$A(x) \to \{x = n\}_n,$$

which consists of first order formulas, and hence there is an l such that

$$A(x) \to x = 0 \vee x = 1 \vee \cdots \vee x = l$$

follows from a finite subset of D. For $m = l+1$, then, $\neg A(m)$. On the other hand, $A(m)$ is provable since $\underset{n}{\curlyvee}\ m = n$ is. So, $\neg A(m)$ and $A(m)$ yield a contradiction.

2) For a variable l of sort N, $l = l$ expresses the "set of all natural numbers," which is of first order.

In the subsequent sections, we shall carry out the syntactical interpretation of non-standard calculus. The propositions there will have the following form:

Let A be a first order sentence which expresses a statement in the calculus, and let A^i be the corresponding expression in terms of the infinitesimal calculus. Then A (is a theorem of *SR*) if and only if A^i (is a theorem of *GR*).

Notice that, although A^i is a statement in the infinitesimal calculus, we do not assume the existence of infinitesimals or infinites.

In order to make propositions assume an outlook of ordinary mathematics, we shall state them in an informal manner.

Let us briefly explain how each of the linkage principles is to be used.

[SPC] relate A and A^i both ways, from A to A^i and vice versa. [CML] and [TRF] are the principles for passing from A to A^i, while [SBT] and [CMP] are those for passing from A^i to A.

Since the techniques are much the same for all propositions, we shall take up a few cases as examples.

§4. Sequences.

Proposition 4.1. Let s be a real number and let $\{s_n\}_n$ be a sequence of reals (in the language *LR*). Then

$$\lim_{n \to \infty} s_n = s \text{ (a theorem of } SR\text{)}$$

if and only if

$$s_n \approx s \text{ for all infinite n (a theorem of } GR\text{).}$$

(The necessary condition can be formally stated as $\forall n(\infty(n) \vdash s_n \approx s)$. That is, n is a variable.)

Proof. Suppose $\lim s_n = s$. Then by [SPC],

$$\bigwedge[\varepsilon > 0]\ \underset{m}{\curlyvee}\ \forall n > m(|s - s_n| < \varepsilon)$$

in *SR*. For any fixed $\varepsilon > 0$,

$$\underset{m}{\curlyvee}\ \forall n > m(|s - s_n| < \varepsilon)$$

is a theorem of *SR*. By [CML],

$$\forall n > m(|s - s_n| < \varepsilon)$$

is a theorem of *SR* for some m, and is hence, by virtue of [TRF], a theorem of *GR*. So,

(1) $n > m \to |s - s_n| < \varepsilon$

in *GR* with n a free variable. On the other hand,

(2) $\infty(n) \to n > m$

is a theorem of *GR* for the m as above. From (1) and (2) follows

$\infty(n) \to |s - s_n| < \varepsilon.$

This is the case for each $\varepsilon > 0$, and hence

$\infty(n) \to \bigwedge[\varepsilon > 0]|s - s_n| < \varepsilon,$

or

$\infty(n) \to s_n \approx s$

is a theorem of *GR* with n a variable.

Conversely, suppose

$\forall n(\infty(n) \vdash s \approx s_n)$

in *GR*; that is,

$\bigwedge_m n > m \to \bigwedge[\varepsilon > 0]|s - s_n| < \varepsilon$

with n a variable, and so for each $\varepsilon > 0$,

$\{n > m\}_m \to |s - s_n| < \varepsilon.$

By [CMP], there is an $m = m_\varepsilon$ such that

$n > m \to |s - s_n| < \varepsilon,$

from which follows

$\exists m \forall n > m(|s - s_n| < \varepsilon)$

in *GR*, and therefore, by [SBT],

$\bigwedge[\varepsilon > 0]\exists m \forall n > m(|s - s_n| < \varepsilon)$

is a theorem of *SR*. From this follows, by [SPC],

$\forall \varepsilon > 0 \exists m \forall n > m(|s - s_n| < \varepsilon).$

Proposition 4.2. $\{s_n\}$ is Cauchy (in *SR*) if and only if $s_n \approx s_m$ for all infinite m and n (in *GR*). (The necessary condition be stated as

$\forall m \forall n(\infty(m) \wedge \infty(n) \vdash \bigwedge[\varepsilon > 0]|s_m - s_n| < \varepsilon.)$

Proof. Assume Cauchy-ness in *SR*. Then for each $\varepsilon > 0$, by [SBT],

$\bigvee_k \forall m \forall n(m, n \geqq k \vdash |s_n - s_m| < \varepsilon);$

and so, by [CML],

(1) $\forall m, n \geqq k(|s_n - s_m| < \varepsilon)$

in *SR* for some k. By [TRF], this is a theorem of *GR*. With m and n variables,

(2) $\infty(m), \infty(n) \to m, n \geqq k.$

By (1) and (2),

$$\infty(m), \infty(n) \rightarrow |s_n - s_m| < \varepsilon$$

for each $\varepsilon > 0$. So,

$$\infty(m), \infty(n) \rightarrow \bigwedge[\varepsilon > 0] |s_n - s_m| < \varepsilon,$$

or

$$\infty(m), \infty(n) \rightarrow s_n \approx s_m.$$

Assume the necessary condition. For each $\varepsilon > 0$,

$$\{m, n > l\}_l \rightarrow |s_n - s_m| < \varepsilon$$

in *GR*. By virtue of [CMP], there is an $l = l_\varepsilon$ such that $m, n > l \rightarrow |s_n - s_m| < \varepsilon$, and hence $\bigwedge[\varepsilon > 0]\exists l \bigvee m, n > l(|s_n - s_m| < \varepsilon)$ in *GR*. By [SBT], this is a theorem of *SR*, and hence, by [SPC],

$$\forall \varepsilon > 0 \exists l \forall m, n > l(|s_n - s_m| < \varepsilon).$$

Proposition 4.3. (Convergence of Cauchy sequences) If $s_m \approx s_n$ for infinite m and n (in *GR*), then there is a standard a such that $a \approx s_n$ for n infinite.

Proof. We first show that for some μ,

(1) $\forall n > \mu(s_n < s_\mu + 1)$.

From the premise,

$$\{m, n > l\}_l \rightarrow |s_m - s_n| < 1$$

with m and n variables. By [CMP] then

$$m, n > l \rightarrow |s_m - s_n| < 1$$

for some l. Putting $\mu = l + 1$, we obtain

$$n > \mu \rightarrow s_n < s_\mu + 1,$$

which is (1).

Now put

$$p = \max\{s_1 + 1,\ s_2 + 1,\ \cdots,\ s_{\mu-1} + 1,\ s_\mu + 1\}$$

From (1) follows

$$\forall n(s_n < p);$$

that is, $\{s_n\}$ is bounded above by p. Similarly we can show $\{s_n\}$ is bounded below, say by q; that is,

$$\forall n(q < s_n < p)$$

in *GR*, and is hence, by [SBT], in *SR*. By the completeness of standard reals ((Group-CMR) in §2), [CML] and [SPC], the following $\bar{a}_n$, $\underline{a}_n$, a and b, where

(2) $\forall n(\bar{a}_n = \sup\{m \geqq n; s_m\}), \forall n(\underline{a}_n = \inf\{m \geqq n; s_m\}), a = \lim_n \bar{a}_n, b = \lim_m \underline{a}_m,$

exist. From (2) we obtain in *GR*, with n a variable,

(3) $\exists m \geqq n(0 \leqq \bar{a}_n - s_m < \varepsilon)$,

(4) $\exists m \geqq n(0 \leqq s_m - \underline{a}_n < \varepsilon)$,

(5) $\infty(n) \rightarrow \bar{a}_n \approx a$,

(6) $\infty(n) \rightarrow \underline{a}_n \approx b$.

Notice that the formulas in (3) and (4) are provable in *SR*, which can be transferred to *GR* since they are of first order (by [TRF]). (5) and (6) follow from (2) by virtue of Proposition 4.1. From (3) and (4),

$$\infty(n) \rightarrow \exists m_1 \geqq n\, \exists m_2 \geqq n$$
$$(|\bar{a}_n - \underline{a}_n| \leqq |\bar{a}_n - s_{m_1}| + |s_{m_1} - s_{m_2}| + |s_{m_2} - \underline{a}_n| < \varepsilon + \varepsilon + \varepsilon < 3\varepsilon)$$

Since $s_{m_1} \approx s_{m_2}$. This holds for each $\varepsilon > 0$, and hence

(7) $\infty(n) \rightarrow \bar{a}_n \approx \underline{a}_n$.

(5) ~ (7) imply

(8) $\infty(n) \rightarrow a \approx b$

But a and b are real constants, and hence (8) is reduced to $a = b$ (Proposition 3.2).

$$\infty(n) \rightarrow s_n \approx a$$

follows from the premise, (3) and (5).

Proposition 4.4. $\{s_n\}$ tends to (positive) infinity if and only if s_n is infinite for any infinite n.

Proposition 4.5. If $\{s_n\}$ is increasing and bounded above (in *SR*), then $s_m \approx s_n$ for all infinite m and n.

§5. Continuous functions.

Proposition 5.1. Let b and c be real constants, and let f be a real function. $\lim f(x) = c(x \rightarrow b)$ if and only if

$$\forall x(x \approx b \wedge \neg x = b \vdash f(x) \approx c).$$

Proof. From the sufficiency and [SPC], we have

$$\bigwedge[\varepsilon > 0] \bigvee[\delta > 0] \forall x(0 < |x - b| < \delta \vdash |f(x) - c| < \varepsilon)$$

in *SR* and so, by [CML], for each $\varepsilon > 0$ and for some $\delta > 0$,

$$\forall x(0 < |x - b| < \delta \vdash |f(x) - c| < \varepsilon)$$

in *SR*. By [TRF], this is a theorem of *GR*. In *GR* we have also

$$\neg x = b,\ x \approx b \rightarrow 0 < |x - b| < \delta,$$

and hence

$$x \approx b \rightarrow |f(x) - c| < \varepsilon.$$

This holds for each $\varepsilon > 0$, and hence the necessity.

Assume the necessity in *GR*. Then, for each $\varepsilon > 0$,

$$\{0 < |x - b| < \delta\}_{\delta > 0} \rightarrow |f(x) - c| < \varepsilon.$$

By [CMP], there is a $\delta > 0$ such that

$$0 < |x - b| < \delta \rightarrow |f(x) - c| < \varepsilon.$$

So,

$$\forall x(0 < |x - b| < \delta \vdash |f(x) - c| < \varepsilon).$$

Passing to *SR* by [SBT] and applying [SPC], we obtain the sufficiency.

Proposition 5.2. f is continuous at a if and only if $\forall x(x \approx a \vdash f(x) \approx f(a))$.

Proposition 5.3. f is uniformly continuous on a domain D if and only if

$$\forall x, y \in D(x \approx y \vdash f(x) \approx f(y)).$$

Proposition 5.4. Suppose f is (uniformly) continuous on [a, b], f(a) < 0 and f(b) >0. Then there is a $c \in (a, b)$ such that f(c) =0. (A nonstandard proof of the standard intermediate value theorem.)

Proof. Define in *SR* (as a function of n and i)

$$\alpha(i, n) = a + i(b - a)/2^n.$$
$$f(\alpha(0, n)) = f(a) < 0,$$
$$f(\alpha(2^n, n)) = f(b) > 0.$$

So,

$$0 \leqq \exists i < 2^n(f(\alpha(i, n)) < 0 < f(\alpha(i+1, n))).$$

Put

$$\iota(n) = \min\{i ; f(\alpha(i, n)) < 0 < f(\alpha(i+1, n))\}.$$

Then

$$(1) \qquad f(\alpha(\iota(n), n)) < 0 < f(\alpha(\iota(n)+1, n)).$$

Put $\gamma(n) = \alpha(\iota(n), n)$. Then "$\{\gamma(n)\}$ is increasing and bounded," and hence by Proposition 4.5,

$$\infty(m), \infty(n) \rightarrow \gamma(m) \approx \gamma(n)$$

in *GR*, and hence by Proposition 4.3 there is a c such that

$$\infty(n) \rightarrow \gamma(n) \approx c.$$
$$\gamma(n) \approx c. \rightarrow f(c) \approx f(\gamma(n)) < 0$$

by Proposition 5.2, and so

$$(2) \qquad \infty(n) \rightarrow f(c) \approx f(\gamma(n)) < 0.$$
$$\infty(n) \rightarrow \alpha(\iota(n)+1, n) = \gamma(n) \approx \alpha(\iota(n)+1, n),$$

and hence

$$(3) \qquad \infty(n) \rightarrow f(\gamma(n)) \approx f(\alpha(\iota(n)+1, n))$$

by Proposition 5.2. By (1) and (2),

$$(4) \qquad \infty(n) \rightarrow f(c) \approx f(\gamma(n)) < 0 < f(\alpha(\iota(n)+1, n)).$$

(3) and (4) imply

$$\infty(n) \rightarrow f(c) \approx 0,$$

or

$$f(c) \approx 0,$$

from which follows $f(c) = 0$.

Proposition 5.5. If f is (uniformly) continuous on [a, b], then $\exists c \forall x (f(x) \leqq f(c))$.

Proof. Define

$$\alpha(i, n) = a + i(b - a)/2^n.$$

$$0 \leqq \exists j \leqq 2^n,\ f(\alpha(j, n)) = \max\{f(\alpha(i, n)); 0 \leqq i \leqq 2^n\}.$$

Define

$$\iota(n) = \min\{j; 0 \leqq \forall i \leqq 2^n,\ f(\alpha(j, n)) \geqq f(\alpha(i, n))\}$$

and

$$\gamma(n) = \alpha(\iota(n), n).$$

$\{\gamma(n)\}$ is bounded. By the completeness of standard reals, there is a subsequence of γ, say $\{d_k\}$, which is convergent; $d = \lim d_k$ for some d. By Proposition 4.1 and 5.1,

$$(1) \quad \infty(n) \rightarrow f(d_k) \approx f(d)$$

in *GR*. $d_k = \gamma(\nu(k))$, where $\{\nu(k)\}$ is increasing. So,

$$(2) \quad 0 \leqq i \leqq 2^{\nu(k)} \rightarrow f(\alpha(i, \nu(k))) \leqq f(d_k)$$

$$= f(\gamma(\nu(k))) = f(\alpha(\iota(\nu(k)), \nu(k))).$$

by the definitions of ι, γ and ν.

$$(3) \quad \infty(k), x \in [a, b]$$

$$\rightarrow \exists i(0 \leqq i \leqq 2^{\nu(k)} \wedge x \approx \alpha(i, \nu(k))).$$

By (1) ~ (3), $f(d_k) \approx f(d)$ and

$$f(x) \leqq f(d_k) \vee f(x) \approx f(d_k)$$

(if $\infty(k)$). So, $f(x) \leqq f(d) \vee f(x) \approx f(d)$. By [SBT], this is carried to *SR*. But in *SR*, $f(x) \approx f(d)$ implies $f(x) = f(d)$, hence the conclusion.

§6. Differential calculus.

Proposition 6.1. 1) $c = f'(a)$ if and only if $\forall x(\neg x = a \wedge x \approx a \vdash (f(x) - f(x))/(x - a) \approx c)$.

2) If f is differentiable at a, then f is continuous at a (a standard statement).

Proof. 1) $c = f'(a)$ is expressed as

$$\lim(f(x) - f(a))/(x - a) = c(x \rightarrow a).$$

By Proposition 5.1, this holds if and only if

$$\forall x(\neg x = a \wedge x \approx a \vdash (f(x) - f(a))/(x - a) \approx c).$$

2) Suppose f is differentiable. (That is, there is a c such that $c = f'(a)$.) Then the necessary condition in 1) holds, from which follows

$$x \approx a \rightarrow f(x) \approx f(a),$$

which means f is continuous at a by Proposition 5.2.

Proposition 6.2. (Rolle's theorem). Suppose f is differentiable in [a, b] and $f(a) = f(b) = 0$. Then for some $c \in (a, b)$, $f'(c) = 0$.

Proof. For some $c \in [a, b]$,

$$\forall x(|f(x)| \leqq |f(c)|)$$

by Proposition 5.5, and $f'(c) (= d)$ exists. By 1) of Proposition 6.1,

$$\neg x = c, x \approx c \rightarrow (f(x) - f(c))/(x - c) \approx d.$$

Suppose $f(x) \leqq f(c)$. Then

$$x > c \rightarrow (f(x) - f(c))/(x - c) \leqq 0$$

and

$$x < c \rightarrow (f(x) - f(c))/(x - c) \geqq 0.$$

When $f(c) < f(x)$, the opposite inequalities are the case. From these follows $d = 0$. If $f(x) \equiv 0$ on [a, b], then we can take, for example, $c = (b - a)/2 \in (a, b)$. Otherwise $c \in (a, b)$ is forced since $f(a) = f(b) = 0$.

§7. Integration.

Definition 7.1. Define the following.

$$\alpha(j, n) = a + j(b - a)/2^n,$$

$$S(n) = \Sigma(j = 0, \cdots, 2^n - 1; f(a_j)(b - a)/2^n).$$

Proposition 7.1. Let f be continuous on [a, b]. Then $\int_a^b f(x)\,dx$ (= c) exists and $\infty(n) \rightarrow S(n) \approx c$.

Proof. By Proposition 5.5, $f(a_j)$ is finite, and hence

$$\infty(m), \infty(n) \rightarrow S(m) \approx S(n).$$

Then by Proposition 4.3 there is a c such that

$$\infty(n) \rightarrow S(n) \approx c,$$

and hence, by Proposition 4.1,

$$c = \lim S(n) = \int_a^b f(x)\,dx$$

exists.

Definition 7.2. Let $\{ \varepsilon_n \}$ be any decreasing sequence of positive numbers converging to zero and let $\pi \equiv \{ \pi(n) \}$ be any sequence of partitions of [a, b] where, if $I_{n,j}$ is the jth interval of $\pi(n)$ and $|I_{n,j}|$ is the length of $I_{n,j}$, then $|I_{n,j}| < \varepsilon_n$. Let $a_{n,j} \in I_{n,j}$.

$$S(\pi, n) = \Sigma(j = 1, \cdots, k_n; |I_{n,j}| f(a_{n,j})),$$

where kn is the number of intervals of $\pi(n)$.

Proposition 7.2. For any partition as above

$\infty(n) \to S(\pi, n) \approx \int_a^b f(x)\,dx$.

Proposition 7.3. (Fundamental theorem of the calculus). Let f be continuous on [a, b]. Then

$F(x) = \int_a^x f(t)\,dt$.

exists and $F'(x_0) = f(x_0)$ if $a < x_0 < b$.

Proof. F(x) exists by Proposition 7.1. Assume $x_0, x_0 + h \in (a, b)$, where $h > 0$. Put

$$F(x_0 + h) - F(x_0) = \int_{x_0}^{x_0+h} f(t)\,dt,$$

$\pi_0(n) = \{x_0, x_1, \cdots, x_{2^n} = x_0 + h\}$,

$x_j = x_0 + jh/2^n$,

$S_0(n) = \Sigma(j = 1, \cdots, 2^n; f(x_{j-1})h/2^n)$.

(1) $\infty(n) \to F(x_0 + h) - F(x_0)$

$$= \int_{x_0}^{x_0+h} f(t)\,dt = S_0(n)$$

$$= (h/2^n)\,\Sigma(j = 1, \cdots, 2^n; f(x_{j-1}))$$

by Proposition 7.1. f is continuous, and hence there are $c, d \in [x_0, x_0 + h]$ such that $f(d) \leqq f(x) \leqq f(c)$ in $[x_0, x_0 + h]$ by Proposition 5.5. So,

$1 \leqq j \leqq 2^n \to f(d) \leqq f(x_{j-1}) \leqq f(c)$,

and hence

$f(d) \leqq (1/2^n)\,\Sigma(j = 1, \cdots, 2^n; f(x_{j-1}) \leqq f(c))$,

or, by (1) and [TRF]

(2) $f(d) \lesssim (F(x_0 + h) - F(x))/h \lesssim f(c)$,

where $u \lesssim v$ abbreviates $u < v \vee u \approx v$. That is,

(3) $\forall h > 0 \exists c, d \in [x_0, x_0 + h]$ (2).

(4) $i(h) \wedge h > 0 \wedge c, d \in [x_0, x_0 + h]$

$\to f(c) \approx f(d) \approx f(x_0)$

by Proposition 5.3. From (3) and (4) follows that

$i(h) \wedge h > 0 \to (F(x_0 + h) - F(x_0))/h \approx f(x_0)$.

By Proposition 6.1, this means that $F'(x_0) = f(x_0)$.

§8. Differentials.

Definition 8.1. Let n be a fixed natural number and let $t_1, \cdots, t_n$ be terms of sort R. Define the following.

$u \in O(t_1, \cdots, t_n): \exists x_1 \cdots \exists x_n$ (finite) $(u = x_1 t_1 + \cdots + x_n t_n)$

$u \in o(t_1, \cdots, t_n): \exists x_1 \cdots \exists x_n$ (infinitesimal) $(u = x_1 t_1 + \cdots + x_n t_n)$

$M_0 \equiv O(1), M_1 \equiv o(1)$.

Proposition 8.1. The following are theorem of *GR*. 1) $u \in M_0$ if and only if u is infinitesimal.

2) Put $t = \max(|t_1|, \cdots, |t_n|)$. Then $O(t_1, \cdots, t_n) = O(t)$ and $o(t_1, \cdots, t_n) = o(t)$.

3) $O(t)$ and $o(t)$ are respectively additive groups and admit M_0 as the domain of multiplicative operators.

4) If t is not infinitesimal, then $O(t)$ and M_0, $o(t)$ and M_1, $O(t)/o(t)$ and R are respectively isomorphic as additive groups.

Proposition 8.2. Let f be continuous in (a, b) (in *SR*). Let dx be a variable of sort R. Define,

$dy = df = f(x + dx) - f(x)$ if $x \in (a, b - dx)$,

$d^2y = d(dy) = f(x + 2dx) - 2f(x + dx) + f(x)$ if $x \in (a, b - 2dx)$.

$d^n(y) = d(d^{n-1}y)$.

If f is differentiable at $x_0 \in (a, b)$ (real constant) and ni(dx), then

$dy / dx \approx f'(x) \pmod{o(dx)}$

by Proposition 6.1 (in *GR*). In general,

$d^n y \approx f^{(n)}(x_0)\, dx^n \pmod{o(dx^n)}$.

References

[Dv] Martin Davis, Applied nonstandard analysis, Wiley (1976).

[Ff] Solomon Feferman, Lectures on proof theory, LNM 70 (1968), Springer-Verlag, 1-108.

[Hs-Ks] Ward Henson and Jerome Keisler, On the strength of nonstandard analysis, JSL 51 (1986), 377-386.

[Kk] Yuzuru Kakuda, Theory of infinitesimals without nonstandard models, Kobe J. Math. 2 (1985),187-213.

[Kw] Toru Kawai, Nonstandard analysis by axiomatic method, Southeast Asian Conference on Logic, Elsevier (1983), 55-76.

[Mt] Nobuyoshi Motohashi, Developing nonstandard analysis in a text of proof theory (in Japanese), Proceedings of RIMS 436 (1980), 135-149.

[Nl] Edward Nelson, The syntax of nonstandard analysis , mimeographed notes.

[Rb] Abraham Robinson, Nonstandard analysis, 2nd ed., American Elsevier, New York (1974).

[St] Masahiko Saito, "What is nonstandard analysis ?" (in Japanese), Sugaku 38 (1986), 133-149.

[Tk] Gaisi Takeuti, Proof theory, North-Holland Publishing Company, Amsterdam (1975).

List of Participants

NAME	AFFILIATION
Yoshihiro Abe	Fukushima Technical College
Kunimasa Aoki	Nagoya University
Kiwamu Aoyama	Kyushu University
Toshiyasu Arai	Nagoya University
Chi-Tat Chong	National University of Singapore
Katsuya Eda	University of Tsukuba
Hiroshi Fujita	Nagoya University
Marcia J. Groszek	Dartmouth College
Masazumi Hanazawa	Tokai University
Mikio Harada	University of Tsukuba
Leo A. Harrington	University of California, Berkeley
Shoji Hatano	Nagoya University
Koichi Hirano	Nagoya University
Sachio Hirokawa	Shizuoka University
Kiyoshi Iseki	Naruto University of Education
Hiromi Ishikawa	Ehime University
Noriya Kadota	Hiroshima University
Yuzuru Kakuda	Kobe University
Takao Kashiwagi	Yamaguchi University
Hiroaki Katsutani	Kobe University
Kenji Kawada	Aichi University
Tsuyoshi Kawaguchi	Kyoto University
Hirotaka Kikyo	Waseda University
Satoshi Kobayashi	University of Tokyo
Heiji Kodera	Aichi University of Education
Masahiro Kumabe	Waseda University
Reijiro Kurata	Kawai Institute of Culture and Education
Eido Kyuma	Nagoya University
Tosio Miyamoto	Toyama Mercantile Marine College
Nobuyoshi Motohashi	University of Tsukuba
Kazuaki Nagaoka	Tsuda College
Takashi Nagashima	Hitotsubashi University
Hayao Nakahara	University of Tokyo
Yoshihiro Nakano	Yamaguchi University
Koji Nakatogawa	Gunma University
Hideki Nakaya	Kanazawa University
Kanji Namba	University of Tokyo
Shigeo Ohama	Toyota Technical College
Masao Ohnishi	Kinki University
Minolu Ohta	Aichi University of Education
Hiroakira Ono	Hiroshima university
Takeshi Oshiba	Nagoya Institute of Technology
Masanao Ozawa	Nagoya University

Mamoru Shimoda	Shimonoseki City College
Tatsuya Shimura	University of Tokyo
Juichi Shinoda	Nagoya University
Kokio Shirai	Shizuoka University
Theodore A. Slaman	The University of Chicago
John Steel	University of California, Los Angeles
Joji Takahashi	Kobe University
Makoto Takahashi	Kobe University
Shuichi Takahashi	Science University of Tokyo
Michio Takano	Niigata University
Hisao Tanaka	Hosei University
Katsumi Tanaka	Kobe University
Akito Tsuboi	University of Tsukuba
Nobutaka Tsukada	University of Tsukuba
Tosiyuki Tugué	Nagoya University
Yoshiaki Uemura	Kyoto Sangyo University
Tadahiro Uesu	Science University of Tokyo
Masamichi Wate	Nihon University
W. Hugh Woodin	California Institute of Technology
Ken-ichi Yamane	Nagoya University
Yutaka Yasuda	Tokai University
Mariko Yasugi	Kyoto Sangyo University
Masahiro Yasumoto	Nagoya University
Yoshimi Yonezawa	Toyota Technical College

Author's Addresses

Chi-Tat Chong
Department of Mathematics, National University of Singapore
Lower Kent Ridge Road, Singapore 0511, SINGAPORE

Yuzuru Kakuda
Department of Mathematics, College of Liberal Arts, Kobe University
Nada, Kobe 657, JAPAN

Hiroaki Katsutani
Division of System Science, The Graduate School of Science and Technology
Kobe University
Nada, Kobe 657, JAPAN

Satoshi Kobayashi
Department of Mathematics, Faculty of Science, University of Tokyo
Hongo, Bunkyo-ku, Tokyo 113, JAPAN

Mamoru Shimoda
Department of Mathematics, Shimonoseki City College, Shimonoseki 751, JAPAN

Juichi Shinoda
Department of Mathematics, College of General Education, Nagoya University
Chikusa-ku, Nagoya 464, JAPAN

Theodore A. Slaman
Department of Mathematics, The University of Chicago
5734 University Avenue, Chicago, IL 60637, U.S.A.

W.Hugh Woodin
Department of Mathematics, The California Institute of Technology
Pasadena, CA 91125, U.S.A.

Tomoyuki Yamakami
Department of Mathematics, Rikkyo University
Nishi-Ikebukuro, Toshima-ku, Tokyo 171, JAPAN

Mariko Yasugi
Department of Mathematics, Faculty of Science, Kyoto Sangyo University
Kita-ku, Kyoto 603, JAPAN

Vol. 1232: P.C. Schuur, Asymptotic Analysis of Soliton Problems. VIII, 180 pages. 1986.

Vol. 1233: Stability Problems for Stochastic Models. Proceedings, 1985. Edited by V.V. Kalashnikov, B. Penkov and V.M. Zolotarev. VI, 223 pages. 1986.

Vol. 1234: Combinatoire énumérative. Proceedings, 1985. Edité par G. Labelle et P. Leroux. XIV, 387 pages. 1986.

Vol. 1235: Séminaire de Théorie du Potentiel, Paris, No. 8. Directeurs: M. Brelot, G. Choquet et J. Deny. Rédacteurs: F. Hirsch et G. Mokobodzki. III, 209 pages. 1987.

Vol. 1236: Stochastic Partial Differential Equations and Applications. Proceedings, 1985. Edited by G. Da Prato and L. Tubaro. V, 257 pages. 1987.

Vol. 1237: Rational Approximation and its Applications in Mathematics and Physics. Proceedings, 1985. Edited by J. Gilewicz, M. Pindor and W. Siemaszko. XII, 350 pages. 1987.

Vol. 1238: M. Holz, K.-P. Podewski and K. Steffens, Injective Choice Functions. VI, 183 pages. 1987.

Vol. 1239: P. Vojta, Diophantine Approximations and Value Distribution Theory. X, 132 pages. 1987.

Vol. 1240: Number Theory, New York 1984–85. Seminar. Edited by D.V. Chudnovsky, G.V. Chudnovsky, H. Cohn and M.B. Nathanson. V, 324 pages. 1987.

Vol. 1241: L. Gårding, Singularities in Linear Wave Propagation. III, 125 pages. 1987.

Vol. 1242: Functional Analysis II, with Contributions by J. Hoffmann-Jørgensen et al. Edited by S. Kurepa, H. Kraljević and D. Butković. VII, 432 pages. 1987.

Vol. 1243: Non Commutative Harmonic Analysis and Lie Groups. Proceedings, 1985. Edited by J. Carmona, P. Delorme and M. Vergne. V, 309 pages. 1987.

Vol. 1244: W. Müller, Manifolds with Cusps of Rank One. XI, 158 pages. 1987.

Vol. 1245: S. Rallis, L-Functions and the Oscillator Representation. XVI, 239 pages. 1987.

Vol. 1246: Hodge Theory. Proceedings, 1985. Edited by E. Cattani, F. Guillén, A. Kaplan and F. Puerta. VII, 175 pages. 1987.

Vol. 1247: Séminaire de Probabilités XXI. Proceedings. Edité par J. Azéma, P.A. Meyer et M. Yor. IV, 579 pages. 1987.

Vol. 1248: Nonlinear Semigroups, Partial Differential Equations and Attractors. Proceedings, 1985. Edited by T.L. Gill and W.W. Zachary. IX, 185 pages. 1987.

Vol. 1249: I. van den Berg, Nonstandard Asymptotic Analysis. IX, 187 pages. 1987.

Vol. 1250: Stochastic Processes – Mathematics and Physics II. Proceedings 1985. Edited by S. Albeverio, Ph. Blanchard and L. Streit. VI, 359 pages. 1987.

Vol. 1251: Differential Geometric Methods in Mathematical Physics. Proceedings, 1985. Edited by P.L. García and A. Pérez-Rendón. VII, 300 pages. 1987.

Vol. 1252: T. Kaise, Représentations de Weil et GL_2 Algèbres de division et GL_n. VII, 203 pages. 1987.

Vol. 1253: J. Fischer, An Approach to the Selberg Trace Formula via the Selberg Zeta-Function. III, 184 pages. 1987.

Vol. 1254: S. Gelbart, I. Piatetski-Shapiro, S. Rallis. Explicit Constructions of Automorphic L-Functions. VI, 152 pages. 1987.

Vol. 1255: Differential Geometry and Differential Equations. Proceedings, 1985. Edited by C. Gu, M. Berger and R.L. Bryant. XII, 243 pages. 1987.

Vol. 1256: Pseudo-Differential Operators. Proceedings, 1986. Edited by H.O. Cordes, B. Gramsch and H. Widom. X, 479 pages. 1987.

Vol. 1257: X. Wang, On the C*-Algebras of Foliations in the Plane. V, 165 pages. 1987.

Vol. 1258: J. Weidmann, Spectral Theory of Ordinary Differential Operators. VI, 303 pages. 1987.

Vol. 1259: F. Cano Torres, Desingularization Strategies for Three-Dimensional Vector Fields. IX, 189 pages. 1987.

Vol. 1260: N.H. Pavel, Nonlinear Evolution Operators and Semigroups. VI, 285 pages. 1987.

Vol. 1261: H. Abels, Finite Presentability of S-Arithmetic Groups. Compact Presentability of Solvable Groups. VI, 178 pages. 1987.

Vol. 1262: E. Hlawka (Hrsg.), Zahlentheoretische Analysis II. Seminar, 1984–86. V, 158 Seiten. 1987.

Vol. 1263: V.L. Hansen (Ed.), Differential Geometry. Proceedings, 1985. XI, 288 pages. 1987.

Vol. 1264: Wu Wen-tsün, Rational Homotopy Type. VIII, 219 pages. 1987.

Vol. 1265: W. Van Assche, Asymptotics for Orthogonal Polynomials. VI, 201 pages. 1987.

Vol. 1266: F. Ghione, C. Peskine, E. Sernesi (Eds.), Space Curves. Proceedings, 1985. VI, 272 pages. 1987.

Vol. 1267: J. Lindenstrauss, V.D. Milman (Eds.), Geometrical Aspects of Functional Analysis. Seminar. VII, 212 pages. 1987.

Vol. 1268: S.G. Krantz (Ed.), Complex Analysis. Seminar, 1986. VII, 195 pages. 1987.

Vol. 1269: M. Shiota, Nash Manifolds. VI, 223 pages. 1987.

Vol. 1270: C. Carasso, P.-A. Raviart, D. Serre (Eds.), Nonlinear Hyperbolic Problems. Proceedings, 1986. XV, 341 pages. 1987.

Vol. 1271: A.M. Cohen, W.H. Hesselink, W.L.J. van der Kallen, J.R. Strooker (Eds.), Algebraic Groups Utrecht 1986. Proceedings. XII, 284 pages. 1987.

Vol. 1272: M.S. Livšic, L.L. Waksman, Commuting Nonselfadjoint Operators in Hilbert Space. III, 115 pages. 1987.

Vol. 1273: G.-M. Greuel, G. Trautmann (Eds.), Singularities, Representation of Algebras, and Vector Bundles. Proceedings, 1985. XIV, 383 pages. 1987.

Vol. 1274: N. C. Phillips, Equivariant K-Theory and Freeness of Group Actions on C*-Algebras. VIII, 371 pages. 1987.

Vol. 1275: C.A. Berenstein (Ed.), Complex Analysis I. Proceedings, 1985–86. XV, 331 pages. 1987.

Vol. 1276: C.A. Berenstein (Ed.), Complex Analysis II. Proceedings, 1985–86. IX, 320 pages. 1987.

Vol. 1277: C.A. Berenstein (Ed.), Complex Analysis III. Proceedings, 1985–86. X, 350 pages. 1987.

Vol. 1278: S.S. Koh (Ed.), Invariant Theory. Proceedings, 1985. V, 102 pages. 1987.

Vol. 1279: D. Ieşan, Saint-Venant's Problem. VIII, 162 Seiten. 1987.

Vol. 1280: E. Neher, Jordan Triple Systems by the Grid Approach. XII, 193 pages. 1987.

Vol. 1281: O.H. Kegel, F. Menegazzo, G. Zacher (Eds.), Group Theory. Proceedings, 1986. VII, 179 pages. 1987.

Vol. 1282: D.E. Handelman, Positive Polynomials, Convex Integral Polytopes, and a Random Walk Problem. XI, 136 pages. 1987.

Vol. 1283: S. Mardešić, J. Segal (Eds.), Geometric Topology and Shape Theory. Proceedings, 1986. V, 261 pages. 1987.

Vol. 1284: B.H. Matzat, Konstruktive Galoistheorie. X, 286 pages. 1987.

Vol. 1285: I.W. Knowles, Y. Saitō (Eds.), Differential Equations and Mathematical Physics. Proceedings, 1986. XVI, 499 pages. 1987.

Vol. 1286: H.R. Miller, D.C. Ravenel (Eds.), Algebraic Topology. Proceedings, 1986. VII, 341 pages. 1987.

Vol. 1287: E.B. Saff (Ed.), Approximation Theory, Tampa. Proceedings, 1985–1986. V, 228 pages. 1987.

Vol. 1288: Yu. L. Rodin, Generalized Analytic Functions on Riemann Surfaces. V, 128 pages, 1987.

Vol. 1289: Yu. I. Manin (Ed.), K-Theory, Arithmetic and Geometry. Seminar, 1984–1986. V, 399 pages. 1987.

Vol. 1290: G. Wüstholz (Ed.), Diophantine Approximation and Transcendence Theory. Seminar, 1985. V, 243 pages. 1987.

Vol. 1291: C. Mœglin, M.-F. Vignéras, J.-L. Waldspurger, Correspondances de Howe sur un Corps p-adique. VII, 163 pages. 1987

Vol. 1292: J.T. Baldwin (Ed.), Classification Theory. Proceedings, 1985. VI, 500 pages. 1987.

Vol. 1293: W. Ebeling, The Monodromy Groups of Isolated Singularities of Complete Intersections. XIV, 153 pages. 1987.

Vol. 1294: M. Queffélec, Substitution Dynamical Systems – Spectral Analysis. XIII, 240 pages. 1987.

Vol. 1295: P. Lelong, P. Dolbeault, H. Skoda (Réd.), Séminaire d'Analyse P. Lelong – P. Dolbeault – H. Skoda. Seminar, 1985/1986. VII, 283 pages. 1987.

Vol. 1296: M.-P. Malliavin (Ed.), Séminaire d'Algèbre Paul Dubreil et Marie-Paule Malliavin. Proceedings, 1986. IV, 324 pages. 1987.

Vol. 1297: Zhu Y.-l., Guo B.-y. (Eds.), Numerical Methods for Partial Differential Equations. Proceedings. XI, 244 pages. 1987.

Vol. 1298: J. Aguadé, R. Kane (Eds.), Algebraic Topology, Barcelona 1986. Proceedings. X, 255 pages. 1987.

Vol. 1299: S. Watanabe, Yu.V. Prokhorov (Eds.), Probability Theory and Mathematical Statistics. Proceedings, 1986. VIII, 589 pages. 1988.

Vol. 1300: G.B. Seligman, Constructions of Lie Algebras and their Modules. VI, 190 pages. 1988.

Vol. 1301: N. Schappacher, Periods of Hecke Characters. XV, 160 pages. 1988.

Vol. 1302: M. Cwikel, J. Peetre, Y. Sagher, H. Wallin (Eds.), Function Spaces and Applications. Proceedings, 1986. VI, 445 pages. 1988.

Vol. 1303: L. Accardi, W. von Waldenfels (Eds.), Quantum Probability and Applications III. Proceedings, 1987. VI, 373 pages. 1988.

Vol. 1304: F.Q. Gouvêa, Arithmetic of *p*-adic Modular Forms. VIII, 121 pages. 1988.

Vol. 1305: D.S. Lubinsky, E.B. Saff, Strong Asymptotics for Extremal Polynomials Associated with Weights on ℝ. VII, 153 pages. 1988.

Vol. 1306: S.S. Chern (Ed.), Partial Differential Equations. Proceedings, 1986. VI, 294 pages. 1988.

Vol. 1307: T. Murai, A Real Variable Method for the Cauchy Transform, and Analytic Capacity. VIII, 133 pages. 1988.

Vol. 1308: P. Imkeller, Two-Parameter Martingales and Their Quadratic Variation. IV, 177 pages. 1988.

Vol. 1309: B. Fiedler, Global Bifurcation of Periodic Solutions with Symmetry. VIII, 144 pages. 1988.

Vol. 1310: O.A. Laudal, G. Pfister, Local Moduli and Singularities. V, 117 pages. 1988.

Vol. 1311: A. Holme, R. Speiser (Eds.), Algebraic Geometry, Sundance 1986. Proceedings. VI, 320 pages. 1988.

Vol. 1312: N.A. Shirokov, Analytic Functions Smooth up to the Boundary. III, 213 pages. 1988.

Vol. 1313: F. Colonius, Optimal Periodic Control. VI, 177 pages. 1988.

Vol. 1314: A. Futaki, Kähler-Einstein Metrics and Integral Invariants. IV, 140 pages. 1988.

Vol. 1315: R.A. McCoy, I. Ntantu, Topological Properties of Spaces of Continuous Functions. IV, 124 pages. 1988.

Vol. 1316: H. Korezlioglu, A.S. Ustunel (Eds.), Stochastic Analysis and Related Topics. Proceedings, 1986. V, 371 pages. 1988.

Vol. 1317: J. Lindenstrauss, V.D. Milman (Eds.), Geometric Aspects of Functional Analysis. Seminar, 1986–87. VII, 289 pages. 1988.

Vol. 1318: Y. Felix (Ed.), Algebraic Topology – Rational Homotopy. Proceedings, 1986. VIII, 245 pages. 1988

Vol. 1319: M. Vuorinen, Conformal Geometry and Quasiregular Mappings. XIX, 209 pages. 1988.

Vol. 1320: H. Jürgensen, G. Lallement, H.J. Weinert (Eds.), Semigroups, Theory and Applications. Proceedings, 1986. X, 416 pages. 1988.

Vol. 1321: J. Azéma, P.A. Meyer, M. Yor (Eds.), Séminaire de Probabilités XXII. Proceedings. IV, 600 pages. 1988.

Vol. 1322: M. Métivier, S. Watanabe (Eds.), Stochastic Analysis. Proceedings, 1987. VII, 197 pages. 1988.

Vol. 1323: D.R. Anderson, H.J. Munkholm, Boundedly Controlled Topology. XII, 309 pages. 1988.

Vol. 1324: F. Cardoso, D.G. de Figueiredo, R. Iório, O. Lopes (Eds.), Partial Differential Equations. Proceedings, 1986. VIII, 433 pages. 1988.

Vol. 1325: A. Truman, I.M. Davies (Eds.), Stochastic Mechanics and Stochastic Processes. Proceedings, 1986. V, 220 pages. 1988.

Vol. 1326: P.S. Landweber (Ed.), Elliptic Curves and Modular Forms in Algebraic Topology. Proceedings, 1986. V, 224 pages. 1988.

Vol. 1327: W. Bruns, U. Vetter, Determinantal Rings. VII,236 pages. 1988.

Vol. 1328: J.L. Bueso, P. Jara, B. Torrecillas (Eds.), Ring Theory. Proceedings, 1986. IX, 331 pages. 1988.

Vol. 1329: M. Alfaro, J.S. Dehesa, F.J. Marcellan, J.L. Rubio de Francia, J. Vinuesa (Eds.): Orthogonal Polynomials and their Applications. Proceedings, 1986. XV, 334 pages. 1988.

Vol. 1330: A. Ambrosetti, F. Gori, R. Lucchetti (Eds.), Mathematical Economics. Montecatini Terme 1986. Seminar. VII, 137 pages. 1988.

Vol. 1331: R. Bamón, R. Labarca, J. Palis Jr. (Eds.), Dynamical Systems, Valparaiso 1986. Proceedings. VI, 250 pages. 1988.

Vol. 1332: E. Odell, H. Rosenthal (Eds.), Functional Analysis. Proceedings, 1986–87. V, 202 pages. 1988.

Vol. 1333: A.S. Kechris, D.A. Martin, J.R. Steel (Eds.), Cabal Seminar 81–85. Proceedings, 1981–85. V, 224 pages. 1988.

Vol. 1334: Yu.G. Borisovich, Yu. E. Gliklikh (Eds.), Global Analysis – Studies and Applications III. V, 331 pages. 1988.

Vol. 1335: F. Guillén, V. Navarro Aznar, P. Pascual-Gainza, F. Puerta, Hyperrésolutions cubiques et descente cohomologique. XII, 192 pages. 1988.

Vol. 1336: B. Helffer, Semi-Classical Analysis for the Schrödinger Operator and Applications. V, 107 pages. 1988.

Vol. 1337: E. Sernesi (Ed.), Theory of Moduli. Seminar, 1985. VIII, 232 pages. 1988.

Vol. 1338: A.B. Mingarelli, S.G. Halvorsen, Non-Oscillation Domains of Differential Equations with Two Parameters. XI, 109 pages. 1988.

Vol. 1339: T. Sunada (Ed.), Geometry and Analysis of Manifolds. Procedings, 1987. IX, 277 pages. 1988.

Vol. 1340: S. Hildebrandt, D.S. Kinderlehrer, M. Miranda (Eds.), Calculus of Variations and Partial Differential Equations. Proceedings, 1986. IX, 301 pages. 1988.

Vol. 1341: M. Dauge, Elliptic Boundary Value Problems on Corner Domains. VIII, 259 pages. 1988.

Vol. 1342: J.C. Alexander (Ed.), Dynamical Systems. Proceedings, 1986–87. VIII, 726 pages. 1988.

Vol. 1343: H. Ulrich, Fixed Point Theory of Parametrized Equivariant Maps. VII, 147 pages. 1988.

Vol. 1344: J. Král, J. Lukeš, J. Netuka, J. Veselý (Eds.), Potentia Theory – Surveys and Problems. Proceedings, 1987. VIII, 271 pages. 1988.

Vol. 1345: X. Gomez-Mont, J. Seade, A. Verjovski (Eds.), Holomorphi Dynamics. Proceedings, 1986. VII, 321 pages. 1988.

Vol. 1346: O. Ya. Viro (Ed.), Topology and Geometry – Rohli Seminar. XI, 581 pages. 1988.

Vol. 1347: C. Preston, Iterates of Piecewise Monotone Mappings o an Interval. V, 166 pages. 1988.

Vol. 1348: F. Borceux (Ed.), Categorical Algebra and its Applications Proceedings, 1987. VIII, 375 pages. 1988.

Vol. 1349: E. Novak, Deterministic and Stochastic Error Bounds Numerical Analysis. V, 113 pages. 1988.